Storage Networks Explained

Storage Networks Explained

Basics and Application of Fibre Channel SAN, NAS, iSCSI, InfiniBand and FCoE, Second Edition

Ulf Troppens, Wolfgang Müller-Friedt, Rainer Wolafka

IBM Storage Software Development, Mainz, Germany

Rainer Erkens, Nils Haustein

IBM Advanced Technical Sales Europe, Mainz, Germany

Translated by Rachel Waddington, Member of the Institute of Translating and Interpreting, UK

New material for this edition translated from the original German version into English by Hedy Jourdan

A John Wiley and Sons, Ltd., Publication

First published under the title *Speichernetze, Grundlagen und Einsatz von Fibre Channel SAN, NAS, iSCSI und InfiniBand.*
ISBN: 3-89864-135-X by dpunkt.verlag GmbH
Copyright © 2003 by dpunkt.verlag GmbH, Heidelberg, Germany

1st edition of the English translation first published 2004.
Translation Copyright © 2004, John Wiley & Sons Ltd.

This edition first published 2009
Translation Copyright © 2009, John Wiley & Sons Ltd.

Registered office
John Wiley & Sons, Ltd, The Atrium, Southern Gate, Chichester, West Sussex PO19 8SQ, United Kingdom

For details of our global editorial offices, for customer services and for information about how to apply for permission to reuse the copyright material in this book please see our website at www.wiley.com.

Library of Congress Cataloging-in-Publication Data:

Storage networks explained : basics and application of Fibre Channel SAN, NAS, iSCSI, InfiniBand, and
FCoE / Ulf Troppens ... [et al.]. – 2nd ed.
        p. cm.
    Rev. ed. of Storage networks explained / Ulf Troppens. c2004.
    ISBN 978-0-470-74143-6 (cloth)
    1. Storage area networks (Computer networks) 2. Information storage and retrieval systems. I.
Troppens, Ulf. II. Troppens, Ulf. Storage networks explained.
    TK5105.86.T78 2009
    004.6–dc22
                                                                            2009014224

A catalogue record for this book is available from the British Library.

ISBN   978-0-470-74143-6

For Silke, Hannah, Nina, and Julia
You keep showing me what really matters in life.

For Christina, Marie and Tom
For your love and support.

For Christel
Only your patience and your understanding have made my contribution to
this book possible.

For Susann
In Love.

For Tineke, Daniel and Marina
For the love, motivation and reassurance you have always given me.

Contents

About the Authors

The authors are employed at IBM's storage competence center in Mainz, Germany. They work at the interface between technology and customers. Their duties cover a wide field of responsibilities. They develop and test new software for storage networks. They present the latest hardware and software products in the field of storage networks to customers and explain their underlying concepts. Last but not least they deploy and support respective hardware and software in customer environments.

Ulf Troppens (centre) studied Computer Science at the University of Karlsruhe. Since 1989 he has been primarily involved in the development and administration of Unix systems, storage systems, data and storage networks and distributed applications.

Rainer Erkens (left) studied Mathematics at the University of Mainz. His experience in the management of computers and distributed applications goes back to 1992. Since 2005 he is a technical support manager in IBM's European Storage Competence Center.

Wolfgang Müller-Friedt (right) studied Computer Science at the FH Darmstadt. He is a software architect focussing on the software development of management applications for storage networks which support open standards such as SMI-S and IEEE 1244.

Nils Haustein (left front) studied Electrical Engineering at the TU Chemnitz. For several years he is with IBM's advanced technical sales support in Europe where he is focussing on digital archiving.

Rainer Wolafka (right front) studied Electrical Engineering at the FH Frankfurt and Software Engineering at the Santa Clara University. Since 1997 he is working in the field of storage networks and the software development of management applications for storage networks.

Foreword to the Second Edition by Hermann Strass

A book on the subject of storage networks is especially important during these fast-moving times. The technology for storage networking is basically bringing with it new structures and procedures that will remain topical in the foreseeable future regardless of incremental differences and changes in products. This book is based on the experience of its authors in their day-to-day work with the material. It provides system administrators and system planners in particular with the tools they need for an optimal selection and cost-effective implementation of this complex technology, the use and operation of which currently seems indispensable in view of the ever-increasing storage quantities in companies. The technology of networked storage provides demonstrable and important cost savings. Growth therefore continues even in an unfavourable economic climate.

Storage quantities are growing because we are now working much more in colour, in three-dimension and digitally than was the case years ago. Furthermore, legal regulations that exist in the European Union and in other countries make the electronic/digital storage of all business data compulsory. The law no longer allows old business documents to be filed in printed form in archives. Data quantities continue to increase in good times as well as bad. Even lost contracts and the related data must be stored digitally. The legal regulations on their own are thus ensuring that a certain amount of growth in data is inevitable.

In the past, data was stored on disk and tape drives that were connected directly to a server. Storage was operated as a peripheral to the computer. Access rights, virus protection and other functions could thus be performed on the relevant computer (server). For reasons that are explained in detail in this book, this mode of operation is no longer practical today. Storage has been detached from the servers and combined to form a separate storage network. This has resulted in a fundamentally different approach to dealing with storage. The new procedures required will continue to be developed into the near future. Data storage therefore has a value of its own. It is no longer a matter of attaching another disk drive to a server.

Today stored data and the information it contains are the crown jewels of a company. The computers (servers) needed for processing data can be purchased by the dozen or in larger quantities – individually as server blades or packed into cabinets – at any time, integrated into a LAN or a WAN or exchanged for defective units. However, if stored data is lost, restore of it is very expensive and time-consuming, assuming that all or some of it can even be recovered. As a rule, data must be available 'around the clock'. Data networks must therefore be designed with redundancy and high availability.

These and related topics are covered in detail in this book. The approach is based upon the current state of technology only to a certain degree. What is more important is the description of the fundamental topics and how they relate to one another. This coverage goes beyond the scope of even lengthy magazine articles and will continue to be topical in the future. This is the only book available in the market today that covers this subject so comprehensively.

The requirements of storage networks are fundamentally different from those of the familiar local networks (LANs). Storage networks have therefore almost exclusively been using Fibre Channel technology, which was specially developed as a connection technology for company-critical applications. Storage networking is not a short-term trend and efforts are therefore currently underway to use other existing (for example, Ethernet-LAN-TCP/IP) network technologies as well as new ones that are coming on the market (for example InfiniBand and FCoE). Under certain circumstances these are totally sensible alternatives. This book highlights which selection criteria play a role here. It is usually not technical details or prejudices that are decisive but rather usage requirements, existing infrastructure and devices, along with a careful assessment of the future development in companies. The aim of this book is to provide valuable help in structural planning and the selection of devices and software.

The importance of networked storage technology has grown substantially since the first edition was printed. For the reasons mentioned in this book and due to regulatory requirements, even medium-sized companies need to manage large quantities of data and make them available for many years. This is why the sections on storage archiving have been considerably expanded in the new edition of this book. In a global economy business continuity is overly important for survival. This second edition devotes extensive coverage to this topic.

Overall this book is an excellent work. It explains the chosen subject comprehensively and in great detail, based on solid technical foundations. It is hoped that it will gain a wide circulation, particularly as it corrects a great many half-truths with its presentation of facts and addresses the usual prejudices.

Hermann Strass

Preface by the Authors

This Preface answers the following main questions:

- What does this book deal with?
- Who should read this book?
- How should this book be read?
- Who has written this book?

WHAT DOES THIS BOOK DEAL WITH?

The technology of storage networks fundamentally changes the architecture of IT systems. In conventional IT systems, storage devices are connected to servers by means of SCSI cables. The idea behind storage networks is that these SCSI cables are replaced by a network, which is installed in addition to the existing LAN. Server and storage devices can exchange data over this new network using the SCSI protocol. Storage networks have long been a known quantity in the world of mainframes. Fibre Channel, iSCSI, FCoE and Network Attached Storage (NAS) are now also taking storage networks into the field of Open Systems (Unix, Windows, OS/400, Novell Netware, MacOS).

Storage networks are a basic technology like databases and LANs. Storage was previously installed in the servers. Now most storage capacity is provided in external devices that are linked to servers over a storage network. As a result, anyone who is involved in the planning or operation of IT systems requires basic knowledge about the fundamentals and the use of storage networks. These networks are almost as widespread as SCSI, SAS and SATA but are more complex than LANs and TCP/IP.

The book is divided into two parts. Part I deals with fundamental technologies relating to storage networks. It guides the reader from the structure and operating method of storage devices through I/O techniques and I/O protocols to the file systems and storage virtualisation.

The second part of this book presents applications that utilise the new functions of storage networks and intelligent disk subsystems. The emphasis here is on the shared use of resources that are available over a storage network, scalable and adaptable storage architectures, network backup and digital archiving. Another important focus of the book is business continuity with strategies for continuous and loss-free operation as protection against small failures and large catastrophes. Further focal points are the discussions on the management of storage networks and the management of removable media. Last but not least, the SNIA Shared Storage Model provides a reference model to describe storage networks.

At the end of the book we have added a glossary, an index and an annotated bibliography, which in addition to further literature also highlights numerous freely available sources on the Internet.

Section 1.4 sets out in detail the structure of the book and the relationships between the individual chapters. Figure 1.7 illustrates the structure of the book. At this point, it is worth casting a glance at this illustration. Note that the illustration also describes the subjects that we will not be covering.

Long before the second edition was printed, many readers of the first edition wanted to know what the differences are between the two editions. Here we want to express that our approach was successful, we aimed at introducing basic concepts rather than presenting actual products and overly technical details. The chapter on I/O techniques was the only one that required some updating on Fibre Channel and iSCSI. The key distinction of the second edition is the addition of two new chapters covering the topics of digital archiving and business continuity. We have also expanded the coverage on the copy services of intelligent disk subsystems.

WHO SHOULD READ THIS BOOK?

Our approach is, first, to explain the basic techniques behind storage networks and, secondly, to show how these new techniques help to overcome problems in current IT systems. The book is equally suitable for beginners with basic IT knowledge and for old hands. It is more an introduction to the basic concepts and techniques than a technical reference work. The target group thus includes:

- System administrators and system architects
- System consultants
- Decision makers
- Users
- Students

After reading the whole book you will be familiar with the following:

- The concepts of storage networks and their basic techniques
- Usage options for storage networks
- Proposed solutions for the support of business processes with the aid of storage networks
- The advantages of storage networks
- New possibilities opened up by storage networks.

HOW SHOULD THIS BOOK BE READ?

There are two options for reading this book. Those readers who are only interested in the concepts and usage options of storage networks should read Chapter 1 (Introduction) and Part II (Application and Management of Storage Networks); they can use Part I as a reference to look up any basic technical information they might require. Readers who are also interested in the technical background of storage networks should read the book through from the beginning.

WHO HAS WRITTEN THIS BOOK?

Ulf Troppens began work on this book in 2001. Rainer Erkens joined him soon after, providing his contributions on the topics of storage virtualisation, management of storage networks and NDMP for the first edition in 2002. In 2004 Wolfgang Müller-Friedt expanded the English translation – which was presented with the 'Editor's Choice Award 2005' by Linux Journal – with his sound knowledge of magnetic tape, tape libraries and their management. Lastly, the second edition has been expanded considerably through contributions by Nils Haustein (digital archiving) and Rainer Wolafka (business continuity).

All five authors have different roles at the Storage Competence Center of IBM in Mainz, Germany. Our responsibilities range from the development and testing of new software for storage networks to providing guidance to customers on the procurement of suitable products and the respective underlying concepts as well as on the installation and support of relevant hardware and software for customer environments. We advise customers on how storage networks can help to solve problems in their current IT systems. This experience has made us familiar with the types of questions customers have in respect of storage networks. Our involvement extends to customers with experience in storage networks as well as to those who are novices in this field. The positive feedback we have received from readers of the first edition show that our work has helped us to structure the content of this book and to choose topics in a way that are important to readers of books on storage networks.

Our intention has been to take off our 'IBM hats' and to write this book from an unbiased viewpoint. As employees of IBM in the area of storage technology, the experience and opinions that have been formed in our day-to-day work have of course had some influence on this book. In this connection, we have to be very familiar with our own company's products as well as with those of our competitors and to position these products so that we inevitably have a view that goes beyond the IBM scope. In the end, this book is our personal work and has no connection with IBM apart from our employee relationship. Most importantly, this book does not represent any of the official opinions of IBM.

ACKNOWLEDGEMENTS FOR THE SECOND EDITION

We would like to give special thanks to our technical advisors on the second edition: Dirk Jahn (Archiving), Hans-Peter Kemptner (Business Continuity), Robert Haas (Limitations of RAID 5) and Hermann Strass for the Foreword. Other contributions were made by Jens-Peter Akelbein. We also appreciate the help we received on the publishing side from Rene Wiegand (copy-editing), Ulrich Kilian (LaTeX) and Rene Schoenfeld (editorial), all who helped to make our manuscript ready for printing.

With regard to the second English edition we would like to thank Birgit Gruber, Tiina Ruonamaa, Brett Wells, Liz Benson, Anna Smart, Sarah Tilley, Mary Lawrence and Sarah Hinton (all Wiley & Sons) as well as Deepthi Unni and her team at Laserwords. Last but not the least we thank Hedy Jourdan for the great translation of the new parts from German to English.

ACKNOWLEDGEMENTS FOR THE FIRST EDITION

We would also like to use this preface to thank some of the people who have made a significant contribution to the first edition of this book. From a chronological point of view, we should start by mentioning the editorial department of *iX* magazine and the copy-editing staff of dpunkt.verlag as they set the whole project in motion in March 2001 with the question 'Could you see yourselves writing a book on the subject of storage in the network?'

Regarding content, our colleagues from the IBM Mainz storage community, especially the former SAN Lab and the current TotalStorage Interoperability Center (meanwhile renamed to Systems Lab Europe), deserve mention: Without the collaboration on storage hardware and software with customers and employees of partner companies, business partners and IBM, and without the associated knowledge exchange, we would lack the experience and knowledge that we have been able to put into this book. The list of people in question is much too long for us to include it here. The cooperation of one of the authors with the students of the BAITI 2000 course of the Berufsakademie Mannheim (University of Applied Science Mannheim), from whom we have learnt that we have to explain subjects such as 'RAID', 'disk subsystems', 'instant copy', 'remote mirroring' and 'file server', was also valuable from a didactic point of view.

With regard to quality control, we thank our proofreaders Axel Köster, Bernd Blaudow, Birgit Bäuerlein, Frank Krämer, Gaetano Bisaz, Hermann Strass, Jürgen Deicke, Julia Neumann, Michael Lindner, Michael Riepe, Peter Münch, René Schönfeldt, Steffen Fischer, Susanne Nolte, Thorsten Schäfer, Uwe Harms and Willi Gardt, as well as our helpers at dpunkt.verlag, whose names we do not know.

We should emphasise in particular the many constructive suggestions for improvement by Susanne Nolte, who also contributed a few paragraphs on 'DAFS', and the numerous comments from our colleagues Axel Köster and Jürgen Deicke and our manuscript reader René Schönfeldt. In this connection, the efforts of Jürgen Deicke and Tom Clark should also be mentioned regarding the 'SNIA Recommended Reading' logo, which is printed on the front cover of the book.

With regard to the first English edition of this book we have to thank even more people: First of all, we would like to thank René Schönfeldt from dpunkt.verlag for convincing Birgit Gruber from Wiley & Sons to invest in the translation. We greatly appreciate Birgit Gruber for taking a risk on the translation project and having so much patience with all our editorial changes. Rachel Waddington did an outstanding job of translating the text and all the figures from German into English. Last but not least, we would like to thank Daniel Gill for leading the production process, including copy-editing and typesetting, and we would like to thank the team at Laserwords for typesetting the whole book.

Closing comments

Finally, the support of our parents, parents-in-law and partners deserves mention. I, Nils Haustein, would like to thank my dear wife Susann who gave me a lot of 'computer time' and the opportunity to make a contribution to this book. I, Rainer Wolafka, would like to thank my dear wife Tineke for her support and her constant encouragement and motivation to work on this book and to my son Daniel for understanding why I did not always have the time he deserved during this time. I, Wolfgang Müller-Friedt, would like to thank my dear wife Christel for her patience, her emotional support and for many more reasons than there is room to list in these notes. I, Ulf Troppens, at this point would like to thank my dear wife Silke for her support and for taking many household and family duties off my hands and thus giving me the time I needed to write this book. And I, Rainer Erkens, would like to thank my dear partner Christina, who never lost sight of worldly things and thus enabled me to travel untroubled through the world of storage networks, for her support. We are pleased that we again have more time for children, our families and friends. May we have many more happy and healthy years together.

Mainz, April 2009

Ulf Troppens
Rainer Erkens
Wolfgang Müller-Friedt
Nils Haustein
Rainer Wolafka

List of Figures and Tables

FIGURES

TABLES

1

Introduction

The purpose of this chapter is to convey the basic idea underlying this book. To this end we will first describe conventional server-centric IT architecture and sketch out its limitations (Section 1.1). We will then introduce the alternative approach of storage-centric IT architecture (Section 1.2), explaining its advantages using the case study 'Replacing a Server with Storage Networks' (Section 1.3). Finally, we explain the structure of the entire book and discuss which subjects are not covered (Section 1.4).

1.1 SERVER-CENTRIC IT ARCHITECTURE AND ITS LIMITATIONS

In conventional IT architectures, storage devices are normally only connected to a single server (Figure 1.1). To increase fault tolerance, storage devices are sometimes connected to two servers, with only one server actually able to use the storage device at any one time. In both cases, the storage device exists only in relation to the server to which it is connected. Other servers cannot directly access the data; they always have to go through the server that is connected to the storage device. This conventional IT architecture is therefore called server-centric IT architecture. In this approach, servers and storage devices are generally connected together by SCSI cables.

As mentioned above, in conventional server-centric IT architecture storage devices exist only in relation to the one or two servers to which they are connected. The failure of both of these computers would make it impossible to access this data. Most companies find

Storage Networks Explained: Basics and Application of Fibre Channel SAN, NAS, iSCSI, InfiniBand and FCoE, Second Edition
U. Troppens R. Erkens W. Müller-Friedt N. Haustein R. Wolafka © 2009 John Wiley & Sons, Ltd

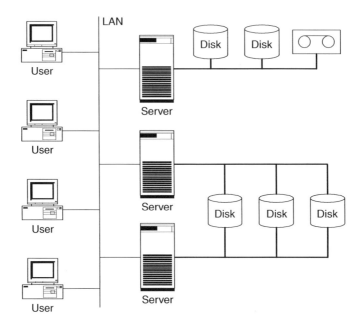

Figure 1.1   In a server-centric IT architecture storage devices exist only in relation to servers.

this unacceptable: at least some of the company data (for example, patient files, websites) must be available around the clock.

Although the storage density of hard disks and tapes is increasing all the time due to ongoing technical development, the need for installed storage is increasing even faster. Consequently, it is necessary to connect ever more storage devices to a computer. This throws up the problem that each computer can accommodate only a limited number of I/O cards (for example, SCSI cards). Furthermore, the length of SCSI cables is limited to a maximum of 25 m. This means that the storage capacity that can be connected to a computer using conventional technologies is limited. Conventional technologies are therefore no longer sufficient to satisfy the growing demand for storage capacity.

In server-centric IT environments the storage device is statically assigned to the computer to which it is connected. In general, a computer cannot access storage devices that are connected to a different computer. This means that if a computer requires more storage space than is connected to it, it is no help whatsoever that another computer still has attached storage space, which is not currently used (Figure 1.2).

Last, but not least, storage devices are often scattered throughout an entire building or branch. Sometimes this is because new computers are set up all over the campus without any great consideration and then upgraded repeatedly. Alternatively, computers may be consciously set up where the user accesses the data in order to reduce LAN data traffic. The result is that the storage devices are distributed throughout many rooms, which are

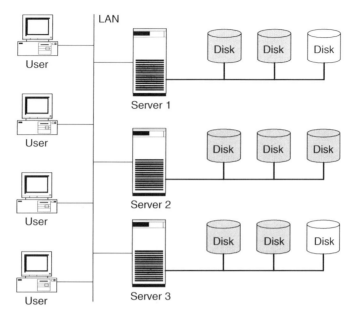

Figure 1.2   The storage capacity on server 2 is full. It cannot make use of the fact that there is still storage space free on server 1 and server 3.

neither protected against unauthorised access nor sufficiently air-conditioned. This may sound over the top, but many system administrators could write a book about replacing defective hard disks that are scattered all over the country.

1.2   STORAGE-CENTRIC IT ARCHITECTURE AND ITS ADVANTAGES

Storage networks can solve the problems of server-centric IT architecture that we have just discussed. Furthermore, storage networks open up new possibilities for data management. The idea behind storage networks is that the SCSI cable is replaced by a network that is installed in addition to the existing LAN and is primarily used for data exchange between computers and storage devices (Figure 1.3).

In contrast to server-centric IT architecture, in storage networks storage devices exist completely independently of any computer. Several servers can access the same storage device directly over the storage network without another server having to be involved. Storage devices are thus placed at the centre of the IT architecture; servers, on the other

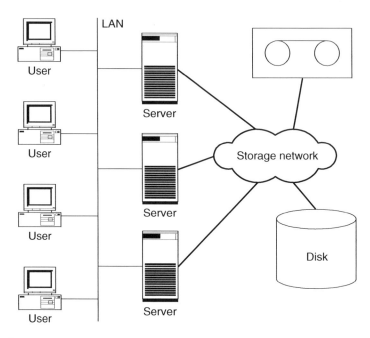

Figure 1.3 In storage-centric IT architecture the SCSI cables are replaced by a network. Storage devices now exist independently of a server.

hand, become an appendage of the storage devices that 'just process data'. IT architectures with storage networks are therefore known as storage-centric IT architectures.

When a storage network is introduced, the storage devices are usually also consolidated. This involves replacing the many small hard disks attached to the computers with a large disk subsystem. Disk subsystems currently (in the year 2009) have a maximum storage capacity of up to a petabyte. The storage network permits all computers to access the disk subsystem and share it. Free storage capacity can thus be flexibly assigned to the computer that needs it at the time. In the same manner, many small tape libraries can be replaced by one big one.

More and more companies are converting their IT systems to a storage-centric IT architecture. It has now become a permanent component of large data centres and the IT systems of large companies. In our experience, more and more medium-sized companies and public institutions are now considering storage networks. Even today, most storage capacity is no longer fitted into the case of a server (internal storage device), but has its own case (external storage device).

1.3   CASE STUDY: REPLACING A SERVER WITH STORAGE NETWORKS

In the following we will illustrate some advantages of storage-centric IT architecture using a case study: in a production environment an application server is no longer powerful enough. The ageing computer must be replaced by a higher-performance device. Whereas such a measure can be very complicated in a conventional, server-centric IT architecture, it can be carried out very elegantly in a storage network.

1. Before the exchange, the old computer is connected to a storage device via the storage network, which it uses partially (Figure 1.4 shows stages 1, 2 and 3).
2. First, the necessary application software is installed on the new computer. The new computer is then set up at the location at which it will ultimately stand. With storage networks it is possible to set up the computer and storage device several kilometres apart.

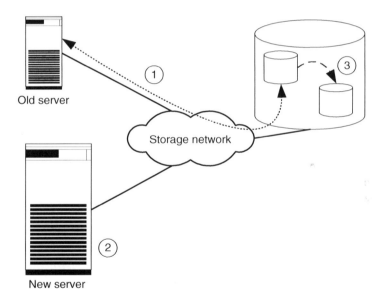

Figure 1.4   The old server is connected to a storage device via a storage network (1). The new server is assembled and connected to the storage network (2). To generate test data the production data is copied within the storage device (3).

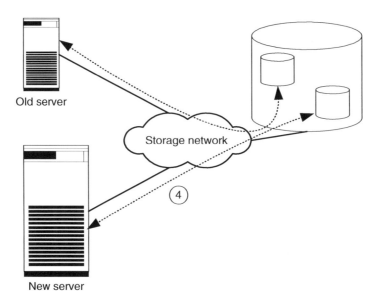

Figure 1.5   Old server and new server share the storage system. The new server is intensively tested using the copied production data (4).

3. Next, the production data for generating test data within the disk subsystem is copied. Modern storage systems can (practically) copy even terabyte-sized data files within seconds. This function is called instant copy and is explained in more detail in Chapter 2.

   To copy data it is often necessary to shut down the applications, so that the copied data is in a consistent state. Consistency is necessary to permit the application to resume operation with the data. Some applications are also capable of keeping a consistent state on the disk during operation (online backup mode of database systems, snapshots of file systems).

4. Then the copied data is assigned to the new computer and the new computer is tested intensively (Figure 1.5). If the storage system is placed under such an extreme load by the tests that its performance is no longer sufficient for the actual application, the data must first be transferred to a second storage system by means of remote mirroring. Remote mirroring is also explained in more detail in Chapter 2.

5. After successful testing, both computers are shut down and the production data assigned to the new server. The assignment of the production data to the new server also takes just a few seconds (Figure 1.6 shows steps 5 and 6).

6. Finally, the new server is restarted with the production data.

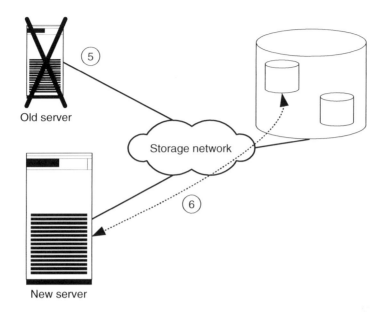

Figure 1.6   Finally, the old server is powered down (5) and the new server is started up with the production data (6).

1.4   THE STRUCTURE OF THE BOOK

One objective of this book is to illustrate the benefits of storage networks. In order to provide an introduction to this subject, this chapter has presented a few fundamental problems of conventional server-centric IT architecture and concluded by mentioning a few advantages of storage-centric IT architecture based upon the upgrade of an application server. The remaining chapters deal with the concepts and techniques that have already been sketched out and discuss further case studies in detail. The book is structured around the path from the storage device to the application (Figure 1.7).

  In modern IT systems, data is normally stored on hard disks and tapes. It is more economical to procure and manage a few large storage systems than several small ones. This means that the individual disk drives are being replaced by disk subsystems. In contrast to a file server, an intelligent disk subsystem can be visualised as a hard disk server; other servers can use these hard disks that are exported via the storage network just as they can use locally connected disk drives. Chapter 2 shows what modern disk subsystems can do in addition to the instant copy and remote mirroring functions mentioned above.

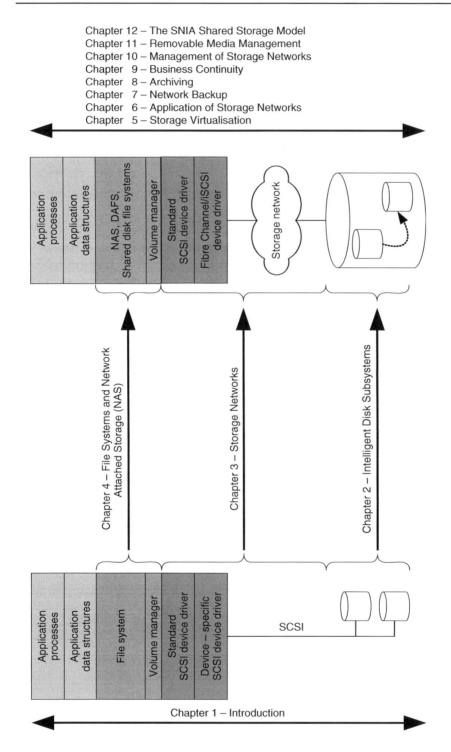

The hardware of tapes and tape libraries changes only slightly as a result of the transition to storage networks, so we only touch upon this subject in the book. In Section 6.2.2 we will discuss the sharing of large tape libraries by several servers and access to these over a storage network and Chapter 11 will present the management of removable media including – among other removable media – tapes and tape libraries.

Fibre Channel has established itself as a technology with which storage networks can be efficiently realised for both open systems (Unix, Windows, Novell Netware, MacOS, OS/400) and mainframes. Where Fibre Channel introduces a new transmission technology, its alternative Internet SCSI (iSCSI) is based upon the proven TCP/IP and Gigabit Ethernet. InfiniBand and Fibre Channel over Ethernet (FCoE) are two additional approaches to consolidate all data traffic (storage, cluster) onto a single transmission technology. All these technologies are subject of Chapter 3.

File systems are of interest in this book for two reasons. First, pre-configured file servers, also known as Network Attached Storage (NAS), have established themselves as an important building block for current IT systems. Storage networks can also be realised using NAS servers. In contrast to the block-oriented data traffic of Fibre Channel, iSCSI and FCoE in this approach whole files or file fragments are transferred.

So-called shared-disk file systems represent the other interesting development in the field of file systems. In shared-disk file systems, several computers can access the same data area in an intelligent disk subsystem over the storage network. The performance of shared-disk file systems is currently significantly better than those of Network File System (NFS), Common Internet File System (CIFS), AppleTalk or the above-mentioned NAS servers. Examples of problems are discussed on the basis of shared-disk file systems that must also be solved in the same manner for comparable applications such as parallel databases. Chapter 4 deals with Network Attached Storage (NAS) and shared-disk file systems.

The first four chapters of the book discuss fundamental components and technologies with regard to storage networks. As storage networks have become more widespread, it has become clear that the implementation of a storage network alone is not sufficient to make efficient use of the resources of ever growing storage networks. Chapter 5 sketches

Figure 1.7 The book is divided into two main parts. The first part discusses the fundamental techniques that underlie storage networks. In particular, these apply to intelligent disk subsystems (Chapter 2), block-oriented storage networks (Chapter 3) and file systems (Chapter 4). We also outline how virtualisation can manage storage more efficiently (Chapter 5). The second part of the book discusses the application of these new technologies. In particular, we discuss standard applications such as storage pooling and clustering (Chapter 6), backup (Chapter 7), archiving (Chapter 8) and business continuity (Chapter 9). These chapters show how storage networks help to develop IT systems that are more flexible, fault-tolerant and powerful than traditional systems. We then discuss the management of storage networks (Chapter 10) and removable media (Chapter 11). Finally, the SNIA Shared Storage Model is presented (Chapter 12).

out the difficulties associated with the use of storage networks and it introduces storage virtualisation – an approach that aims to reduce the total cost of ownership (TCO) for accessing and managing huge amounts of data. It further discusses possible locations for the realisation of storage virtualisation and discusses various alternative approaches to storage virtualisation such as virtualisation on block level and virtualisation on file level or symmetric and asymmetric storage virtualisation.

The first chapters introduce a whole range of new technologies. In Chapter 6 we turn our attention to the application of these new techniques. This chapter uses many case studies to show how storage networks help in the design of IT systems that are more flexible and more fault-tolerant than conventional server-centric IT systems.

Data protection (Backup) is a central application in every IT system. Using network backup systems it is possible to back up heterogeneous IT environments with several thousands of computers largely automatically. Chapter 7 explains the fundamentals of network backup and shows how these new techniques help to back up data even more efficiently. Once again, this clarifies the limitations of server-centric IT architecture and the benefits of the storage-centric IT architecture.

Digital archiving is another important application in storage networks. The law requires that more and more data is kept for years, decades and even longer under strictly regulated conditions. For example, none of the archived data is allowed to be changed or deleted prior to the expiration of the retention times. Due to the long retention times and technical progress, data is required to be copied periodically to new storage media or systems. Chapter 8 discusses the fundamental requirements for digital archiving and presents a number of techniques and solutions that are based on them.

Continuous access to business-critical data and applications, even in a crisis situation, is essential for a company's ability to exist. This does not only apply to those areas one thinks of automatically in this context, such as stock broking, air traffic control, patient data, or Internet companies like Amazon and eBay. An increasing number of smaller and medium-sized companies are now delivering their products to customers worldwide or are tightly integrated into the production processes of larger companies, such as automobile manufacturers, through just-in-time production and contractually agreed delivery times. Chapter 9 introduces the area of business continuity with special consideration of storage networks and discusses different techniques along with possible solutions.

Storage networks are complex systems made up of numerous individual components. As one of the first steps in the management of storage networks it is necessary to understand the current state. This calls for tools that help to answer such questions as 'Which server occupies how much space on which disk subsystem?', 'Which servers are connected to my storage network at all?', 'Which hardware components are in use and how great is the load upon the network?'. In this connection the monitoring of the storage network with regard to faults and performance and capacity bottlenecks of file systems is also important. The second step relates to the automation of the management of storage networks: important subjects are rule-based error handling and the automatic allocation of free storage capacity. Chapter 10 deals in detail with the management of storage networks and in this connection also discusses standards such as Simple Network Management Protocol

(SNMP), Common Information Model/Web-based Enterprise Management (CIM/WBEM) and Storage Management Initiative Specification (SMI-S).

Removable media represent a central component of the storage architecture of large data centres. Storage networks allow several servers, and thus several different applications, to share media and libraries. Therefore, the management of removable media in storage networks is becoming increasingly important. Chapter 11 deals with the requirements of removable media management and it introduces the IEEE 1244 Standard for Removable Media Management.

Storage networks are a complex subject area. There is still a lack of unified terminology, with different manufacturers using the same term to refer to different features and, conversely, describing the same feature using different terms. As a result, it is often unclear what kind of a product is being offered by a manufacturer and which functions a customer can ultimately expect from this product. It is thus difficult for the customer to compare the products of the individual manufacturers and to work out the differences between the alternatives on offer. For this reason, the Technical Council of the Storage Networking Industry Association (SNIA) has introduced the so-called Shared Storage Model in 2001 in order to unify the terminology and descriptive models used by the storage network industry. We introduce this model in Chapter 12.

What doesn't this book cover?

In order to define the content it is also important to know which subjects are not covered:

- *Specific products*
  Product lifecycles are too short for specific products to be discussed in a book. Products change, concepts do not.

- *Economic aspects*
  This book primarily deals with the technical aspects of storage networks. It discusses concepts and approaches to solutions. Prices change very frequently, concepts do not.

- *Excessively technical details*
  The book is an introduction to storage networks. It does not deal with the details necessary for the development of components for storage networks. The communication of the overall picture is more important to us.

- *The planning and implementation of storage networks*
  Planning and implementation require knowledge of specific products, but products change very frequently. Planning and implementation require a great deal of experience. This book, on the other hand, is designed as an introduction. Inexperienced readers should consult experts when introducing a storage network. Furthermore, a specific implementation must always take into account the specific environment in question. It is precisely this that this book cannot do.

Part I

Technologies
for Storage Networks

2

Intelligent Disk Subsystems

Hard disks and tapes are currently the most important media for the storage of data. When storage networks are introduced, the existing small storage devices are replaced by a few large storage systems (storage consolidation). For example, individual hard disks and small disk stacks are replaced by large disk subsystems that can store between a few hundred gigabytes and several ten petabytes of data, depending upon size. Furthermore, they have the advantage that functions such as high availability, high performance, instant copies and remote mirroring are available at a reasonable price even in the field of open systems (Unix, Windows, OS/400, Novell Netware, MacOS). The administration of a few large storage systems is significantly simpler, and thus cheaper, than the administration of many small disk stacks. However, the administrator must plan what he is doing more precisely when working with large disk subsystems. This chapter describes the functions of such modern disk subsystems.

This chapter begins with an overview of the internal structure of a disk subsystem (Section 2.1). We then go on to consider the hard disks used inside the system and the configuration options for the internal I/O channels (Section 2.2). The controller represents the control centre of a disk subsystem. Disk subsystems without controllers are called JBODs (Just a Bunch of Disks); JBODs provide only an enclosure and a common power supply for several hard disks (Section 2.3). So-called RAID (Redundant Array of Independent Disks) controllers bring together several physical hard disks to form virtual hard disks that are faster and more fault-tolerant than individual physical hard disks (Sections 2.4 and 2.5). Some RAID controllers use a cache to further accelerate write and read access to the server (Section 2.6). In addition, intelligent controllers provide services such as instant copy and remote mirroring (Section 2.7). The conclusion to this chapter

Storage Networks Explained: Basics and Application of Fibre Channel SAN, NAS, iSCSI, InfiniBand and FCoE, Second Edition
U. Troppens R. Erkens W. Müller-Friedt N. Haustein R. Wolafka © 2009 John Wiley & Sons, Ltd

summarises the measures discussed for increasing the fault-tolerance of intelligent disk subsystems (Section 2.8).

2.1 ARCHITECTURE OF INTELLIGENT DISK SUBSYSTEMS

In contrast to a file server, a disk subsystem can be visualised as a hard disk server. Servers are connected to the connection port of the disk subsystem using standard I/O techniques such as Small Computer System Interface (SCSI), Fibre Channel or Internet SCSI (iSCSI) and can thus use the storage capacity that the disk subsystem provides (Figure 2.1). The internal structure of the disk subsystem is completely hidden from the server, which sees only the hard disks that the disk subsystem provides to the server.

The connection ports are extended to the hard disks of the disk subsystem by means of internal I/O channels (Figure 2.2). In most disk subsystems there is a controller between the connection ports and the hard disks. The controller can significantly increase the data availability and data access performance with the aid of a so-called RAID procedure. Furthermore, some controllers realise the copying services instant copy and remote mirroring and further additional services. The controller uses a cache in an attempt to accelerate read and write accesses to the server.

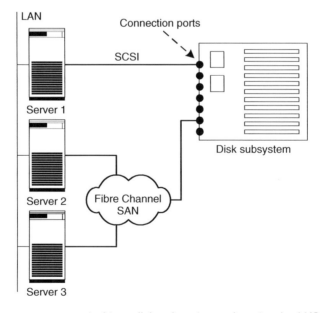

Figure 2.1   Servers are connected to a disk subsystem using standard I/O techniques. The figure shows a server that is connected by SCSI. Two others are connected by Fibre Channel SAN.

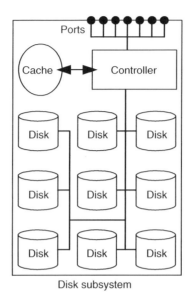

Disk subsystem

Figure 2.2   Servers are connected to the disk subsystems via the ports. Internally, the disk subsystem consists of hard disks, a controller, a cache and internal I/O channels.

Disk subsystems are available in all sizes. Small disk subsystems have one to two connection ports for servers or storage networks, six to eight hard disks and, depending on the disk capacity, storage capacity of a few terabytes. Large disk subsystems have multiple ten connection ports for servers and storage networks, redundant controllers and multiple I/O channels. A considerably larger number of servers can access a subsystem through a connection over a storage network. Large disk subsystems can store up to a petabyte of data and, depending on the supplier, can weigh well over a tonne. The dimensions of a large disk subsystem are comparable to those of a wardrobe.

Figure 2.2 shows a simplified schematic representation. The architecture of real disk subsystems is more complex and varies greatly. Ultimately, however, it will always include the components shown in Figure 2.2. The simplified representation in Figure 2.2 provides a sufficient basis for the further discussion in the book.

Regardless of storage networks, most disk subsystems have the advantage that free disk space can be flexibly assigned to each server connected to the disk subsystem (storage pooling). Figure 2.3 refers back once again to the example of Figure 1.2. In Figure 1.2 it is not possible to assign more storage to server 2, even though free space is available on servers 1 and 3. In Figure 2.3 this is not a problem. All servers are either directly connected to the disk subsystem or indirectly connected via a storage network. In this configuration each server can be assigned free storage. Incidentally, free storage capacity should be understood to mean both hard disks that have already been installed and have not yet been used and also free slots for hard disks that have yet to be installed.

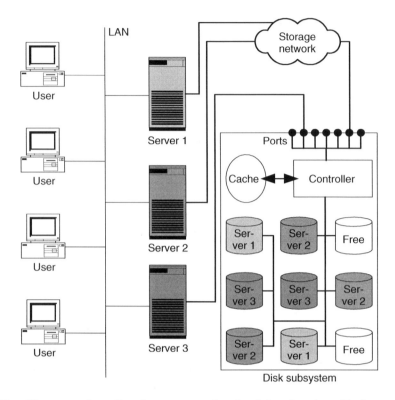

Figure 2.3   All servers share the storage capacity of a disk subsystem. Each server can be assigned free storage more flexibly as required.

2.2   HARD DISKS AND INTERNAL I/O CHANNELS

The controller of the disk subsystem must ultimately store all data on physical hard disks. Standard hard disks that range in size from 36 GB to 1 TB are currently (2009) used for this purpose. Since the maximum number of hard disks that can be used is often limited, the size of the hard disk used gives an indication of the maximum capacity of the overall disk subsystem.

When selecting the size of the internal physical hard disks it is necessary to weigh the requirements of maximum performance against those of the maximum capacity of the overall system. With regard to performance it is often beneficial to use smaller hard disks at the expense of the maximum capacity: given the same capacity, if more hard disks are available in a disk subsystem, the data is distributed over several hard disks and thus the overall load is spread over more arms and read/write heads and usually over more I/O channels (Figure 2.4). For most applications, medium-sized hard disks are sufficient. Only

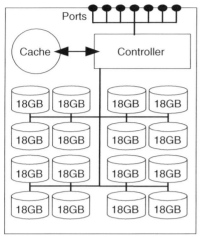
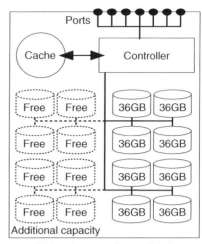

Disk subsystem (small disks)   Disk subsystem (large disks)

Figure 2.4 If small internal hard disks are used, the load is distributed over more hard disks and thus over more read and write heads. On the other hand, the maximum storage capacity is reduced, since in both disk subsystems only 16 hard disks can be fitted.

for applications with extremely high performance requirements should smaller hard disks be considered. However, consideration should be given to the fact that more modern, larger hard disks generally have shorter seek times and larger caches, so it is necessary to carefully weigh up which hard disks will offer the highest performance for a certain load profile in each individual case.

Standard I/O techniques such as SCSI, Fibre Channel, increasingly Serial ATA (SATA) and Serial Attached SCSI (SAS) and, still to a degree, Serial Storage Architecture (SSA) are being used for internal I/O channels between connection ports and controller as well as between controller and internal hard disks. Sometimes, however, proprietary – i.e., manufacturer-specific – I/O techniques are used. Regardless of the I/O technology used, the I/O channels can be designed with built-in redundancy in order to increase the fault-tolerance of a disk subsystem. The following cases can be differentiated here:

- *Active*
  In active cabling the individual physical hard disks are only connected via one I/O channel (Figure 2.5, left). If this access path fails, then it is no longer possible to access the data.

- *Active/passive*
  In active/passive cabling the individual hard disks are connected via two I/O channels (Figure 2.5, right). In normal operation the controller communicates with the hard disks via the first I/O channel and the second I/O channel is not used. In the event of the

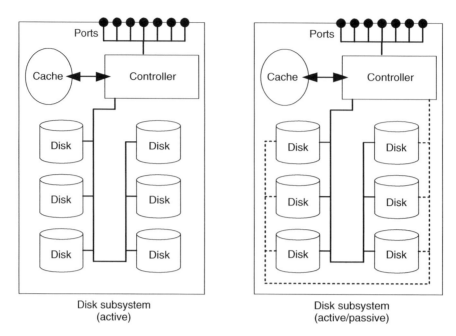

Figure 2.5   In active cabling all hard disks are connected by just one I/O channel. In active/passive cabling all hard disks are additionally connected by a second I/O channel. If the primary I/O channel fails, the disk subsystem switches to the second I/O channel.

failure of the first I/O channel, the disk subsystem switches from the first to the second I/O channel.

- *Active/active (no load sharing)*
  In this cabling method the controller uses both I/O channels in normal operation (Figure 2.6, left). The hard disks are divided into two groups: in normal operation the first group is addressed via the first I/O channel and the second via the second I/O channel. If one I/O channel fails, both groups are addressed via the other I/O channel.

- *Active/active (load sharing)*
  In this approach all hard disks are addressed via both I/O channels in normal operation (Figure 2.6, right). The controller divides the load dynamically between the two I/O channels so that the available hardware can be optimally utilised. If one I/O channel fails, then the communication goes through the other channel only.

Active cabling is the simplest and thus also the cheapest to realise but offers no protection against failure. Active/passive cabling is the minimum needed to protect against failure, whereas active/active cabling with load sharing best utilises the underlying hardware.

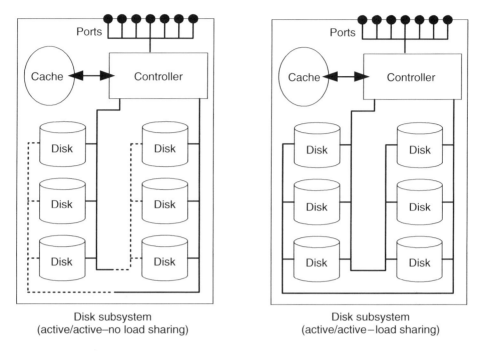

Disk subsystem
(active/active–no load sharing)
                                                Disk subsystem
                                                (active/active–load sharing)

Figure 2.6   Active/active cabling (no load sharing) uses both I/O channels at the same time. However, each disk is addressed via one I/O channel only, switching to the other channel in the event of a fault. In active/active cabling (load sharing) hard disks are addressed via both I/O channels.

2.3   JBOD: JUST A BUNCH OF DISKS

If we compare disk subsystems with regard to their controllers we can differentiate between three levels of complexity: (1) no controller; (2) RAID controller (Sections 2.4 and 2.5); and (3) intelligent controller with additional services such as instant copy and remote mirroring (Section 2.7).

If the disk subsystem has no internal controller, it is only an enclosure full of disks (JBODs). In this instance, the hard disks are permanently fitted into the enclosure and the connections for I/O channels and power supply are taken outwards at a single point. Therefore, a JBOD is simpler to manage than a few loose hard disks. Typical JBOD disk subsystems have space for 8 or 16 hard disks. A connected server recognises all these hard disks as independent disks. Therefore, 16 device addresses are required for a JBOD disk subsystem incorporating 16 hard disks. In some I/O techniques such as SCSI (Section 3.2) and Fibre Channel arbitrated loop (Section 3.3.6), this can lead to a bottleneck at device addresses.

In contrast to intelligent disk subsystems, a JBOD disk subsystem in particular is not capable of supporting RAID or other forms of virtualisation. If required, however, these can be realised outside the JBOD disk subsystem, for example, as software in the server (Section 5.1) or as an independent virtualisation entity in the storage network (Section 5.6.3).

2.4 STORAGE VIRTUALISATION USING RAID

A disk subsystem with a RAID controller offers greater functional scope than a JBOD disk subsystem. RAID was originally developed at a time when hard disks were still very expensive and less reliable than they are today. RAID was originally called 'Redundant Array of Inexpensive Disks'. Today RAID stands for 'Redundant Array of Independent Disks'. Disk subsystems that support RAID are sometimes also called RAID arrays.

RAID has two main goals: to increase performance by striping and to increase fault-tolerance by redundancy. Striping distributes the data over several hard disks and thus distributes the load over more hardware. Redundancy means that additional information is stored so that the operation of the application itself can continue in the event of the failure of a hard disk. You cannot increase the performance of an individual hard disk any more than you can improve its fault-tolerance. Individual physical hard disks are slow and have a limited life-cycle. However, through a suitable combination of physical hard disks it is possible to significantly increase the fault-tolerance and performance of the system as a whole.

The bundle of physical hard disks brought together by the RAID controller are also known as virtual hard disks. A server that is connected to a RAID system sees only the virtual hard disk; the fact that the RAID controller actually distributes the data over several physical hard disks is completely hidden to the server (Figure 2.7). This is only visible to the administrator from outside.

A RAID controller can distribute the data that a server writes to the virtual hard disk amongst the individual physical hard disks in various manners. These different procedures are known as RAID levels. Section 2.5 explains various RAID levels in detail.

One factor common to almost all RAID levels is that they store redundant information. If a physical hard disk fails, its data can be reconstructed from the hard disks that remain intact. The defective hard disk can even be replaced by a new one during operation if a disk subsystem has the appropriate hardware. Then the RAID controller reconstructs the data of the exchanged hard disk. This process remains hidden to the server apart from a possible reduction in performance: the server can continue to work uninterrupted on the virtual hard disk.

Modern RAID controllers initiate this process automatically. This requires the definition of so-called hot spare disks (Figure 2.8). The hot spare disks are not used in normal operation. If a disk fails, the RAID controller immediately begins to copy the data of the remaining intact disk onto a hot spare disk. After the replacement of the defective disk,

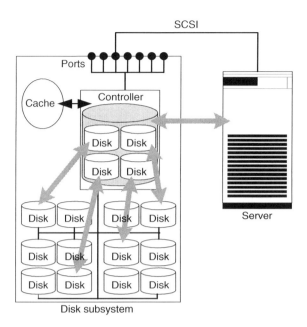

Figure 2.7 The RAID controller combines several physical hard disks to create a virtual hard disk. The server sees only a single virtual hard disk. The controller hides the assignment of the virtual hard disk to the individual physical hard disks.

this is included in the pool of hot spare disks. Modern RAID controllers can manage a common pool of hot spare disks for several virtual RAID disks. Hot spare disks can be defined for all RAID levels that offer redundancy.

The recreation of the data from a defective hard disk takes place at the same time as write and read operations of the server to the virtual hard disk, so that from the point of view of the server, performance reductions at least can be observed. Modern hard disks come with self-diagnosis programs that report an increase in write and read errors to the system administrator in plenty of time: 'Caution! I am about to depart this life. Please replace me with a new disk. Thank you!' To this end, the individual hard disks store the data with a redundant code such as the Hamming code. The Hamming code permits the correct recreation of the data, even if individual bits are changed on the hard disk. If the system is looked after properly you can assume that the installed physical hard disks will hold out for a while. Therefore, for the benefit of higher performance, it is generally an acceptable risk to give access by the server a higher priority than the recreation of the data of an exchanged physical hard disk.

A further side-effect of the bringing together of several physical hard disks to form a virtual hard disk is the higher capacity of the virtual hard disks. As a result, less device addresses are used up in the I/O channel and thus the administration of the server is also simplified, because less hard disks (drive letters or volumes) need to be used.

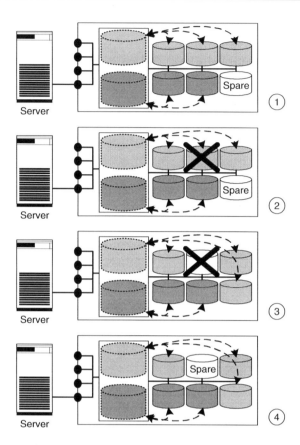

Figure 2.8   Hot spare disk: The disk subsystem provides the server with two virtual disks for which a common hot spare disk is available (1). Due to the redundant data storage the server can continue to process data even though a physical disk has failed, at the expense of a reduction in performance (2). The RAID controller recreates the data from the defective disk on the hot spare disk (3). After the defective disk has been replaced a hot spare disk is once again available (4).

2.5   DIFFERENT RAID LEVELS IN DETAIL

RAID has developed since its original definition in 1987. Due to technical progress some RAID levels are now practically meaningless, whilst others have been modified or added at a later date. This section introduces the RAID levels that are currently the most significant in practice. We will not introduce RAID levels that represent manufacturer-specific variants and variants that only deviate slightly from the basic forms mentioned in the following.

2.5.1   RAID 0: block-by-block striping

RAID 0 distributes the data that the server writes to the virtual hard disk onto one physical hard disk after another block-by-block (block-by-block striping). Figure 2.9 shows a RAID array with four physical hard disks. In Figure 2.9 the server writes the blocks A, B, C, D, E, etc. onto the virtual hard disk one after the other. The RAID controller distributes the sequence of blocks onto the individual physical hard disks: it writes the first block, A, to the first physical hard disk, the second block, B, to the second physical hard disk, block C to the third and block D to the fourth. Then it begins to write to the first physical hard disk once again, writing block E to the first disk, block F to the second, and so on.

RAID 0 increases the performance of the virtual hard disk as follows: the individual hard disks can exchange data with the RAID controller via the I/O channel significantly more quickly than they can write to or read from the rotating disk. In Figure 2.9 the RAID controller sends the first block, block A, to the first hard disk. This takes some time to write the block to the disk. Whilst the first disk is writing the first block to the physical hard disk, the RAID controller is already sending the second block, block B, to the second hard disk and block C to the third hard disk. In the meantime, the first two physical hard disks are still engaged in depositing their respective blocks onto the physical hard disk. If the RAID controller now sends block E to the first hard disk, then this has written block A at least partially, if not entirely, to the physical hard disk.

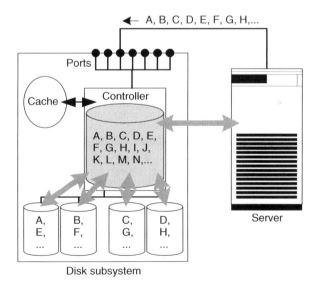

Figure 2.9   RAID 0 (striping): As in all RAID levels, the server sees only the virtual hard disk. The RAID controller distributes the write operations of the server amongst several physical hard disks. Parallel writing means that the performance of the virtual hard disk is higher than that of the individual physical hard disks.

In the example, it was possible to increase the throughput fourfold in 2002: Individual hard disks were able to achieve a throughput of around 50 MB/s. The four physical hard disks achieve a total throughput of around 4×50 MB/s ≈ 200 MB/s. In those days I/O techniques such as SCSI or Fibre Channel achieve a throughput of 160 MB/s or 200 MB/s. If the RAID array consisted of just three physical hard disks the total throughput of the hard disks would be the limiting factor. If, on the other hand, the RAID array consisted of five physical hard disks the I/O path would be the limiting factor. With five or more hard disks, therefore, performance increases are only possible if the hard disks are connected to different I/O paths so that the load can be striped not only over several physical hard disks, but also over several I/O paths.

RAID 0 increases the performance of the virtual hard disk, but not its fault-tolerance. If a physical hard disk is lost, all the data on the virtual hard disk is lost. To be precise, therefore, the 'R' for 'Redundant' in RAID is incorrect in the case of RAID 0, with 'RAID 0' standing instead for 'zero redundancy'.

2.5.2 RAID 1: block-by-block mirroring

In contrast to RAID 0, in RAID 1 fault-tolerance is of primary importance. The basic form of RAID 1 brings together two physical hard disks to form a virtual hard disk by mirroring the data on the two physical hard disks. If the server writes a block to the virtual hard disk, the RAID controller writes this block to both physical hard disks (Figure 2.10). The individual copies are also called mirrors. Normally, two or sometimes three copies of the data are kept (three-way mirror).

In a normal operation with pure RAID 1, performance increases are only possible in read operations. After all, when reading the data the load can be divided between the two disks. However, this gain is very low in comparison to RAID 0. When writing with RAID 1 it tends to be the case that reductions in performance may even have to be taken into account. This is because the RAID controller has to send the data to both hard disks. This disadvantage can be disregarded for an individual write operation, since the capacity of the I/O channel is significantly higher than the maximum write speed of the two hard disks put together. However, the I/O channel is under twice the load, which hinders other data traffic using the I/O channel at the same time.

2.5.3 RAID 0+1/RAID 10: striping and mirroring combined

The problem with RAID 0 and RAID 1 is that they increase either performance (RAID 0) or fault-tolerance (RAID 1). However, it would be nice to have both performance and fault-tolerance. This is where RAID 0+1 and RAID 10 come into play. These two RAID levels combine the ideas of RAID 0 and RAID 1.

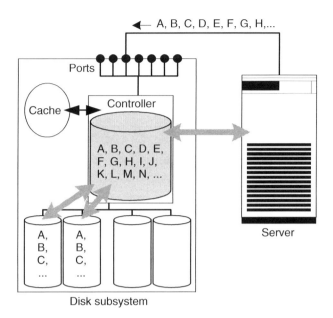

← A, B, C, D, E, F, G, H,...

Ports

Cache

Controller

A, B, C, D, E, F, G, H, I, J, K, L, M, N, ...

A, B, C, ...

A, B, C, ...

Server

Disk subsystem

Figure 2.10   RAID 1 (mirroring): As in all RAID levels, the server sees only the virtual hard disk. The RAID controller duplicates each of the server's write operations onto two physical hard disks. After the failure of one physical hard disk the data can still be read from the other disk.

RAID 0+1 and RAID 10 each represent a two-stage virtualisation hierarchy. Figure 2.11 shows the principle behind RAID 0+1 (mirrored stripes). In the example, eight physical hard disks are used. The RAID controller initially brings together each four physical hard disks to form a total of two virtual hard disks that are only visible within the RAID controller by means of RAID 0 (striping). In the second level, it consolidates these two virtual hard disks into a single virtual hard disk by means of RAID 1 (mirroring); only this virtual hard disk is visible to the server.

In RAID 10 (striped mirrors) the sequence of RAID 0 (striping) and RAID 1 (mirroring) is reversed in relation to RAID 0+1 (mirrored stripes). Figure 2.12 shows the principle underlying RAID 10 based again on eight physical hard disks. In RAID 10 the RAID controller initially brings together the physical hard disks in pairs by means of RAID 1 (mirroring) to form a total of four virtual hard disks that are only visible within the RAID controller. In the second stage, the RAID controller consolidates these four virtual hard disks into a virtual hard disk by means of RAID 0 (striping). Here too, only this last virtual hard disk is visible to the server.

In both RAID 0+1 and RAID 10 the server sees only a single hard disk, which is larger, faster and more fault-tolerant than a physical hard disk. We now have to ask the question: which of the two RAID levels, RAID 0+1 or RAID 10, is preferable?

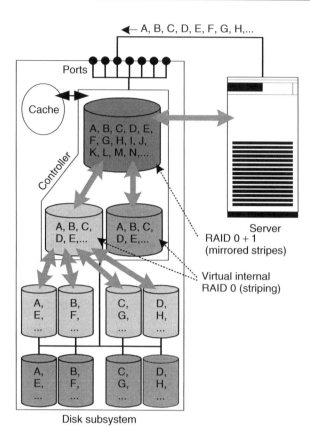

Disk subsystem

Figure 2.11 RAID 0+1 (mirrored stripes): As in all RAID levels, the server sees only the virtual hard disk. Internally, the RAID controller realises the virtual disk in two stages: in the first stage it brings together every four physical hard disks into one virtual hard disk that is only visible within the RAID controller by means of RAID 0 (striping); in the second stage it consolidates these two virtual hard disks by means of RAID 1 (mirroring) to form the hard disk that is visible to the server.

The question can be answered by considering that when using RAID 0 the failure of a hard disk leads to the loss of the entire virtual hard disk. In the example relating to RAID 0+1 (Figure 2.11) the failure of a physical hard disk is thus equivalent to the effective failure of four physical hard disks (Figure 2.13). If one of the other four physical hard disks is lost, then the data is lost. In principle it is sometimes possible to reconstruct the data from the remaining disks, but the RAID controllers available on the market cannot do this particularly well.

In the case of RAID 10, on the other hand, after the failure of an individual physical hard disk, the additional failure of a further physical hard disk – with the exception of the

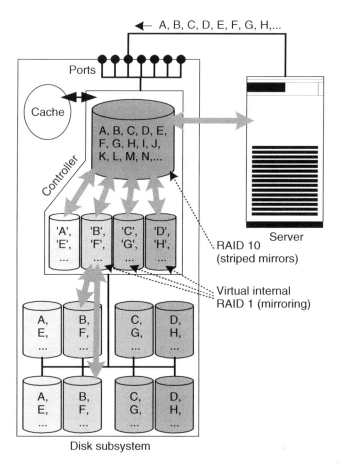

Figure 2.12   RAID 10 (striped mirrors): As in all RAID levels, the server sees only the virtual hard disk. Here too, we proceed in two stages. The sequence of striping and mirroring is reversed in relation to RAID 0+1. In the first stage the controller links every two physical hard disks by means of RAID 1 (mirroring) to a virtual hard disk, which it unifies by means of RAID 0 (striping) in the second stage to form the hard disk that is visible to the server.

corresponding mirror – can be withstood (Figure 2.14). RAID 10 thus has a significantly higher fault-tolerance than RAID 0+1. In addition, the cost of restoring the RAID system after the failure of a hard disk is much lower in the case of RAID 10 than RAID 0+1. In RAID 10 only one physical hard disk has to be recreated. In RAID 0+1, on the other hand, a virtual hard disk must be recreated that is made up of four physical disks. However, the cost of recreating the defective hard disk can be significantly reduced because a physical hard disk is exchanged as a preventative measure when the number of read errors start to increase. In this case it is sufficient to copy the data from the old disk to the new.

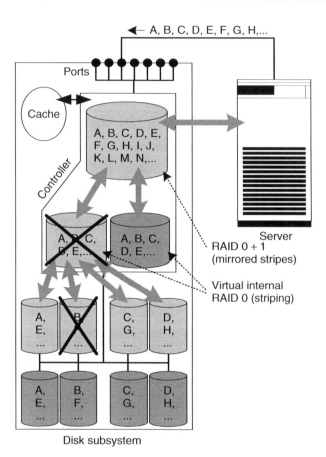

Figure 2.13 The consequences of the failure of a physical hard disk in RAID 0+1 (mirrored stripes) are relatively high in comparison to RAID 10 (striped mirrors). The failure of a physical hard disk brings about the failure of the corresponding internal RAID 0 disk, so that in effect half of the physical hard disks have failed. The recovery of the data from the failed disk is expensive.

However, things look different if the performance of RAID 0+1 is compared with the performance of RAID 10. In Section 5.1 we discuss a case study in which the use of RAID 0+1 is advantageous.

With regard to RAID 0+1 and RAID 10 it should be borne in mind that the two RAID procedures are often confused. Therefore the answer 'We use RAID 10!' or 'We use RAID 0+1' does not always provide the necessary clarity. In discussions it is better to ask if mirroring takes place first and the mirror is then striped or if striping takes place first and the stripes are then mirrored.

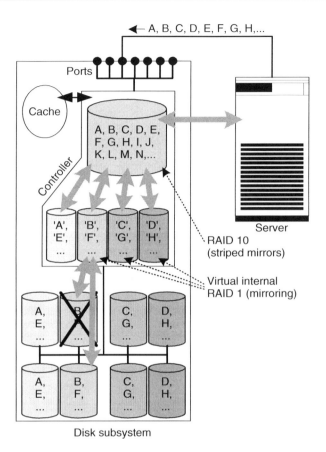

Figure 2.14   In RAID 10 (striped mirrors) the consequences of the failure of a physical hard disk are not as serious as in RAID 0+1 (mirrored stripes). All virtual hard disks remain intact. The recovery of the data from the failed hard disk is simple.

2.5.4   RAID 4 and RAID 5: parity instead of mirroring

RAID 10 provides excellent performance at a high level of fault-tolerance. The problem with this is that mirroring using RAID 1 means that all data is written to the physical hard disk twice. RAID 10 thus doubles the required storage capacity.

The idea of RAID 4 and RAID 5 is to replace all mirror disks of RAID 10 with a single parity hard disk. Figure 2.15 shows the principle of RAID 4 based upon five physical hard disks. The server again writes the blocks A, B, C, D, E, etc. to the virtual hard disk sequentially. The RAID controller stripes the data blocks over the first four physical hard disks. Instead of mirroring all data onto the further four physical hard disks, as in

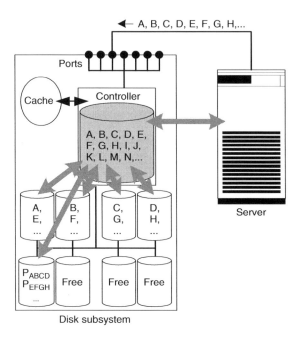

Figure 2.15   RAID 4 (parity disk) is designed to reduce the storage requirement of RAID 0+1 and RAID 10. In the example, the data blocks are distributed over four physical hard disks by means of RAID 0 (striping). Instead of mirroring all data once again, only a parity block is stored for each four blocks.

RAID 10, the RAID controller calculates a parity block for every four blocks and writes this onto the fifth physical hard disk. For example, the RAID controller calculates the parity block P_{ABCD} for the blocks A, B, C and D. If one of the four data disks fails, the RAID controller can reconstruct the data of the defective disks using the three other data disks and the parity disk. In comparison to the examples in Figures 2.11 (RAID 0+1) and 2.12 (RAID 10), RAID 4 saves three physical hard disks. As in all other RAID levels, the server again sees only the virtual disk, as if it were a single physical hard disk.

From a mathematical point of view the parity block is calculated with the aid of the logical XOR operator (Exclusive OR). In the example from Figure 2.15, for example, the equation $P_{ABCD} = A$ XOR B XOR C XOR D applies.

The space saving offered by RAID 4 and RAID 5, which remains to be discussed, comes at a price in relation to RAID 10. Changing a data block changes the value of the associated parity block. This means that each write operation to the virtual hard disk requires (1) the physical writing of the data block, (2) the recalculation of the parity block and (3) the physical writing of the newly calculated parity block. This extra cost for write operations in RAID 4 and RAID 5 is called the write penalty of RAID 4 or the write penalty of RAID 5.

The cost for the recalculation of the parity block is relatively low due to the mathematical properties of the XOR operator. If the block A is overwritten by block $\tilde{A}$ and Δ is the difference between the old and new data block, then $\Delta = A$ XOR $\tilde{A}$. The new parity block $\tilde{P}$ can now simply be calculated from the old parity block P and Δ, i.e. $\tilde{P} = P$ XOR Δ. Proof of this property can be found in Appendix A. Therefore, if P_{ABCD} is the parity block for the data blocks A, B, C and D, then after the data block A has been changed, the new parity block can be calculated without knowing the remaining blocks B, C and D. However, the old block A must be read in before overwriting the physical hard disk in the controller, so that this can calculate the difference Δ.

When processing write commands for RAID 4 and RAID 5 arrays, RAID controllers use the above-mentioned mathematical properties of the XOR operation for the recalculation of the parity block. Figure 2.16 shows a server that changes block D on the virtual hard disk. The RAID controller reads the data block and the associated parity block from the disk in question into its cache. Then it uses the XOR operation to calculate the difference

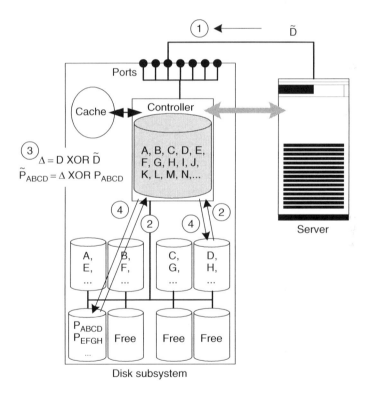

Figure 2.16 Write penalty of RAID 4 and RAID 5: The server writes a changed data block (1). The RAID controller reads in the old data block and the associated old parity block (2) and calculates the new parity block (3). Finally it writes the new data block and the new parity block onto the physical hard disk in question (4).

between the old and the new parity block, i.e. $\Delta = D \text{ XOR } \tilde{D}$, and from this the new parity block $\tilde{P}_{ABCD}$ by means of $\tilde{P}_{ABCD} = P_{ABCD} \text{ XOR } \Delta$. Therefore it is not necessary to read in all four associated data blocks to recalculate the parity block. To conclude the write operation to the virtual hard disk, the RAID controller writes the new data block and the recalculated parity block onto the physical hard disks in question.

Advanced RAID 4 and RAID 5 implementations are capable of reducing the write penalty even further for certain load profiles. For example, if large data quantities are written sequentially, then the RAID controller can calculate the parity blocks from the data flow without reading the old parity block from the disk. If, for example, the blocks E, F, G and H in Figure 2.15 are written in one go, then the controller can calculate the parity block P_{EFGH} from them and overwrite this without having previously read in the old value. Likewise, a RAID controller with a suitably large cache can hold frequently changed parity blocks in the cache after writing to the disk, so that the next time one of the data blocks in question is changed there is no need to read in the parity block. In both cases the I/O load is now lower than in the case of RAID 10. In the example only five physical blocks now need to be written instead of eight as is the case with RAID 10.

RAID 4 saves all parity blocks onto a single physical hard disk. For the example in Figure 2.15 this means that the write operations for the data blocks are distributed over four physical hard disks. However, the parity disk has to handle the same number of write operations all on its own. Therefore, the parity disk become the performance bottleneck of RAID 4 if there are a high number of write operations.

To get around this performance bottleneck, RAID 5 distributes the parity blocks over all hard disks. Figure 2.17 illustrates the procedure. As in RAID 4, the RAID controller writes the parity block P_{ABCD} for the blocks A, B, C and D onto the fifth physical hard disk. Unlike RAID 4, however, in RAID 5 the parity block P_{EFGH} moves to the fourth physical hard disk for the next four blocks E, F, G, H.

RAID 4 and RAID 5 distribute the data blocks over many physical hard disks. Therefore, the read performance of RAID 4 and RAID 5 is as good as that of RAID 0 and almost as good as that of RAID 10. As discussed, the write performance of RAID 4 and RAID 5 suffers from the write penalty; in RAID 4 there is an additional bottleneck caused by the parity disk. Therefore, RAID 4 is seldom used in practice because RAID 5 accomplishes more than RAID 4 with the same amount of physical resources (see also Section 2.5.6).

RAID 4 and RAID 5 can withstand the failure of a physical hard disk. Due to the use of parity blocks, the data on the defective hard disk can be restored with the help of other hard disks. In contrast to RAID 10, the failure of an individual sector of the remaining physical hard disks always results in data loss. This is compensated for with RAID 6, whereby a second parity hard disk is kept so that data is protected twice (Section 2.5.5).

In RAID 4 and RAID 5 the recovery of a defective physical hard disk is significantly more expensive than is the case for RAID 1 and RAID 10. In the latter two RAID levels only the mirror of the defective disk needs to be copied to the replaced disk. In RAID 4 and RAID 5, on the other hand, the RAID controller has to read the data from all disks, use this to recalculate the lost data blocks and parity blocks, and then write these blocks to the

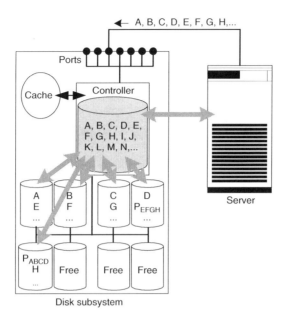

Figure 2.17 RAID 5 (striped parity): In RAID 4 each write access by the server is associated with a write operation to the parity disk for the update of parity information. RAID 5 distributes the load of the parity disk over all physical hard disks.

replacement disk. As in RAID 0+1 this high cost can be avoided by replacing a physical hard disk as a precaution as soon as the rate of read errors increases. If this is done, it is sufficient to copy the data from the hard disk to be replaced onto the new hard disk.

If the fifth physical hard disk has to be restored in the examples from Figure 2.15 (RAID 4) and Figure 2.17 (RAID 5), the RAID controller must first read the blocks A, B, C and D from the physical hard disks, recalculate the parity block P_{ABCD} and then write to the exchanged physical hard disk. If a data block has to be restored, only the calculation rule changes. If, in the example, the third physical hard disk is to be recreated, the controller would first have to read in the blocks A, B, D and P_{ABCD}, use these to reconstruct block C and write this to the replaced disk.

2.5.5 RAID 6: double parity

'How can it be that I am losing data even though it is protected by RAID 5?' Unfortunately, there are many storage administrators who do not deal with this question until *after* they have lost data despite using RAID 5 and not *before* this loss has taken place. With the size of the hard disks available today large quantities of data can quickly be lost. The main cause of this unpleasant surprise is that RAID 5 offers a high availability solution for the failure of individual physical hard disks, whereas many people expect RAID 5 to provide a business continuity solution for the failure of multiple hard disks and, consequently, the entire RAID 5 array. Section 9.4.2 discusses this difference in detail.

Table 2.1 shows the failure probability for some RAID 5 configurations. The calculations are based on the technical data for the type of Fibre Channel hard disk commercially used in 2007. For example, the average serviceability of the disk was assumed to be 1,400,000 hours (approximately 160 years). In the model only the size of the hard disks has been varied. The first four columns of the table show the number of physical disks in a RAID array, the size of the individual disks, the net capacity of the RAID array and the ratio of net capacity to gross capacity (storage efficiency).

The fifth column shows the bit error rate (BER). The Fibre Channel hard disks commonly used in 2007 have a BER of 10^{-15} and the SATA hard disks a BER of 10^{-14}. A BER of 10^{-15} means that a sector of 100 TB of read-in data cannot be read, and a BER of 10^{-14} means that a sector of 10 TB of read-in data cannot be read. Hard disks with lower BER rates can be produced today. However, the advances in disk technology have mainly been focussed on increasing the capacity of individual hard disks and not on improving BER rates.

Table 2.1 The table shows the failure probability for different RAID 5 configurations (see text). In 2007, for example, a typical SATA disk had a size of 512 GB and a BER of 1E-15. At 93.8% a RAID array consisting of 16 such hard disks has a high level of storage efficiency, whereas the probability of data loss due to a stripe-kill error is around 22 years (line 9). This means that anyone who operates ten such arrays will on average lose one array every 2 years.

No. of disks	Disk size GB	Array size GB	Efficiency	BER	MTTDL BER	MTTDL DD	MTTDL Total
8	512	3.584	87.5%	1E-15	706	73.109	700
8	256	1.792	87.5%	1E-15	1.403	146.217	1.389
8	1.024	7.168	87.5%	1E-15	358	36.554	355
4	512	1.536	75.0%	1E-15	3.269	341.173	3.238
16	512	7.680	93.8%	1E-15	168	17.059	166
8	256	1.792	87.5%	1E-14	149	146.217	149
8	512	3.584	87.5%	1E-14	80	73.109	80
8	1.024	7.168	87.5%	1E-14	46	36.554	46
16	512	3.584	93.8%	1E-14	22	17.059	22
16	1.024	15.360	93.8%	1E-14	14	8.529	14
8	256	1.792	87.5%	1E-16	13.937	146.217	12.724
8	512	3.584	87.5%	1E-16	6.973	73.109	6.336
8	1.024	7.168	87.5%	1E-16	3.492	36.554	3.187
16	1.024	15.360	93.8%	1E-16	817	8.529	746

The last three columns show the probability of data loss in RAID 5 arrays. If a hard disk fails in a RAID 5 array, the data of all the intact disks is read so that the data of the faulty disk can be reconstructed (Section 2.5.4). This involves large quantities of data being read, thus increasing the probability of a corrupt sector surfacing during the reconstruction. This means that two blocks belonging to the same parity group are lost and therefore cannot be reconstructed again (mean time to data loss due to BER, MTTDL BER). Sometimes a second hard disk fails during the reconstruction. This too will lead to data loss (mean time to data loss due to double disk failure, MTTDL DD). The last column shows the overall probability of data loss in a RAID 5 array (mean time to data loss, MTTDL).

Line 1 shows the reference configuration for the comparison: a RAID 5 array with eight hard disks at 512 GB. The helper lines in table 2.1 show how different disk sizes or different array sizes have an effect on the probability of failure. The next two blocks show the probability of failure with a BER of 10^{-14}, which was common for SATA hard disks in 2007. In the model a life expectancy of 1,400,000 hours was assumed for these disks. In comparison, real SATA hard disks have a noticeably lower life expectancy and, consequently, an even higher probability of failure. Furthermore, SATA hard disks are often designed for office use (8×5 operations $= 5$ days $\times 8$ hours). Hence, the likelihood of failures occurring earlier increases when these hard disks are used on a continuous basis in data centres (24×7 operations). Finally, the last block shows how the probability of failure would fall if hard disks were to have a BER of 10^{-16}.

RAID 5 arrays cannot correct the double failures described above. The table shows that there is a clear rise in the probability of data loss in RAID 5 arrays due to the increasing capacity of hard disks. RAID 6 offers a compromise between RAID 5 and RAID 10 by adding a second parity hard disk to extend RAID 5, which then uses less storage capacity than RAID 10. There are different approaches available today for calculating the two parity blocks of a parity group. However, none of these procedures has been adopted yet as an industry standard. Irrespective of an exact procedure, RAID 6 has a poor write performance because the write penalty for RAID 5 strikes twice (Section 2.5.4).

2.5.6 RAID 2 and RAID 3

When introducing the RAID levels we are sometimes asked: 'and what about RAID 2 and RAID 3?'. The early work on RAID began at a time when disks were not yet very reliable: bit errors were possible that could lead to a written 'one' being read as 'zero' or a written 'zero' being read as 'one'. In RAID 2 the Hamming code is used, so that redundant information is stored in addition to the actual data. This additional data permits the recognition of read errors and to some degree also makes it possible to correct them. Today, comparable functions are performed by the controller of each individual hard disk, which means that RAID 2 no longer has any practical significance.

Like RAID 4 or RAID 5, RAID 3 stores parity data. RAID 3 distributes the data of a block amongst all the disks of the RAID 3 system so that, in contrast to RAID 4 or RAID 5, all disks are involved in every read or write access. RAID 3 only permits the reading and writing of whole blocks, thus dispensing with the write penalty that occurs in RAID 4 and RAID 5. The writing of individual blocks of a parity group is thus not possible. In addition, in RAID 3 the rotation of the individual hard disks is synchronised so that the data of a block can truly be written simultaneously. RAID 3 was for a long time called the recommended RAID level for sequential write and read profiles such as data mining and video processing. Current hard disks come with a large cache of their own, which means that they can temporarily store the data of an entire track, and they have significantly higher rotation speeds than the hard disks of the past. As a result of these innovations, other RAID levels are now suitable for sequential load profiles, meaning that RAID 3 is becoming less and less important.

2.5.7 A comparison of the RAID levels

The various RAID levels raise the question of which RAID level should be used when. Table 2.2 compares the criteria of fault-tolerance, write performance, read performance and space requirement for the individual RAID levels. The evaluation of the criteria can be found in the discussion in the previous sections.

CAUTION PLEASE: The comparison of the various RAID levels discussed in this section is only applicable to the theoretical basic forms of the RAID level in question. In practice, manufacturers of disk subsystems have design options in

- *the selection of the internal physical hard disks;*
- *the I/O technique used for the communication within the disk subsystem;*
- *the use of several I/O channels;*
- *the realisation of the RAID controller;*
- *the size of the cache;*
- *the cache algorithms themselves;*
- *the behaviour during rebuild; and*
- *the provision of advanced functions such as data scrubbing and preventive rebuild.*

Table 2.2   The table compares the theoretical basic forms of different RAID levels. In practice, huge differences exist in the quality of the implementation of RAID controllers.

RAID level	Fault-tolerance	Read performance	Write performance	Space requirement
RAID 0	None	Good	Very good	Minimal
RAID 1	High	Poor	Poor	High
RAID 10	Very high	Very good	Good	High
RAID 4	High	Good	Very very poor	Low
RAID 5	High	Good	Very poor	Low
RAID 6	Very high	Good	Very very poor	Low

The performance data of the specific disk subsystem must be considered very carefully for each individual case. For example, in the previous chapter measures were discussed that greatly reduce the write penalty of RAID 4 and RAID 5. Specific RAID controllers may implement these measures, but they do not have to.

Another important difference is the behaviour during rebuild. The reconstruction of a defective hard disk puts a heavy burden on the internal I/O buses of a disk subsystem and on the disks of the affected array. The disk subsystem must cope with rebuild in addition to its normal workload. In reality, RAID controllers offer no possibility for controlling rebuild behaviour, such as through a higher prioritisation of the application workload compared to the rebuild. The disk subsystem must be accepted as it is. This means that during capacity planning the behaviour of a disk subsystem in failure situations must be taken into account, such as in a RAID rebuild or in the failure of a redundant RAID controller.

Additional functions such as data scrubbing and preventive rebuild are also important. With data scrubbing, a RAID controller regularly reads all blocks and in the process detects defective blocks before they are actively accessed by the applications. This clearly reduces the probability of data loss due to bit errors (see Table 2.1).

Another measure is preventive rebuild. Hard disks become slower before they fail, because the controller of a hard disk tries to correct read errors – for example, through a repetitive read of the block. High-end RAID controllers detect this kind of performance degradation in an individual hard disk and replace it before it fails altogether. The time required to restore a defective hard disk can be substantially reduced if it is preventively replaced as soon as the correctible read error is detected. It is then sufficient to copy the data from the old hard disk onto the new one.

Subject to the above warning, RAID 0 is the choice for applications for which the maximum write performance is more important than protection against the failure of a disk. Examples are the storage of multimedia data for film and video production and the recording of physical experiments in which the entire series of measurements has no value if all measured values cannot be recorded. In this case it is more beneficial to record all of the measured data on a RAID 0 array first and then copy it after the experiment, for example on a RAID 5 array. In databases, RAID 0 is used as a fast store for segments in which intermediate results for complex requests are to be temporarily stored. However, as a rule hard disks tend to fail at the most inconvenient moment so database administrators only use RAID 0 if it is absolutely necessary, even for temporary data.

With RAID 1, performance and capacity are limited because only two physical hard disks are used. RAID 1 is therefore a good choice for small databases for which the configuration of a virtual RAID 5 or RAID 10 disk would be too large. A further important field of application for RAID 1 is in combination with RAID 0.

RAID 10 is used in situations where high write performance and high fault-tolerance are called for. For a long time it was recommended that database log files be stored on RAID 10. Databases record all changes in log files so this application has a high write component. After a system crash the restarting of the database can only be guaranteed if all log files are fully available. Manufacturers of storage systems disagree as to whether this recommendation is still valid as there are now fast RAID 4 and RAID 5 implementations.

RAID 4 and RAID 5 save disk space at the expense of a poorer write performance. For a long time the rule of thumb was to use RAID 5 where the ratio of read operations to write operations is 70:30. At this point we wish to repeat that there are now storage systems on the market with excellent write performance that store the data internally using RAID 4 or RAID 5.

RAID 6 is a compromise between RAID 5 and RAID 10. It offers large hard disks considerably more protection from failure than RAID 5, albeit at the cost of clearly poorer write performance. What is particularly interesting is the use of RAID 6 for archiving, such as with hard disk WORM storage, where read access is prevalent (Section 8.3.1). With the appropriate cache algorithms, RAID 6 can be as effective as RAID 5 for workloads with a high proportion of sequential write access.

2.6   CACHING: ACCELERATION OF HARD DISK ACCESS

In all fields of computer systems, caches are used to speed up slow operations by operating them from the cache. Specifically in the field of disk subsystems, caches are designed to accelerate write and read accesses to physical hard disks. In this connection we can differentiate between two types of cache: (1) cache on the hard disk (Section 2.6.1) and (2) cache in the RAID controller. The cache in the RAID controller is subdivided into write cache (Section 2.6.2) and read cache (Section 2.6.3).

2.6.1   Cache on the hard disk

Each individual hard disk comes with a very small cache. This is necessary because the transfer rate of the I/O channel to the disk controller is significantly higher than the speed at which the disk controller can write to or read from the physical hard disk. If a server or a RAID controller writes a block to a physical hard disk, the disk controller stores this in its cache. The disk controller can thus write the block to the physical hard disk in its own time whilst the I/O channel can be used for data traffic to the other hard disks. Many RAID levels use precisely this state of affairs to increase the performance of the virtual hard disk.

Read access is accelerated in a similar manner. If a server or an intermediate RAID controller wishes to read a block, it sends the address of the requested block to the hard disk controller. The I/O channel can be used for other data traffic while the hard disk controller copies the complete block from the physical hard disk into its cache at a slower data rate. The hard disk controller transfers the block from its cache to the RAID controller or to the server at the higher data rate of the I/O channel.

2.6.2   Write cache in the disk subsystem controller

In addition to the cache of the individual hard drives many disk subsystems come with their own cache, which in some models is gigabytes in size. As a result it can buffer much greater data quantities than the cache on the hard disk. The write cache should have a battery backup and ideally be mirrored. The battery backup is necessary to allow the data in the write cache to survive a power cut. A write cache with battery backup can significantly reduce the write penalty of RAID 4 and RAID 5, particularly for sequential write access (cf. Section 2.5.4 'RAID 4 and RAID 5: parity instead of mirroring'), and smooth out load peaks.

Many applications do not write data at a continuous rate, but in batches. If a server sends several data blocks to the disk subsystem, the controller initially buffers all blocks into a write cache with a battery backup and immediately reports back to the server that all data has been securely written to the drive. The disk subsystem then copies the data from the write cache to the slower physical hard disk in order to make space for the next write peak.

2.6.3   Read cache in the disk subsystem controller

The acceleration of read operations is difficult in comparison to the acceleration of write operations using cache. To speed up read access by the server, the disk subsystem's controller must copy the relevant data blocks from the slower physical hard disk to the fast cache before the server requests the data in question.

The problem with this is that it is very difficult for the disk subsystem's controller to work out in advance what data the server will ask for next. The controller in the disk subsystem knows neither the structure of the information stored in the data blocks nor the access pattern that an application will follow when accessing the data. Consequently, the controller can only analyse past data access and use this to extrapolate which data blocks the server will access next. In sequential read processes this prediction is comparatively simple, in the case of random access it is almost impossible. As a rule of thumb, good RAID controllers manage to provide around 40% of the requested blocks from the read cache in mixed read profiles.

The disk subsystem's controller cannot further increase the ratio of read access provided from the cache (pre-fetch hit rate), because it does not have the necessary application knowledge. Therefore, it is often worthwhile realising a further cache within applications. For example, after opening a file, file systems can load all blocks of the file into the main memory (RAM); the file system knows the structures that the files are stored in. File systems can thus achieve a pre-fetch hit rate of 100%. However, it is impossible to know whether the expense for the storage of the blocks is worthwhile in an individual case, since the application may not actually request further blocks of the file.

2.7 INTELLIGENT DISK SUBSYSTEMS

Intelligent disk subsystems represent the third level of complexity for controllers after JBODs and RAID arrays. The controllers of intelligent disk subsystems offer additional functions over and above those offered by RAID. In the disk subsystems that are currently available on the market these functions are usually instant copies (Section 2.7.1), remote mirroring (Section 2.7.2) and LUN masking (Section 2.7.3).

2.7.1 Instant copies

Instant copies can virtually copy data sets of several terabytes within a disk subsystem in a few seconds. Virtual copying means that disk subsystems fool the attached servers into believing that they are capable of copying such large data quantities in such a short space of time. The actual copying process takes significantly longer. However, the same server, or a second server, can access the virtually copied data after a few seconds (Figure 2.18).

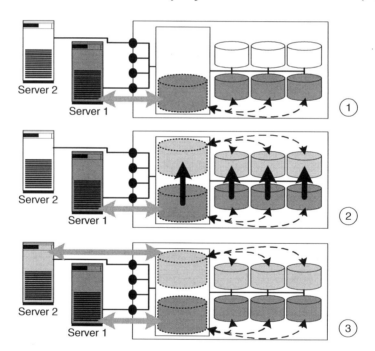

Figure 2.18   Instant copies can virtually copy several terabytes of data within a disk subsystem in a few seconds: server 1 works on the original data (1). The original data is virtually copied in a few seconds (2). Then server 2 can work with the data copy, whilst server 1 continues to operate with the original data (3).

Instant copies are used, for example, for the generation of test data, for the backup of data and for the generation of data copies for data mining. Based upon the case study in Section 1.3 it was shown that when copying data using instant copies, attention should be paid to the consistency of the copied data. Sections 7.8.4 and 7.10.3 discuss in detail the interaction of applications and storage systems for the generation of consistent instant copies.

There are numerous alternative implementations for instant copies. One thing that all implementations have in common is that the pretence of being able to copy data in a matter of seconds costs resources. All realisations of instant copies require controller computing time and cache and place a load on internal I/O channels and hard disks. The different implementations of instant copy force the performance down at different times. However, it is not possible to choose the most favourable implementation alternative depending upon the application used because real disk subsystems only ever realise one implementation alternative of instant copy.

In the following, two implementation alternatives will be discussed that function in very different ways. At one extreme the data is permanently mirrored (RAID 1 or RAID 10). Upon the copy command both mirrors are separated: the separated mirrors can then be used independently of the original. After the separation of the mirror the production data is no longer protected against the failure of a hard disk. Therefore, to increase data protection, three mirrors are often kept prior to the separation of the mirror (three-way mirror), so that the production data is always mirrored after the separation of the copy.

At the other extreme, no data at all is copied prior to the copy command, only after the instant copy has been requested. To achieve this, the controller administers two data areas, one for the original data and one for the data copy generated by means of instant copy. The controller must ensure that during write and read access operations to original data or data copies the blocks in question are written to or read from the data areas in question. In some implementations it is permissible to write to the copy, in some it is not. Some implementations copy just the blocks that have actually changed (partial copy), others copy all blocks as a background process until a complete copy of the original data has been generated (full copy).

In the following, the case differentiations of the controller will be investigated in more detail based upon the example from Figure 2.18. We will first consider access by server 1 to the original data. Read operations are completely unproblematic; they are always served from the area of the original data. Handling write operations is trickier. If a block is changed for the first time since the generation of the instant copy, the controller must first copy the old block to the data copy area so that server 2 can continue to access the old data set. Only then may it write the changed block to the original data area. If a block that has already been changed in this manner has to be written again, it must be written to the original data area. The controller may not even back up the previous version of the block to the data copy area because otherwise the correct version of the block would be overwritten.

The case differentiations for access by server 2 to the data copy generated by means of instant copy are somewhat simpler. In this case, write operations are unproblematic: the controller always writes all blocks to the data copy area. On the other hand, for read operations it has to distinguish whether the block in question has already been copied or not. This determines whether it has to read the block from the original data area or read it from the data copy area and forward it to the server.

The subsequent copying of the blocks offers the basis for important variants of instant copy. Space-efficient instant copy only copies the blocks that were changed (Figure 2.19). These normally require considerably less physical storage space than the entire copy. Yet the exported virtual hard disks of the original hard disk and the copy created through space-efficient instant copy are of the same size. From the view of the server both virtual disks continue to have the same size. Therefore, less physical storage space is needed overall and, therefore, the cost of using instant copy can be reduced.

Incremental instant copy is another important variant of instant copy. In some situations such a heavy burden is placed on the original data and the copy that the performance within

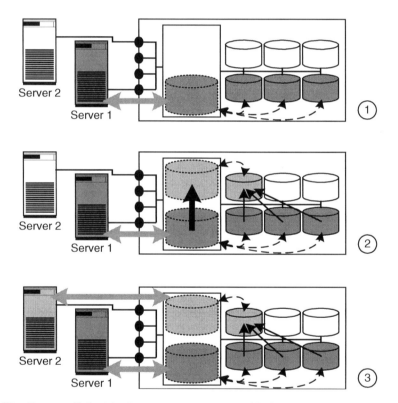

Figure 2.19   Space-efficient instant copy manages with fewer storage systems than the basic form of instant copy (Figure 2.18). Space-efficient instant copy actually only copies the changed blocks before they have been overwritten into the separate area (2). From the view of server 2 the hard disk that has been copied in this way is just as large as the source disk (3). The link between source and copy remains as the changed blocks are worthless on their own.

the disk subsystem suffers unless the data has been copied completely onto the copy. An example of this is backup when data is completely backed up through instant copy (Section 7.8.4). On the other side, the background process for copying all the data of an instant copy requires many hours when very large data volumes are involved, making this not a viable alternative. One remedy is incremental instant copy where data is only copied in its entirety the first time around. Afterwards the instant copy is repeated – for example, perhaps daily – whereby only those changes since the previous instant copy are copied.

A reversal of instant copy is yet another important variant. If data is backed up through instant copy, then the operation should be continued with the copy if a failure occurs. A simple approach is to shut down the application, copy back the data on the productive hard disks per a second instant copy from the copy onto the productive hard disks and restart the application. In this case, the disk subsystem must enable a reversal of the instant copy. If this function is not available, if a failure occurs the data either has to be copied back to the productive disks by different means or the operation continues directly with the copy. Both of these approaches are coupled with major copying operations or major configuration changes, and, consequently the recovery takes considerably longer than a reversal of the instant copy.

2.7.2   Remote mirroring

Instant copies are excellently suited for the copying of data sets within disk subsystems. However, they can only be used to a limited degree for data protection. Although data copies generated using instant copy protect against application errors (accidental deletion of a file system) and logical errors (errors in the database program), they do not protect against the failure of a disk subsystem. Something as simple as a power failure can prevent access to production data and data copies for several hours. A fire in the disk subsystem would destroy original data and data copies. For data protection, therefore, the proximity of production data and data copies is fatal.

Remote mirroring offers protection against such catastrophes. Modern disk subsystems can now mirror their data, or part of their data, independently to a second disk subsystem, which is a long way away. The entire remote mirroring operation is handled by the two participating disk subsystems. Remote mirroring is invisible to application servers and does not consume their resources. However, remote mirroring requires resources in the two disk subsystems and in the I/O channel that connects the two disk subsystems together, which means that reductions in performance can sometimes make their way through to the application.

Figure 2.20 shows an application that is designed to achieve high availability using remote mirroring. The application server and the disk subsystem, plus the associated data, are installed in the primary data centre. The disk subsystem independently mirrors the application data onto the second disk subsystem that is installed 50 kilometres away in the backup data centre by means of remote mirroring. Remote mirroring ensures that the application data in the backup data centre is always kept up-to-date with the time

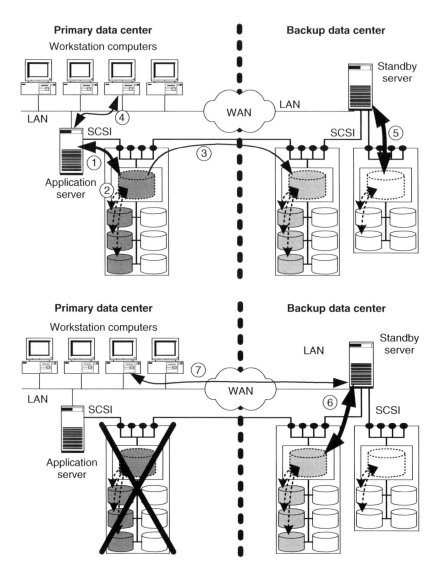

Figure 2.20 High availability with remote mirroring: (1) The application server stores its data on a local disk subsystem. (2) The disk subsystem saves the data to several physical drives by means of RAID. (3) The local disk subsystem uses remote mirroring to mirror the data onto a second disk subsystem located in the backup data centre. (4) Users use the application via the LAN. (5) The stand-by server in the backup data centre is used as a test system. The test data is located on a further disk subsystem. (6) If the first disk subsystem fails, the application is started up on the stand-by server using the data of the second disk subsystem. (7) Users use the application via the WAN.

interval for updating the second disk subsystem being configurable. If the disk subsystem in the primary data centre fails, the backup application server in the backup data centre can be started up using the data of the second disk subsystem and the operation of the application can be continued. The I/O techniques required for the connection of the two disk subsystems will be discussed in the next chapter.

We can differentiate between synchronous and asynchronous remote mirroring. In synchronous remote mirroring the first disk subsystem sends the data to the second disk subsystem first before it acknowledges a server's write command. By contrast, asynchronous remote mirroring acknowledges a write command immediately; only then does it send the copy of the block to the second disk subsystem.

Figure 2.21 illustrates the data flow of synchronous remote mirroring. The server writes block A to the first disk subsystem. This stores the block in its write cache and immediately sends it to the second disk subsystem, which also initially stores the block in its write cache. The first disk subsystem waits until the second reports that it has written the block. The question of whether the block is still stored in the write cache of the second disk subsystem or has already been written to the hard disk is irrelevant to the first disk subsystem. It does not acknowledge to the server that the block has been written until it has received confirmation from the second disk subsystem that this has written the block.

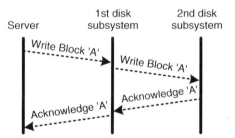

Figure 2.21   In synchronous remote mirroring a disk subsystem does not acknowledge write operations until it has saved a block itself and received write confirmation from the second disk subsystem.

Synchronous remote mirroring has the advantage that the copy of the data held by the second disk subsystem is always up-to-date. This means that if the first disk subsystem fails, the application can continue working with the most recent data set by utilising the data on the second disk subsystem.

The disadvantage is that copying the data from the first disk subsystem to the second and sending the write acknowledgement back from the second to the first increases the response time of the first disk subsystem to the server. However, it is precisely this response time that determines the throughput of applications such as databases and file systems. An important factor for the response time is the signal transit time between the two disk subsystems. After all, their communication is encoded in the form of physical signals, which propagate at a certain speed. The propagation of the signals from one disk subsystem to another simply costs time. As a rule of thumb, it is worth using synchronous remote mirroring if the cable lengths from the server to the second disk subsystem via the first are a maximum of 6–10 kilometres. However, many applications can deal with noticeably longer distances. Although performance may then not be optimal, it is still good enough.

If we want to mirror the data over longer distances, then we have to switch to asynchronous remote mirroring. Figure 2.22 illustrates the data flow in asynchronous remote mirroring. In this approach the first disk subsystem acknowledges the receipt of data as soon as it has been temporarily stored in the write cache. The first disk subsystem does not send the copy of the data to the second disk subsystem until later. The write confirmation of the second disk subsystem to the first is not important to the server that has written the data.

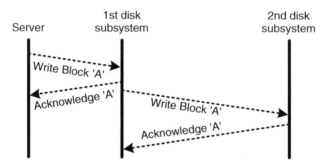

Figure 2.22   In asynchronous remote mirroring one disk subsystem acknowledges a write operation as soon as it has saved the block itself.

The price of the rapid response time achieved using asynchronous remote mirroring is obvious. In contrast to synchronous remote mirroring, in asynchronous remote mirroring there is no guarantee that the data on the second disk subsystem is up-to-date. This is precisely the case if the first disk subsystem has sent the write acknowledgement to the server but the block has not yet been saved to the second disk subsystem.

If we wish to mirror data over long distances but do not want to use only asynchronous remote mirroring it is necessary to use three disk subsystems (Figure 2.23). The first two may be located just a few kilometres apart, so that synchronous remote mirroring can be used between the two. In addition, the data of the second disk subsystem is mirrored onto a third by means of asynchronous remote mirroring. However, this solution comes at a price: for most applications the cost of data protection would exceed the costs that would be incurred after data loss in the event of a catastrophe. This approach would therefore only be considered for very important applications.

An important aspect of remote mirroring is the duration of the initial copying of the data. With large quantities of data it can take several hours until all data is copied from

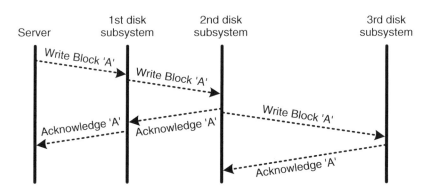

Figure 2.23 The combination of synchronous and asynchronous remote mirroring means that rapid response times can be achieved in combination with mirroring over long distances.

the first disk subsystem to the second one. This is completely acceptable the first time remote mirroring is established. However, sometimes the connection between both disk subsystems is interrupted later during operations – for example, due to a fault in the network between both systems or during maintenance work on the second disk subsystem. After the appropriate configuration the application continues operation on the first disk subsystem without the changes having been transferred to the second disk subsystem. Small quantities of data can be transmitted in their entirety again after a fault has been resolved. However, with large quantities of data a mechanism should exist that allows only those blocks that were changed during the fault to be transmitted. This is also referred to as suspending (or freezing) remote mirroring and resuming it later on. Sometimes there is a deliberate reason for suspending a remote mirroring relationship. Later in the book we will present a business continuity solution that suspends remote mirroring relationships at certain points in time for the purposes of creating consistent copies in backup data centres (Section 9.5.6).

For these same reasons there is a need for a reversal of remote mirroring. In this case, if the first disk subsystem fails, the entire operation is completely switched over to the second disk subsystem and afterwards the data is only changed on that second system. The second disk subsystem logs all changed blocks so that only those blocks that were changed during the failure are transmitted to the first disk subsystem once it is operational again. This ensures that the data on both disk subsystems is synchronised once again.

2.7.3 Consistency groups

Applications such as databases normally stripe their data over multiple virtual hard disks. Depending on data quantities and performance requirements, the data is sometimes even distributed over multiple disk subsystems. Sometimes, as with the web architecture, multiple applications that are running on different operating systems manage common related data sets (Section 6.4.2). The copies created through instant copy and remote mirroring must also be consistent for these types of distributed data sets so that they can be used if necessary to restart operation.

The problem in this case is that, unless other measures are taken, the copying from multiple virtual hard disks through instant copy and remote mirroring will not be consistent. If, for example, a database with multiple virtual hard disks is copied using instant copy, the copies are created at almost the same time, but not exactly at the same time. However, databases continuously write time stamps into their virtual hard disks. At restart the database then checks the time stamp of the virtual hard disks and aborts the start if the time stamps of all the disks do not match 100%. This means that the operation cannot be restarted with the copied data and the use of instant copy has been worthless.

Consistency groups provide help in this situation. A consistency group for instant copy combines multiple instant copy pairs into one unit. If an instant copy is then requested for a consistency group, the disk subsystem makes sure that all virtual hard disks of the consistency group are copied at exactly the same point in time. Due to this simultaneous copying of all the virtual hard disks of a consistency group, the copies are given a consistent set of time stamps. An application can therefore restart with the hard disks that have been copied in this way. When a consistency group is copied via instant copy, attention of course has to be paid to the consistency of the data on each individual virtual hard disk – the same as when an individual hard disk is copied (Section 1.3). It is also important that the instant copy pairs of a consistency group can span multiple disk subsystems as for large databases and large file systems.

The need to combine multiple remote mirroring pairs into a consistency group also exists with remote mirroring (Figure 2.24). Here too the consistency group should be able to span multiple disk subsystems. If the data of an application is distributed over multiple virtual hard disks or even over multiple disk subsystems, and, if the remote mirroring of this application has been deliberately suspended so that a consistent copy of the data can be created in the backup data centre (Section 9.5.6), then all remote mirroring pairs must be suspended at exactly the same point in time. This will ensure that the time stamps on the copies are consistent and the application can restart smoothly from the copy.

Write-order consistency for asynchronous remote mirroring pairs is another important feature of consistency groups for remote mirroring. A prerequisite for the functions referred to as journaling of file systems (Section 4.1.2) and log mechanisms of databases (Section 7.10.1), which are presented later in this book, is that updates of data sets are executed in a very specific sequence. If the data is located on multiple virtual hard disks that are mirrored asynchronously, the changes can get ahead of themselves during the mirroring so that the data in the backup data centre is updated in a different sequence than in the primary data centre (Figure 2.25). This means that the consistency of the data in the backup centre is then at risk. Write-order consistency ensures that, despite asynchronous mirroring, the data on the primary hard disks and on the target hard disks is updated in the same sequence – even when it spans multiple virtual hard disks or multiple disk subsystems.

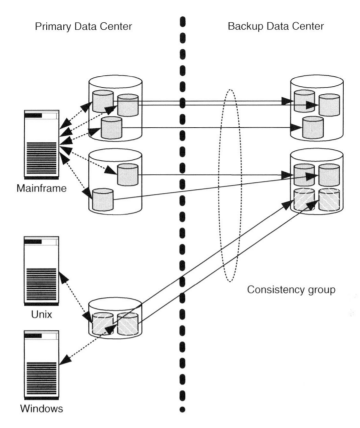

Figure 2.24 The example shows a distributed application that manages shared data over multiple servers. A large database needing two disk subsystems is running on one of the servers. For data consistency it is important that changes in the database are synchronised over all the servers and disk subsystems. Consistency groups help to maintain this synchronisation during copying using instant copy and remote mirroring.

2.7.4 LUN masking

So-called LUN masking brings us to the third important function – after instant copy and remote mirroring – that intelligent disk subsystems offer over and above that offered by RAID. LUN masking limits the access to the hard disks that the disk subsystem exports to the connected server.

A disk subsystem makes the storage capacity of its internal physical hard disks available to servers by permitting access to individual physical hard disks, or to virtual hard disks created using RAID, via the connection ports. Based upon the SCSI protocol, all hard disks – physical and virtual – that are visible outside the disk subsystem are also known as LUN.

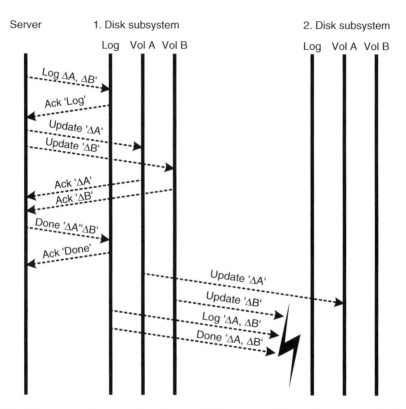

Figure 2.25   The consistency mechanism of databases and file systems records changes to data in a log, requiring adherence to a very specific sequence for writing in the log and the data. With asynchronous remote mirroring this sequence can be changed so that the start-up of an application will fail if a copy is used. The write-order consistency for asynchronous remote mirroring ensures that the copy is changed in the same sequence as the original on all the disks.

Without LUN masking every server would see all hard disks that the disk subsystem provides. Figure 2.26 shows a disk subsystem without LUN masking to which three servers are connected. Each server sees all hard disks that the disk subsystem exports outwards. As a result, considerably more hard disks are visible to each server than is necessary.

In particular, on each server those hard disks that are required by applications that run on a different server are visible. This means that the individual servers must be very carefully configured. In Figure 2.26 an erroneous formatting of the disk LUN 3 of server 1 would destroy the data of the application that runs on server 3. In addition, some operating systems are very greedy: when booting up they try to draw to them each hard disk that is written with the signature (label) of a foreign operating system.

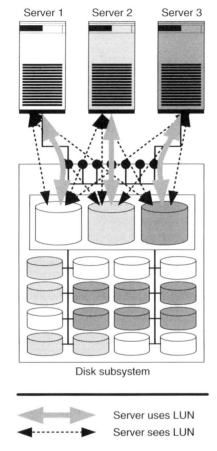

Server uses LUN

Server sees LUN

Figure 2.26   Chaos: Each server works to its own virtual hard disk. Without LUN masking each server sees all hard disks. A configuration error on server 1 can destroy the data on the other two servers. The data is thus poorly protected.

Without LUN masking, therefore, the use of the hard disk must be very carefully configured in the operating systems of the participating servers. LUN masking brings order to this chaos by assigning the hard disks that are externally visible to servers. As a result, it limits the visibility of exported disks within the disk subsystem. Figure 2.27 shows how LUN masking brings order to the chaos of Figure 2.26. Each server now sees only the hard disks that it actually requires. LUN masking thus acts as a filter between the exported hard disks and the accessing servers.

It is now no longer possible to destroy data that belongs to applications that run on another server. Configuration errors are still possible, but the consequences are no longer so devastating. Furthermore, configuration errors can now be more quickly traced since

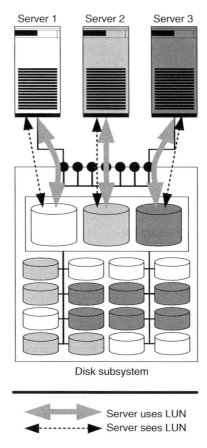

Figure 2.27   Order: each server works to its own virtual hard disk. With LUN masking, each server sees only its own hard disks. A configuration error on server 1 can no longer destroy the data of the two other servers. The data is now protected.

the information is bundled within the disk subsystem instead of being distributed over all servers.

We differentiate between port-based LUN masking and server-based LUN masking. Port-based LUN masking is the 'poor man's LUN masking', it is found primarily in low-end disk subsystems. In port-based LUN masking the filter only works using the granularity of a port. This means that all servers connected to the disk subsystem via the same port see the same disks.

Server-based LUN masking offers more flexibility. In this approach every server sees only the hard disks assigned to it, regardless of which port it is connected via or which other servers are connected via the same port.

2.8   AVAILABILITY OF DISK SUBSYSTEMS

Disk subsystems are assembled from standard components, which have a limited fault-tolerance. In this chapter we have shown how these standard components are combined in order to achieve a level of fault-tolerance for the entire disk subsystem that lies significantly above the fault-tolerance of the individual components. Today, disk subsystems can be constructed so that they can withstand the failure of any component without data being lost or becoming inaccessible. We can also say that such disk subsystems have no 'single point of failure'.

The following list describes the individual measures that can be taken to increase the availability of data:

- The data is distributed over several hard disks using RAID processes and supplemented by further data for error correction. After the failure of a physical hard disk, the data of the defective hard disk can be reconstructed from the remaining data and the additional data.

- Individual hard disks store the data using the so-called Hamming code. The Hamming code allows data to be correctly restored even if individual bits are changed on the hard disk. Self-diagnosis functions in the disk controller continuously monitor the rate of bit errors and the physical variables (e.g., temperature, spindle vibration). In the event of an increase in the error rate, hard disks can be replaced before data is lost.

- Each internal physical hard disk can be connected to the controller via two internal I/O channels. If one of the two channels fails, the other can still be used.

- The controller in the disk subsystem can be realised by several controller instances. If one of the controller instances fails, one of the remaining instances takes over the tasks of the defective instance.

- Other auxiliary components such as power supplies, batteries and fans can often be duplicated so that the failure of one of the components is unimportant. When connecting the power supply it should be ensured that the various power cables are at least connected through various fuses. Ideally, the individual power cables would be supplied via different external power networks; however, in practice this is seldom realisable.

- Server and disk subsystem are connected together via several I/O channels. If one of the channels fails, the remaining ones can still be used.

- Instant copies can be used to protect against logical errors. For example, it would be possible to create an instant copy of a database every hour. If a table is 'accidentally' deleted, then the database could revert to the last instant copy in which the database is still complete.

- Remote mirroring protects against physical damage. If, for whatever reason, the original data can no longer be accessed, operation can continue using the data copy that was generated using remote mirroring.

- Consistency groups and write-order consistency synchronise the copying of multiple virtual hard disks. This means that instant copy and remote mirroring can even guarantee the consistency of the copies if the data spans multiple virtual hard disks or even multiple disk subsystems.
- LUN masking limits the visibility of virtual hard disks. This prevents data being changed or deleted unintentionally by other servers.

This list shows that disk subsystems can guarantee the availability of data to a very high degree. Despite everything it is in practice sometimes necessary to shut down and switch off a disk subsystem. In such cases, it can be very tiresome to co-ordinate all project groups to a common maintenance window, especially if these are distributed over different time zones.

Further important factors for the availability of an entire IT system are the availability of the applications or the application server itself and the availability of the connection between application servers and disk subsystems. Chapter 6 shows how multipathing can improve the connection between servers and storage systems and how clustering can increase the fault-tolerance of applications.

2.9 SUMMARY

Large disk subsystems have a storage capacity of up to a petabyte that is often shared by several servers. The administration of a few large disk subsystems that are used by several servers is more flexible and cheaper than the administration of many individual disks or many small disk stacks. Large disk subsystems are assembled from standard components such as disks, RAM and CPU. Skilful combining of standard components and additional software can make the disk subsystem as a whole significantly more high-performance and more fault-tolerant than its individual components.

A server connected to a disk subsystem only sees those physical and virtual hard disks that the disk subsystem exports via the connection ports and makes available to it by LUN masking. The internal structure is completely hidden to the server. The controller is the control centre of the disk subsystem. The sole advantage of disk subsystems without controllers (JBODs) is that they are easier to handle than the corresponding number of separate hard disks without a common enclosure. RAID controllers offer clear advantages over JBODs. They combine several physical hard disks to form a virtual hard disk that can perform significantly faster, is more fault-tolerant and is larger than an individual physical hard disk. In addition to RAID, intelligent controllers realise the copying services instant copy and remote mirroring, consistency groups for copying services and LUN masking. Instant copy enables large databases of several terabytes to be virtually copied in a few seconds within a disk subsystem. Remote mirroring mirrors the data of a disk subsystem to a second disk subsystem without the need for server resources. Consistency groups synchronise the simultaneous copying of multiple related virtual hard disks.

Disk subsystems can thus take on many tasks that were previously performed within the operating system. As a result, more and more functions are being moved from the operating system to the storage system, meaning that intelligent storage systems are moving to the centre of the IT architecture (storage-centric IT architecture). Storage systems and servers are connected together by means of block-oriented I/O techniques such as SCSI, Fibre Channel, iSCSI and FCoE. These techniques are handled in detail in the next chapter. In the second part of the book we will be explaining the use of instant copy, remote mirroring and consistency groups for backup (Chapter 7) and for business continuity (Chapter 9). It is particularly clear from the discussion on business continuity that these three techniques can be combined together in different ways to optimise protection from small and large outages.

3

I/O Techniques

Computers generate, process and delete data. However, they can only store data for very short periods. Therefore, computers move data to storage devices such as tape libraries and the disk subsystems discussed in the previous chapter for long-term storage and fetch it back from these storage media for further processing. So-called I/O techniques realise the data transfer between computers and storage devices. This chapter describes I/O techniques that are currently in use or that the authors believe will most probably be used in the coming years.

This chapter first considers the I/O path from the CPU to the storage system (Section 3.1). An important technique for the realisation of the I/O path is Small Computer System Interface (SCSI, Section 3.2). To be precise, SCSI defines a medium (SCSI cable) and a communication protocol (SCSI protocol). The idea of Fibre Channel SAN is to replace the SCSI cable by a network that is realised using Fibre Channel technology: servers and storage devices exchange data as before using SCSI commands, but the data is transmitted via a Fibre Channel network and not via a SCSI cable (Sections 3.3, 3.4). An alternative to Fibre Channel SAN is IP storage. Like Fibre Channel, IP storage connects several servers and storage devices via a network on which data exchange takes place using the SCSI protocol. In contrast to Fibre Channel, however, the devices are connected by TCP/IP and Ethernet (Section 3.5). InfiniBand (Section 3.6) and Fibre Channel over Ethernet (FCoE, Section 3.7) are two additional approaches to consolidate all data traffic (LAN, storage, cluster) onto a single transmission technology.

Storage Networks Explained: Basics and Application of Fibre Channel SAN, NAS, iSCSI, InfiniBand and FCoE, Second Edition
U. Troppens R. Erkens W. Müller-Friedt N. Haustein R. Wolafka © 2009 John Wiley & Sons, Ltd

3.1   THE PHYSICAL I/O PATH FROM THE CPU TO THE STORAGE SYSTEM

In the computer, one or more CPUs process data that is stored in the CPU cache or in the random access memory (RAM). CPU cache and RAM are very fast; however, their data is lost when the power is been switched off. Furthermore, RAM is expensive in comparison to disk and tape storage. Therefore, the data is moved from the RAM to storage devices such as disk subsystems and tape libraries via system bus, host bus and I/O bus (Figure 3.1). Although storage devices are slower than CPU cache and RAM,

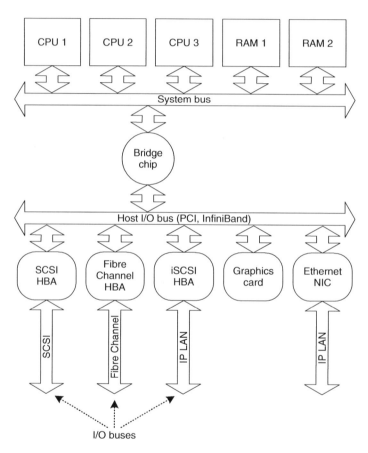

Figure 3.1   The physical I/O path from the CPU to the storage system consists of system bus, host I/O bus and I/O bus. More recent technologies such as InfiniBand, Fibre Channel and Internet SCSI (iSCSI) replace individual buses with a serial network. For historic reasons the corresponding connections are still called host I/O bus or I/O bus.

they compensate for this by being cheaper and by their ability to store data even when the power is switched off. Incidentally, the same I/O path also exists within a disk subsystem between the connection ports and the disk subsystem controller and between the controller and the internal hard disk (Figure 3.2).

At the heart of the computer, the system bus ensures the rapid transfer of data between CPUs and RAM. The system bus must be clocked at a very high frequency so that it can supply the CPU with data sufficiently quickly. It is realised in the form of printed conductors on the main circuit board. Due to physical properties, high clock rates require short printed conductors. Therefore, the system bus is kept as short as possible and thus connects only CPUs and main memory.

In modern computers as many tasks as possible are moved to special purpose processors such as graphics processors in order to free up the CPU for the processing of the application. These cannot be connected to the system bus due to the physical limitations mentioned above. Therefore, most computer architectures realise a second bus, the so-called host I/O bus. So-called bridge communication chips provide the connection between system bus and host I/O bus. Peripheral Component Interconnect (PCI) is currently the most widespread technology for the realisation of host I/O buses.

Device drivers are responsible for the control of and communication with peripheral devices of all types. The device drivers for storage devices are partially realised in the form of software that is processed by the CPU. However, part of the device driver for the communication with storage devices is almost always realised by firmware that is processed by special processors (Application Specific Integrated Circuits, ASICs). These ASICs are currently partially integrated into the main circuit board, such as on-board SCSI controllers, or connected to the main board via add-on cards (PCI cards). These add-on cards are usually called network cards (Network Interface Controller, NIC) or simply controllers. Storage devices are connected to the server via the host bus adapter (HBA) or via the on-board controller. The communication connection between controller and peripheral device is called the I/O bus.

The most important technologies for I/O buses are currently SCSI and Fibre Channel. SCSI defines a parallel bus that can connect up to 16 servers and storage devices with one another. Fibre Channel, on the other hand, defines different topologies for storage

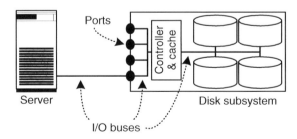

Figure 3.2   The same I/O techniques are used within a disk subsystem as those used between server and disk subsystem.

networks that can connect several millions of servers and storage devices. As an alternative to Fibre Channel, the industry is currently experimenting with different options for the realisation of storage networks by means of TCP/IP and Ethernet such as IP storage and FCoE. It is worth noting that all new technologies continue to use the SCSI protocol for device communication.

The Virtual Interface Architecture (VIA) is a further I/O protocol. VIA permits rapid and CPU-saving data exchange between two processes that run on two different servers or storage devices. In contrast to the I/O techniques discussed previously VIA defines only a protocol. As a medium it requires the existence of a powerful and low-error communication path, which is realised, for example, by means of Fibre Channel, Gigabit Ethernet or InfiniBand. VIA could become an important technology for storage networks and server clusters.

There are numerous other I/O bus technologies on the market that will not be discussed further in this book, for example, Serial Storage Architecture (SSA), IEEE 1394 (Apple's Firewire, Sony's i.Link), High-Performance Parallel Interface (HIPPI), Advanced Technology Attachment (ATA)/Integrated Drive Electronics (IDE), Serial ATA (SATA), Serial Attached SCSI (SAS) and Universal Serial Bus (USB). All have in common that they are either used by very few manufacturers or are not powerful enough for the connection of servers and storage devices. Some of these technologies can form small storage networks. However, none is anywhere near as flexible and scalable as the Fibre Channel and IP storage technologies described in this book.

3.2   SCSI

Small Computer System Interface (SCSI) was for a long time *the* technology for I/O buses in Unix and PC servers, is still very important today and will presumably remain so for a good many years to come. The first version of the SCSI standard was released in 1986. Since then SCSI has been continuously developed in order to keep it abreast with technical progress.

3.2.1   SCSI basics

As a medium, SCSI defines a parallel bus for the transmission of data with additional lines for the control of communication. The bus can be realised in the form of printed conductors on the circuit board or as a cable. Over time, numerous cable and plug types have been defined that are not directly compatible with one another (Table 3.1). A so-called daisy chain can connect up to 16 devices together (Figure 3.3).

The SCSI protocol defines how the devices communicate with each other via the SCSI bus. It specifies how the devices reserve the SCSI bus and in which format data is

Table 3.1 SCSI: maximum cable lengths, transmission speeds.

SCSI version	MByte/s MByte/s	Bus width	Max. no. of devices	Single ended (SE)	High Voltage Differential (HVD)	Low Voltage Differential (LVD)
SCSI-2	5	8	8	6 m	25 m	–
Wide Ultra SCSI	40	16	16	–	25 m	–
Wide Ultra SCSI	40	16	8	1.5 m	–	–
Wide Ultra SCSI	40	16	4	3 m	–	–
Ultra2 SCSI	40	8	8	–	25 m	12 m
Wide Ultra2 SCSI	80	16	16	–	25 m	12 m
Ultra3 SCSI	160	16	16	–	–	12 m
Ultra320 SCSI	320	16	16	–	–	12 m

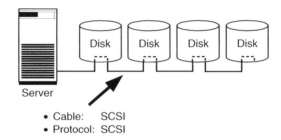

- Cable: SCSI
- Protocol: SCSI

Figure 3.3 An SCSI bus connects one server to several peripheral devices by means of a daisy chain. SCSI defines both the characteristics of the connection cable and also the transmission protocol.

transferred. The SCSI protocol has been further developed over the years. For example, a server could originally only begin a new SCSI command when the previous SCSI command had been acknowledged by the partner; however, precisely this overlapping of SCSI commands is the basis for the performance increase achieved by RAID (Section 2.4). Today it is even possible using asynchronous I/O to initiate multiple concurrently write or read commands to a storage device at the same time.

The SCSI protocol introduces SCSI IDs (sometimes also called target ID or just ID) and Logical Unit Numbers (LUNs) for the addressing of devices. Each device in the SCSI bus must have an unambiguous ID, with the HBA in the server requiring its own ID. Depending upon the version of the SCSI standard, a maximum of 8 or 16 IDs are permitted per SCSI bus. Storage devices such as RAID disk subsystems, intelligent disk subsystems or tape libraries can include several subdevices, such as virtual hard disks, tape drives or a media changer to insert the tapes, which means that the IDs would be used up very quickly. Therefore, so-called LUNs were introduced in order to address subdevices within larger devices (Figure 3.4). A server can be equipped with several SCSI controllers. Therefore, the operating system must note three things for the differentiation of devices – controller ID, SCSI ID and LUN.

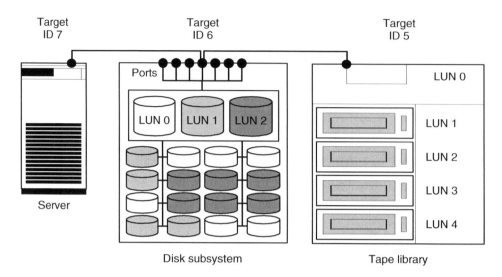

Figure 3.4 Devices on the SCSI bus are differentiated by means of target IDs. Components within devices (virtual hard disks, tape drives and the robots in the tape library) by LUNs.

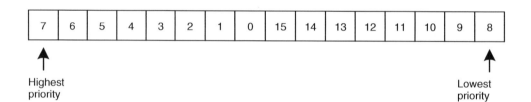

Figure 3.5 SCSI Target IDs with a higher priority win the arbitration of the SCSI bus.

The priority of SCSI IDs is slightly trickier. Originally, the SCSI protocol permitted only eight IDs, with the ID '7' having the highest priority. More recent versions of the SCSI protocol permit 16 different IDs. For reasons of compatibility, the IDs '7' to '0' should retain the highest priority so that the IDs '15' to '8' have a lower priority (Figure 3.5).

Devices (servers and storage devices) must reserve the SCSI bus (arbitrate) before they may send data through it. During the arbitration of the bus, the device that has the highest priority SCSI ID always wins. In the event that the bus is heavily loaded, this can lead to devices with lower priorities never being allowed to send data. The SCSI arbitration procedure is therefore 'unfair'.

3.2.2   SCSI and storage networks

SCSI is only suitable for the realisation of storage networks to a limited degree. First, a SCSI daisy chain can only connect a very few devices with each other. Although it is theoretically possible to connect several servers to a SCSI bus, this does not work very well in practice. Clusters with so-called twin-tailed SCSI cables and a stand-by server have proved their worth in increasing the availability of data and the applications based upon it (Figure 3.6). Both servers can access the shared storage devices, with only one server having active access to the data at any time. If this server fails, then the stand-by server actively accesses the storage device and continues to operate the application.

Second, the maximum lengths of SCSI buses greatly limit the construction of storage networks. Large disk subsystems have over 30 connection ports for SCSI cables so that several dozen servers can access them (Figure 3.7), and many of the advantages of storage-centric IT architectures can be achieved with this layout. However, due to the dimensions of disk subsystems, tape libraries and servers and the length limits of SCSI buses, constructing the configuration shown in Figure 3.7 using real devices is a challenge. Although it is possible to extend the length of the SCSI buses with so-called link extenders, the use of a large number of link extenders is unwieldy.

Despite these limitations, SCSI is of great importance even for storage-centric IT systems. Techniques such as Fibre Channel SAN, iSCSI and FCoE merely replace the SCSI bus by a network; the SCSI protocol is still used for communication over this network. The advantage of continuing to use the SCSI protocol is that the transition of SCSI cables to storage networks remains hidden from applications and higher layers of the operating

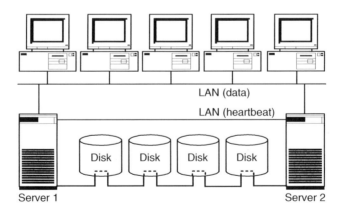

Figure 3.6   In twin-tailed SCSI cabling only one server is active. The second server takes over the devices if the first server fails.

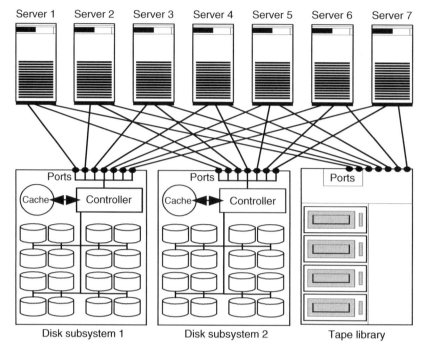

Figure 3.7 SCSI SANs can be built up using multiport storage systems. The dimensions of servers and storage devices and the length restrictions of the SCSI cable make construction difficult. SCSI SANs are difficult to administer, less flexible and only scalable to a limited degree.

system. SCSI also turns up within the disk subsystems and Network Attached Storage (NAS) servers used in storage networks.

3.3 THE FIBRE CHANNEL PROTOCOL STACK

Fibre Channel is currently (2009) the technique most frequently used for implementing storage networks. Interestingly, Fibre Channel was originally developed as a backbone technology for the connection of LANs. The original development objective for Fibre Channel was to supersede Fast-Ethernet (100 Mbit/s) and Fibre Distributed Data Interface (FDDI). Meanwhile Gigabit Ethernet and 10-Gigabit Ethernet have become prevalent or will become prevalent in this market segment.

By coincidence, the design goals of Fibre Channel are covered by the requirements of a transmission technology for storage networks such as:

- Serial transmission for high speed and long distances;
- Low rate of transmission errors;
- Low delay (latency) of the transmitted data;
- Implementation of the Fibre Channel Protocol (FCP) in hardware on HBA cards to free up the server CPUs

In the early 1990s, Seagate was looking for a technology that it could position against IBM's SSA. With the support of the Fibre Channel industry, Fibre Channel was expanded by the arbitrated loop topology, which is cheaper than the originally developed fabric topology. This led to the breakthrough of Fibre Channel for the realisation of storage networks.

Fibre Channel is only one of the transmission technologies with which storage area networks (SANs) can be realised. Nevertheless, the terms 'Storage Area Network' and 'SAN' are often used synonymously with Fibre Channel technology. In discussions, newspaper articles and books the terms 'storage area network' and SAN are often used to mean a storage area network that is built up using Fibre Channel. The advantages of storage area networks and server-centric IT architectures can, however, also be achieved using other technologies for storage area networks, for example, iSCSI and FCoE.

In this book we have taken great pains to express ourselves precisely. We do not use the terms 'storage area network' and 'SAN' on their own. For unambiguous differentiation we always also state the technology, for example, 'Fibre Channel SAN' or 'iSCSI SAN'. In statements about storage area networks in general that are independent of a specific technology we use the term 'storage network'. We use the term 'Fibre Channel' without the suffix 'SAN' when we are referring to the transmission technology that underlies a Fibre Channel SAN.

For the sake of completeness we should also mention that the three letters 'SAN' are also used as an abbreviation for 'System Area Network'. A System Area Network is a network with a high bandwidth and a low latency that serves as a connection between computers in a distributed computer system. In this book we have never used the abbreviation SAN in this manner. However, it should be noted that the VIA standard, for example, does use this second meaning of the abbreviation 'SAN'.

The Fibre Channel protocol stack is subdivided into five layers (Figure 3.8). The lower four layers, FC-0 to FC-3 define the fundamental communication techniques, i.e. the physical levels, the transmission and the addressing. The upper layer, FC-4, defines how application protocols (upper layer protocols, ULPs) are mapped on the underlying Fibre Channel network. The use of the various ULPs decides, for example, whether a real Fibre Channel network is used as an IP network, a Fibre Channel SAN (i.e. as a storage network) or both at the same time. The link services and fabric services are located quasi-adjacent to the Fibre Channel protocol stack. These services will be required in order to administer and operate a Fibre Channel network.

Basic knowledge of the Fibre Channel standard helps to improve understanding of the possibilities for the use of Fibre Channel for a Fibre Channel SAN. This section (Section 3.3) explains technical details of the Fibre Channel protocol. We will restrict

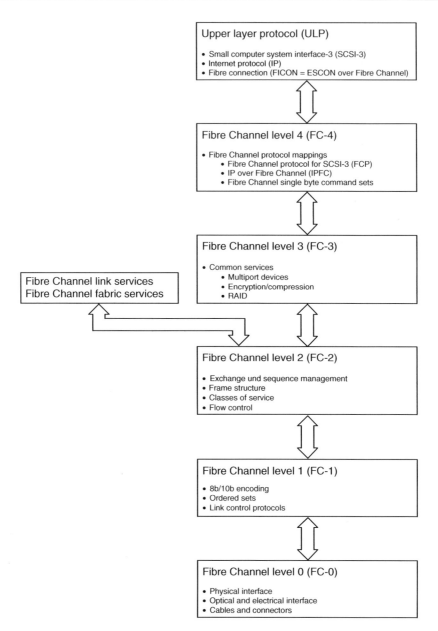

Figure 3.8 The Fibre Channel protocol stack is divided into two parts: the lower four layers (FC-0 to FC-3) realise the underlying Fibre Channel transmission technology. The link services and the fabric services help to administer and configure the Fibre Channel network. The upper layer (FC-4) defines how the application protocols (for example, SCSI and IP) are mapped on a Fibre Channel network.

the level of detail to the parts of the Fibre Channel standard that are helpful in the administration or the design of a Fibre Channel SAN. Building upon this, the next section (Section 3.4) explains the use of Fibre Channel for storage networks.

3.3.1 Links, ports and topologies

The Fibre Channel standard defines three different topologies: fabric, arbitrated loop and point-to-point (Figure 3.9). Point-to-point defines a bi-directional connection between two devices. Arbitrated loop defines a unidirectional ring in which only two devices can ever exchange data with one another at any one time. Finally, fabric defines a network in which several devices can exchange data simultaneously at full bandwidth. A fabric basically requires one or more Fibre Channel switches connected together to form a control centre between the end devices. Furthermore, the standard permits the connection of one or more arbitrated loops to a fabric. The fabric topology is the most frequently used of all topologies, and this is why more emphasis is placed upon the fabric topology than on the two other topologies in the following sections.

Common to all topologies is that devices (servers, storage devices and switches) must be equipped with one or more Fibre Channel ports. In servers, the port is generally realised by means of so-called HBAs (for example, PCI cards) that are also fitted in the server. A port always consists of two channels, one input and one output channel.

The connection between two ports is called a link. In the point-to-point topology and in the fabric topology the links are always bi-directional: in this case the input channel and the output channel of the two ports involved in the link are connected together by a cross, so that every output channel is connected to an input channel. On the other hand, the links of the arbitrated loop topology are unidirectional: each output channel is connected to the input channel of the next port until the circle is closed. The cabling of an arbitrated loop can be simplified with the aid of a hub. In this configuration the end

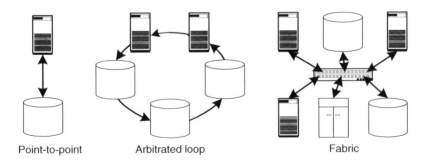

| Point-to-point | Arbitrated loop | Fabric |

Figure 3.9 The fabric topology is the most flexible and scalable Fibre Channel topology.

devices are bi-directionally connected to the hub; the wiring within the hub ensures that the unidirectional data flow within the arbitrated loop is maintained.

The fabric and arbitrated loop topologies are realised by different, incompatible protocols. We can differentiate between the following port types with different capabilities:

- N-Port (Node_Port): Originally the communication of Fibre Channel was developed around N-Ports and F-Ports, with 'N' standing for 'node' and 'F' for 'fabric'. An N-Port describes the capability of a port as an end device (server, storage device), also called node, to participate in the fabric topology or to participate in the point-to-point topology as a partner.
- F-Port (Fabric_Port): F-Ports are the counterpart to N-Ports in the Fibre Channel switch. The F-Port knows how it can pass a frame that an N-Port sends to it through the Fibre Channel network on to the desired end device.
- L-Port (Loop_Port): The arbitrated loop uses different protocols for data exchange than the fabric. An L-Port describes the capability of a port to participate in the arbitrated loop topology as an end device (server, storage device). More modern devices are now fitted with NL-Ports instead of L-Ports. Nevertheless, old devices that are fitted with an L-Port are still encountered in practice.
- NL_Port (Node_Loop_Port): An NL-Port has the capabilities of both an N-Port and an L-Port. An NL-Port can thus be connected both in a fabric and in an arbitrated loop. Most modern HBA cards are equipped with NL-Ports.
- FL-Port (Fabric_Loop_Port): An FL-Port allows a fabric to connect to a loop. However, this is far from meaning that end devices in the arbitrated loop can communicate with end devices in the fabric. More on the subject of connecting fabric and arbitrated loop can be found in Section 3.4.3.
- E-Port (Expansion_Port): Two Fibre Channel switches are connected together by E-Ports. E-Ports transmit the data from end devices that are connected to two different Fibre Channel switches. In addition, Fibre Channel switches smooth out information over the entire Fibre Channel network via E-ports.
- G-Port (Generic_Port): Modern Fibre Channel switches configure their ports automatically. Such ports are called G-Ports. If, for example, a Fibre Channel switch is connected to a further Fibre Channel switch via a G-Port, the G-Port configures itself as an E-Port.
- B-Port (Bridge_Port): B-Ports serve to connect two Fibre Channel switches together via Asynchronous Transfer Mode (ATM), SONET/SDH (Synchronous Optical Networking/Synchronous Digital Hierarchy) as well as Ethernet and IP. Thus Fibre Channel SANs that are a long distance apart can be connected together using classical Wide Area Network (WAN) techniques.

Some Fibre Channel switches have further, manufacturer-specific port types over and above those in the Fibre Channel standard: these port types provide additional functions. When using such port types, it should be noted that you can sometimes bind yourself to the Fibre Channel switches of a certain manufacturer, which cannot subsequently be replaced by Fibre Channel switches of a different manufacturer.

3.3.2   FC-0: cables, plugs and signal encoding

FC-0 defines the physical transmission medium (cable, plug) and specifies which physical signals are used to transmit the bits '0' and '1'. In contrast to the SCSI bus, in which each bit has its own data line plus additional control lines, Fibre Channel transmits the bits sequentially via a single line. In general, buses come up against the problem that the signals have a different transit time on the different data lines (skew), which means that the clock rate can only be increased to a limited degree in buses. The different signal transit times can be visualised as the hand rail in an escalator that runs faster or slower than the escalator stairs themselves.

Fibre Channel therefore transmits the bits serially. This means that, in contrast to the parallel bus, a high transfer rate is possible even over long distances. The high transfer rate of serial transmission more than compensates for the parallel lines of a bus. The transmission rate of actual components increases every few years. Table 3.2 depicts the market entry and the roadmap for new higher transfer rates as of 2009. Fibre Channel components are distinguished as Base2 components and Base10 components. Base2

Table 3.2   Fibre Channel components are distinguished in Base2 components (upper table) and Base10 components (lower table). The table shows the roadmap as of 2009.

Product Naming	MByte/s (per direction)	T11 Spec Completed	Market Availability
1GFC	100	1996	1997
2GFC	200	2000	2001
4GFC	400	2003	2005
8GFC	800	2006	2008
16GFC	1,600	2009	2011
32GFC	3,200	2012	Market demand
64GFC	6,400	2016	Market demand
128GFC	12,800	2020	Market demand
Product Naming	MByte/s (per direction)	T11 Spec Completed	Market Availability
10GFC	1,200	2003	2004
20GFC	2,400	2008	2008
40GFC	4,800	2009	2011
80GFC	9,600	future	Market demand
100GFC	12,000	future	Market demand
160GFC	19,200	future	Market demand

components have to maintain backward compatibility of at least two previous Base2 generations. For instance 8GFC components must be interoperable with 4GFC and 2GFC components. Base10 components have to maintain backward compatibility of at least one Base10 generation, in which case, as an exception for 100GFC no backward compatibility is expected.

When considering the transfer rate it should be noted that in the fabric and point-to-point topologies the transfer is bi-directional and full duplex, which means that, for instance for 2GFC components the transfer rate of 200 MByte/s is available in each direction.

The Fibre Channel standard demands that a single bit error may occur at most once in every 10^{12} transmitted bits. On average, this means that for a 100 Mbit/s connection under full load a bit error may occur only every 16.6 minutes. The error recognition and handling mechanisms of the higher protocol layers are optimised for the maintenance of this error rate. Therefore, when installing a Fibre Channel network it is recommended that the cable is properly laid so that the bit error rate of 10^{-12} is, where possible, also achieved for connections from end device to end device, i.e. including all components connected in between such as repeaters and switches.

Different cable and plug types are defined (Figure 3.10). Fiber-optic cables are more expensive than copper cables. They do, however, have some advantages:

- Greater distances possible than with copper cable;
- Insensitivity to electromagnetic interference;
- No electromagnetic radiation;
- No electrical connection between the devices;
- No danger of 'cross-talking';
- Greater transmission rates possible than with copper cable

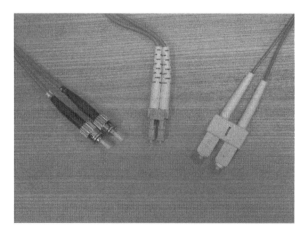

Figure 3.10   Three different plug types for fiber-optic cable.

Cables for long distances are more expensive than those for short distances. The definition of various cables makes it possible to choose the most economical technology for each distance to be bridged.

Typically the Fibre Channel standard specifies for a given medium a minimum distance which must be supported. It is assumed that when respective components enter the market, the state of technology can ensure the error rate for the specified minimum distance. Over the time, further technical improvements and the proper laying of cable enable even larger distances to be bridged in actual installations.

The reduction in the supported distances could present a problem when the equipment of an existing Fibre Channel SAN is upgraded from one generation to the next generation components. Due to technical limitations next generation components typically support for the same medium shorter distances than current generation components. This is especially an issue for longer instances when cables are installed between two collocated data centers which are several 100 meters apart.

3.3.3   FC-1: 8b/10b encoding, ordered sets and link control protocol

FC-1 defines how data is encoded before it is transmitted via a Fibre Channel cable (8b/10b encoding). FC-1 also describes certain transmission words (ordered sets) that are required for the administration of a Fibre Channel connection (link control protocol).

8b/10b Encoding

In all digital transmission techniques, transmitter and receiver must synchronise their clock-pulse rates. In parallel buses the clock rate is transmitted via an additional data line. By contrast, in the serial transmission used in Fibre Channel only one data line is available through which the data is transmitted. This means that the receiver must regenerate the clock rate from the data stream.

The receiver can only synchronise the rate at the points where there is a signal change in the medium. In simple binary encoding (Figure 3.11) this is only the case if the signal changes from '0' to '1' or from '1' to '0'. In Manchester encoding there is a signal change for every bit transmitted. Manchester encoding therefore creates two physical signals for each bit transmitted. It therefore requires a transfer rate that is twice as high as that for binary encoding. Therefore, Fibre Channel – like many other transmission techniques – uses binary encoding, because at a given rate of signal changes more bits can be transmitted than is the case for Manchester encoding.

The problem with this approach is that the signal steps that arrive at the receiver are not always the same length (jitter). This means that the signal at the receiver is sometimes a little longer and sometimes a little shorter (Figure 3.12). In the escalator analogy this means that the escalator bucks. Jitter can lead to the receiver losing synchronisation with

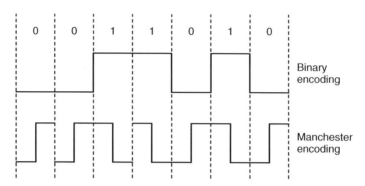

Figure 3.11   In Manchester encoding at least one signal change takes place for every bit transmitted.

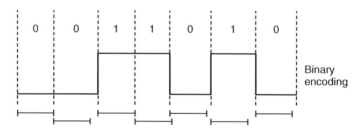

Figure 3.12   Due to physical properties the signals are not always the same length at the receiver (jitter).

the received signal. If, for example, the transmitter sends a sequence of ten zeros, the receiver cannot decide whether it is a sequence of nine, ten or eleven zeros.

If we nevertheless wish to use binary encoding, then we have to ensure that the data stream generates a signal change frequently enough that jitter cannot strike. The so-called 8b/10b encoding represents a good compromise. 8b/10b encoding converts an 8-bit byte to be transmitted into a 10-bit character, which is sent via the medium instead of the 8-bit byte. For Fibre Channel this means, for example, that a useful transfer rate of 100 MByte/s requires a raw transmission rate of 1 Gbit/s instead of 800 Mbit/s. Incidentally, 8b/10b encoding is also used for the Enterprise System Connection Architecture (ESCON), SSA, Gigabit Ethernet and InfiniBand. Finally, it should be noted that 1 Gigabyte Fibre Channel uses the 64b/66b encoding variant for certain cable types.

Expanding the 8-bit data bytes to 10-bit transmission character gives rise to the following advantages:

- In 8b/10b encoding, of all available 10-bit characters, only those that generate a bit sequence that contains a maximum of five zeros one after the other or five ones one after the other for any desired combination of the 10-bit character are selected. Therefore, a signal change takes place at the latest after five signal steps, so that the clock synchronisation of the receiver is guaranteed.

- A bit sequence generated using 8b/10b encoding has a uniform distribution of zeros and ones. This has the advantage that only small direct currents flow in the hardware that processes the 8b/10b encoded bit sequence. This makes the realisation of Fibre Channel hardware components simpler and cheaper.

- Further 10-bit characters are available that do not represent 8-bit data bytes. These additional characters can be used for the administration of a Fibre Channel link.

Ordered sets

Fibre Channel aggregates four 10-bit transmission characters to form a 40-bit transmission word. The Fibre Channel standard differentiates between two types of transmission word: data words and ordered sets. Data words represent a sequence of four 8-bit data bytes. Data words may only stand between a Start-of-Frame (SOF) delimiter and an End-of-Frame (EOF) delimiter.

Ordered sets may only stand between an EOF delimiter and a SOF delimiter, with SOFs and EOFs themselves being ordered sets. All ordered sets have in common that they begin with a certain transmission character, the so-called K28.5 character. The K28.5 character includes a special bit sequence that does not occur elsewhere in the data stream. The input channel of a Fibre Channel port can therefore use the K28.5 character to divide the continuous incoming bit stream into 40-bit transmission words when initialising a Fibre Channel link or after the loss of synchronisation on a link.

Link control protocol

With the aid of ordered sets, FC-1 defines various link level protocols for the initialisation and administration of a link. The initialisation of a link is the prerequisite for data exchange by means of frames. Examples of link level protocols are the initialisation and arbitration of an arbitrated loop.

3.3.4 FC-2: data transfer

FC-2 is the most comprehensive layer in the Fibre Channel protocol stack. It determines how larger data units (for example, a file) are transmitted via the Fibre Channel network. It regulates the flow control that ensures that the transmitter only sends the data at a speed that the receiver can process it. And it defines various service classes that are tailored to the requirements of various applications.

Exchange, sequence and frame

FC-2 introduces a three-layer hierarchy for the transmission of data (Figure 3.13). At the top layer a so-called exchange defines a logical communication connection between two end devices. For example, each process that reads and writes data could be assigned its own exchange. End devices (servers and storage devices) can simultaneously maintain several exchange relationships, even between the same ports. Different exchanges help

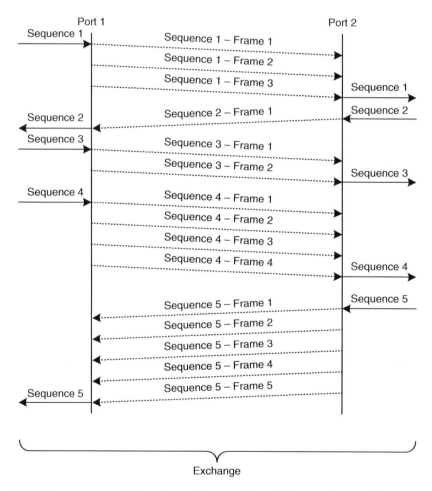

Figure 3.13 One sequence is transferred after another within an exchange. Large sequences are broken down into several frames prior to transmission. On the receiver side, a sequence is not delivered to the next highest protocol layer (FC-3) until all the frames of the sequence have arrived.

the FC-2 layer to deliver the incoming data quickly and efficiently to the correct receiver in the higher protocol layer (FC-3).

A sequence is a larger data unit that is transferred from a transmitter to a receiver. Only one sequence can be transferred after another within an exchange. FC-2 guarantees that sequences are delivered to the receiver in the same order they were sent from the transmitter; hence the name 'sequence'. Furthermore, sequences are only delivered to the next protocol layer up when all frames of the sequence have arrived at the receiver (Figure 3.13). A sequence could represent the writing of a file or an individual database transaction.

A Fibre Channel network transmits control frames and data frames. Control frames contain no useful data, they signal events such as the successful delivery of a data frame. Data frames transmit up to 2,112 bytes of useful data. Larger sequences therefore have to be broken down into several frames. Although it is theoretically possible to agree upon different maximum frame sizes, this is hardly ever done in practice.

A Fibre Channel frame consists of a header, useful data (payload) and a Cyclic Redundancy Checksum (CRC) (Figure 3.14). In addition, the frame is bracketed by an start-of-frame (SOF) delimiter and an end-of-frame (EOF) delimiter. Finally, six filling words must be transmitted by means of a link between two frames. In contrast to Ethernet and TCP/IP, Fibre Channel is an integrated whole: the layers of the Fibre Channel protocol stack are so well harmonised with one another that the ratio of payload to protocol overhead is very efficient at up to 98%. The CRC checking procedure is designed to recognise all transmission errors if the underlying medium does not exceed the specified error rate of 10^{-12}.

Error correction takes place at sequence level: if a frame of a sequence is wrongly transmitted, the entire sequence is re-transmitted. At gigabit speed it is more efficient to

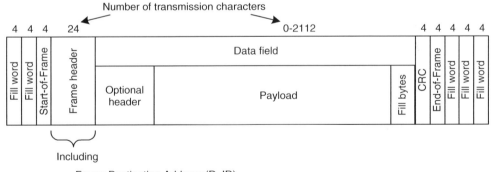

Figure 3.14   The Fibre Channel frame format.

resend a complete sequence than to extend the Fibre Channel hardware so that individual lost frames can be resent and inserted in the correct position. The underlying protocol layer must maintain the specified maximum error rate of 10^{-12} so that this procedure is efficient.

Flow control

Flow control ensures that the transmitter only sends data at a speed that the receiver can receive it. Fibre Channel uses the so-called credit model for this. Each credit represents the capacity of the receiver to receive a Fibre Channel frame. If the receiver awards the transmitter a credit of '4', the transmitter may only send the receiver four frames. The transmitter may not send further frames until the receiver has acknowledged the receipt of at least some of the transmitted frames.

FC-2 defines two different mechanisms for flow control: end-to-end flow control and link flow control (Figure 3.15). In end-to-end flow control two end devices negotiate the end-to-end credit before the exchange of data. The end-to-end flow control is realised on the HBA cards of the end devices. By contrast, link flow control takes place at each physical connection. This is achieved by two communicating ports negotiating the buffer-to-buffer credit. This means that the link flow control also takes place at the Fibre Channel switches.

Service classes

The Fibre Channel standard defines six different service classes for exchange of data between end devices. Three of these defined classes (Class 1, Class 2 and Class 3) are

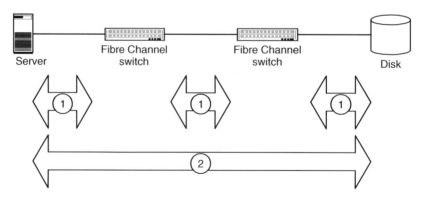

Figure 3.15   In link flow control the ports negotiate the buffer-to-buffer credit at each link (1). By contrast, in end-to-end flow control the end-to-end credit is only negotiated between the end devices (2).

realised in products available on the market, with hardly any products providing the connection-oriented Class 1. Almost all new Fibre Channel products (HBAs, switches, storage devices) support the service classes Class 2 and Class 3, which realise a packet-oriented service (datagram service). In addition, Class F serves for the data exchange between the switches within a fabric.

Class 1 defines a connection-oriented communication connection between two node ports: a Class 1 connection is opened before the transmission of frames. This specifies a route through the Fibre Channel network. Thereafter, all frames take the same route through the Fibre Channel network so that frames are delivered in the sequence in which they were transmitted. A Class 1 connection guarantees the availability of the full bandwidth. A port thus cannot send any other frames while a Class 1 connection is open.

Class 2 and Class 3, on the other hand, are packet-oriented services (datagram services): no dedicated connection is built up, instead the frames are individually routed through the Fibre Channel network. A port can thus maintain several connections at the same time. Several Class 2 and Class 3 connections can thus share the bandwidth.

Class 2 uses end-to-end flow control and link flow control. In Class 2 the receiver acknowledges each received frame (acknowledgement, Figure 3.16). This acknowledgement is used both for end-to-end flow control and for the recognition of lost frames. A missing acknowledgement leads to the immediate recognition of transmission errors by FC-2, which are then immediately signalled to the higher protocol layers. The higher protocol layers can thus initiate error correction measures straight away (Figure 3.17). Users of a Class 2 connection can demand the delivery of the frames in the correct order.

Class 3 achieves less than Class 2: frames are not acknowledged (Figure 3.18). This means that only link flow control takes place, not end-to-end flow control. In addition, the higher protocol layers must notice for themselves whether a frame has been lost. The loss of a frame is indicated to higher protocol layers by the fact that an expected sequence is not delivered because it has not yet been completely received by the FC-2 layer. A switch may dispose off Class 2 and Class 3 frames if its buffer is full. Due to greater time-out values in the higher protocol layers it can take much longer to recognise the loss of a frame than is the case in Class 2 (Figure 3.19).

We have already stated that in practice only Class 2 and Class 3 are important. In practice the service classes are hardly ever explicitly configured, meaning that in current Fibre Channel SAN implementations the end devices themselves negotiate whether they communicate by Class 2 or Class 3. From a theoretical point of view the two service classes differ in that Class 3 sacrifices some of the communication reliability of Class 2 in favour of a less complex protocol. Class 3 is currently the most frequently used service class. This may be because the current Fibre Channel SANs are still very small, so that frames are very rarely lost or overtake each other. The linking of current Fibre Channel SAN islands to a large storage network could lead to Class 2 playing a greater role in future due to its faster error recognition.

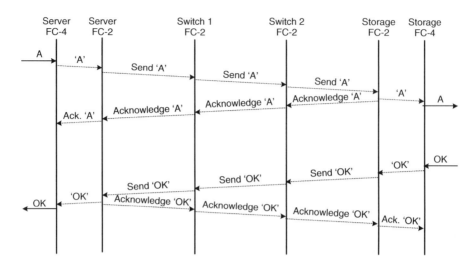

Figure 3.16 Class 2: Each Fibre Channel frame transmitted is acknowledged within the FC-2 layer. The acknowledgement aids the recognition of lost frames (see Figure 3.17) and the end-to-end flow control. The link flow control and the conversion of sequences to frames are not shown.

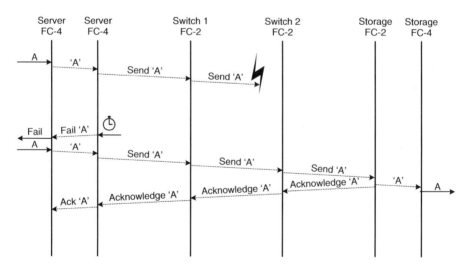

Figure 3.17 Transmission error in Class 2: The time-outs for frames are relatively short on the FC-2 layer. Missing acknowledgements are thus quickly recognised within the FC-2 layer of the transmitter and signalled to the higher protocol levels. The higher protocol layers are responsible for the error processing. In the figure the lost frame is simply resent. The link flow control and the conversion of sequences to frames are not shown.

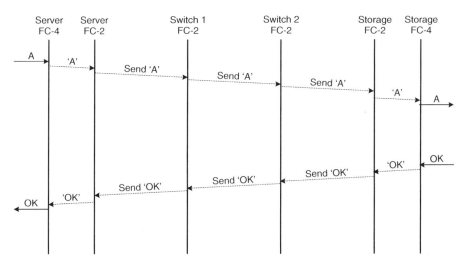

Figure 3.18 Class 3: Transmitted frames are not acknowledged in the FC-2 layer. Lost frames must be recognised in the higher protocol layers (see Figure 3.19). The link flow control and the conversion of sequences to frames are not shown.

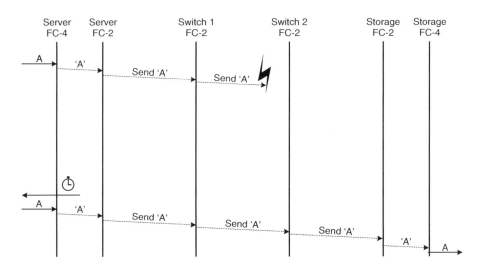

Figure 3.19 Transmission errors in Class 3: Here too the higher protocol layers are responsible for error processing. The time-outs in the higher protocol layers are relatively long in comparison to the time-outs in the FC-2 layer. In Class 3 it thus takes significantly longer before there is a response to a lost frame. In the figure the lost frame is simply resent. The link flow control and the conversion of sequences to frames are not shown.

3.3.5   FC-3: common services

FC-3 has been in its conceptual phase since 1988; in currently available products FC-3 is empty. The following functions are being discussed for FC-3:

- Striping manages several paths between multiport end devices. Striping could distribute the frames of an exchange over several ports and thus increase the throughput between the two devices.
- Multipathing combines several paths between two multiport end devices to form a logical path group. Failure or overloading of a path can be hidden from the higher protocol layers.
- Compressing the data to be transmitted, preferably realised in the hardware on the HBA.
- Encryption of the data to be transmitted, preferably realised in the hardware on the HBA.
- Finally, mirroring and other RAID levels are the last example that are mentioned in the Fibre Channel standard as possible functions of FC-3.

However, the fact that these functions are not realised within the Fibre Channel protocol does not mean that they are not available at all. For example, multipathing functions are currently provided both by suitable additional software in the operating system (Section 6.3.1) and also by some more modern Fibre Channel switches (ISL trunking).

3.3.6   Link services: login and addressing

Link services and the fabric services discussed in the next section stand next to the Fibre Channel protocol stack. They are required to operate data traffic over a Fibre Channel network. Activities of these services do not result from the data traffic of the application protocols. Instead, these services are required to manage the infrastructure of a Fibre Channel network and thus the data traffic on the level of the application protocols. For example, at any given time the switches of a fabric know the topology of the whole network.

Login

Two ports have to get to know each other before application processes can exchange data over them. To this end the Fibre Channel standard provides a three-stage login mechanism (Figure 3.20):

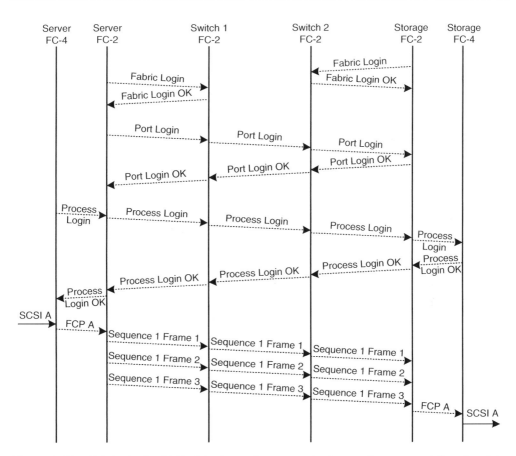

Figure 3.20 Fabric login, N-Port login and process login are the prerequisites for data exchange.

- *Fabric login (FLOGI)*

  The fabric login establishes a session between an N-Port and a corresponding F-Port. The fabric login takes place after the initialisation of the link and is an absolute prerequisite for the exchange of further frames. The F-Port assigns the N-Port a dynamic address. In addition, service parameters such as the buffer-to-buffer credit are negotiated. The fabric login is crucial for the point-to-point topology and for the fabric topology. An N-Port can tell from the response of the corresponding port whether it is a fabric topology or a point-to-point topology. In arbitrated loop topology the fabric login is optional.

- *N-Port login (PLOGI)*
  N-Port login establishes a session between two N-ports. The N-Port login takes place after the fabric login and is a compulsory prerequisite for the data exchange at FC-4 level. N-Port login negotiates service parameters such as end-to-end credit. N-Port login is optional for Class 3 communication and compulsory for all other service classes.

- *Process login (PRLI)*
  Process login establishes a session between two FC-4 processes that are based upon two different N-Ports. These could be system processes in Unix systems and system partitions in mainframes. Process login takes place after the N-Port login. Process login is optional from the point of view of FC-2. However, some FC-4 protocol mappings call for a process login for the exchange of FC-4-specific service parameters.

Addressing

Fibre Channel differentiates between addresses and names. Fibre Channel devices (servers, switches, ports) are differentiated by a 64-bit identifier. The Fibre Channel standard defines different name formats for this. Some name formats guarantee that such a 64-bit identifier will only be issued once worldwide. Such identifiers are thus also known as World Wide Names (WWNs). On the other hand, 64-bit identifiers that can be issued several times in separate networks are simply called Fibre Channel Names (FCNs).

In practice this fine distinction between WWN and FCN is hardly ever noticed, with all 64-bit identifiers being called WWNs. In the following we comply with the general usage and use only the term WWN.

WWNs are differentiated into World Wide Port Names (WWPNs) and World Wide Node Names (WWNNs). As the name suggests, every port is assigned its own WWN in the form of a WWPN and in addition the entire device is assigned its own WWN in the form of a WWNN. The differentiation between WWNN and WWPN allows us to determine which ports belong to a common multiport device in the Fibre Channel network. Examples of multiport devices are intelligent disk subsystems with several Fibre Channel ports or servers with several Fibre Channel HBA cards. WWNNs could also be used to realise services such as striping over several redundant physical paths within the Fibre Channel protocol. As discussed above (Section 3.3.5), the Fibre Channel standard unfortunately does not support these options, so that such functions are implemented in the operating system or by manufacturer-specific expansions of the Fibre Channel standard.

In the fabric, each 64-bit WWPN is automatically assigned a 24-bit port address (N-Port identifier, N-Port_ID) during fabric login. The 24-bit port addresses are used within a Fibre Channel frame for the identification of transmitter and receiver of the frame. The port address of the transmitter is called the Source Identifier (S_ID) and that of the receiver the Destination Identifier (D_ID). The 24-bit addresses are hierarchically structured and mirror the topology of the Fibre Channel network. As a result, it is a simple matter for a Fibre Channel switch to recognise which port it must send an incoming frame to from the destination ID (Figure 3.21). Some of the 24-bit addresses are reserved

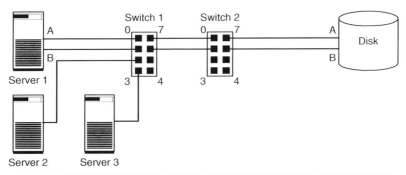

Port_ID	WWPN		WWNN		Device
010000	20000003	EAFE2C31	2100000C	EAFE2C31	Server 1, Port A
010100	20000003	C10E8CC2	2100000C	EAFE2C31	Server 1, Port B
010200	10000007	FE667122	10000007	FE667122	Server 2
010300	20000003	3CCD4431	2100000A	EA331231	Server 3
020600	20000003	EAFE4C31	50000003	214CC4EF	Disk, Port B
020700	20000003	EAFE8C31	50000003	214CC4EF	Disk, Port A

Figure 3.21 Fibre Channel differentiates end devices using World Wide Node Names (WWNN). Each connection port is assigned its own World Wide Port Name (WWPN). For addressing in the fabric WWNNs or WWPNs are converted into shorter Port_IDs that reflect the network topology.

for special purposes, so that 'only' 15.5 million addresses remain for the addressing of devices.

In the arbitrated loop every 64-bit WWPN is even assigned only an 8-bit address, the so-called Arbitrated Loop Physical Address (AL_PA). Of the 256 possible 8-bit addresses, only those for which the 8b/10b encoded transmission word contains an equal number of zeros and ones may be used. Some ordered sets for the configuration of the arbitrated loop are parametrised using AL_PAs. Only by limiting the values for AL_PAs is it possible to guarantee a uniform distribution of zeros and ones in the whole data stream. After the deduction of a few of these values for the control of the arbitrated loop, 127 addresses of the 256 possible addresses remain. One of these addresses is reserved for a Fibre Channel switch so only 126 servers or storage devices can be connected in the arbitrated loop.

3.3.7 Fabric services: name server and co

In a fabric topology the switches manage a range of information that is required for the operation of the fabric. This information is managed by the so-called fabric services. All services have in common that they are addressed via FC-2 frames and can be reached by well-defined addresses (Table 3.3). In the following paragraphs, we introduce the fabric login server, the fabric controller and the name server.

Table 3.3 The Fibre Channel standard specifies the addresses at which the auxiliary services for the administration and configuration of the Fibre Channel network can be addressed.

Address	Description
$0 \times$ FF FF FF	Broadcast addresses
$0 \times$ FF FF FE	Fabric Login Server
$0 \times$ FF FF FD	Fabric Controller
$0 \times$ FF FF FC	Name Server
$0 \times$ FF FF FB	Time Server
$0 \times$ FF FF FA	Management Server
$0 \times$ FF FF F9	Quality of Service Facilitator
$0 \times$ FF FF F8	Alias Server
$0 \times$ FF FF F7	Security Key Distribution Server
$0 \times$ FF FF F6	Clock Synchronisation Server
$0 \times$ FF FF F5	Multicast Server
$0 \times$ FF FF F4	Reserved
$0 \times$ FF FF F3	Reserved
$0 \times$ FF FF F2	Reserved
$0 \times$ FF FF F1	Reserved
$0 \times$ FF FF F0	Reserved

The fabric login server processes incoming fabric login requests under the address '$0 \times$FF FF FE'. All switches must support the fabric login under this address.

The fabric controller manages changes to the fabric under the address '$0 \times$FF FF FD'. N-Ports can register for state changes in the fabric controller (State Change Registration, SCR). The fabric controller then informs registered N-Ports of changes to the fabric (Registered State Change Notification, RSCN). Servers can use this service to monitor their storage devices.

The name server (Simple Name Server to be precise) administers a database on N-Ports under the address '$0 \times$FF FF FC'. It stores information such as port WWN, node WWN, port address, supported service classes, supported FC-4 protocols, etc. N-Ports can register their own properties with the name server and request information on other N-Ports. Like all services, the name server appears as an N-Port to the other ports. N-Ports must log on with the name server by means of port login before they can use its services.

3.3.8 FC-4 and ULPs: application protocols

The layers FC-0 to FC-3 discussed previously serve solely to connect end devices together by means of a Fibre Channel network. However, the type of data that end devices exchange via Fibre Channel connections remains open. This is where the application protocols (Upper Layer Protocols, ULPs) come into play. A specific Fibre Channel network can serve as a medium for several application protocols, for example, SCSI and IP.

The task of the FC-4 protocol mappings is to map the application protocols onto the underlying Fibre Channel network. This means that the FC-4 protocol mappings support the Application Programming Interface (API) of existing protocols upwards in the direction of the operating system and realise these downwards in the direction of the medium via the Fibre Channel network (Figure 3.22). The protocol mappings determine how the mechanisms of Fibre Channel are used in order to realise the application protocol by means of Fibre Channel. For example, they specify which service classes will be used and how the data flow in the application protocol will be projected onto the exchange sequence frame mechanism of Fibre Channel. This mapping of existing protocols aims to ease the transition to Fibre Channel networks: ideally, no further modifications are necessary to the operating system except for the installation of a new device driver.

The application protocol for SCSI is simply called Fibre Channel Protocol (FCP). FCP maps the SCSI protocol onto the underlying Fibre Channel network. For the connection of storage devices to servers the SCSI cable is therefore replaced by a Fibre Channel network. The SCSI protocol operates as before via the new Fibre Channel medium to exchange data between server and storage. It is therefore precisely at this point that the transition from server-centric IT architecture to storage-centric IT architecture takes place. Thus it is here that the Fibre Channel network becomes a Fibre Channel SAN.

The idea of the FCP protocol is that the system administrator merely installs a new device driver on the server and this realises the FCP protocol. The operating system recognises storage devices connected via Fibre Channel as SCSI devices, which it addresses like 'normal' SCSI devices. This emulation of traditional SCSI devices should make it

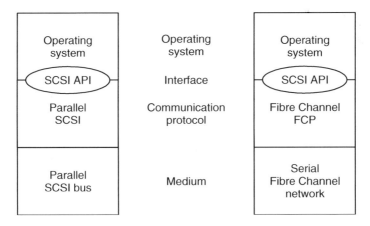

Figure 3.22   Fibre Channel FCP makes its services available to the operating system via the SCSI API. The purpose of this is to ease the transition from SCSI to Fibre Channel SAN.

possible for Fibre Channel SANs to be simply and painlessly integrated into existing hardware and software.

The FCP device driver has to achieve a great deal: SCSI uses parallel cables; daisy chain connects several devices together via a SCSI bus. By contrast, in Fibre Channel the data transmission takes place serially. The parallel transmission via the SCSI bus must therefore be serialised for the Fibre Channel SAN so that the bits are transferred one after the other. Likewise, FCP must map the daisy chain of the SCSI bus onto the underlying Fibre Channel topology. For example, the scanning for devices on a SCSI bus or the arbitration of the SCSI bus requires a totally different logic compared to the same operations in a Fibre Channel network.

A further application protocol is IPFC: IPFC uses a Fibre Channel connection between two servers as a medium for IP data traffic. To this end, IPFC defines how IP packets will be transferred via a Fibre Channel network. Like all application protocols, IPFC is realised as a device driver in the operating system. The connection into the local IP configuration takes place using 'ipconfig' or 'ifconfig'. The IPFC driver then addresses the Fibre Channel HBA card in order to transmit IP packets over Fibre Channel. The IP data traffic over Fibre Channel plays a less important role both in comparison to SCSI over Fibre Channel and in comparison to IP data traffic over Ethernet.

Fibre Connection (FICON) is a further important application protocol. FICON maps the ESCON protocol (Enterprise System Connection) used in the world of mainframes onto Fibre Channel networks. Using ESCON, it has been possible to realise storage networks in the world of mainframes since the 1990s. Fibre Channel is therefore taking the old familiar storage networks from the world of mainframes into the Open System world (Unix, Windows, OS/400, Novell, MacOS) and both worlds can even realise their storage networks on a common infrastructure.

The Fibre Channel standard also defines a few more application protocols. Particularly worth a mention is the Virtual Interface Architecture (VIA, Section 3.6.2). VIA describes a very lightweight protocol that is tailored to the efficient communication within server clusters. VIA intends to construct systems of servers and storage devices in which the boundaries between servers and storage devices disappear to an ever greater degree.

3.4   FIBRE CHANNEL SAN

The previous section introduced the fundamentals of the Fibre Channel protocol stack. This section expands our view of Fibre Channel with the aim of realising storage networks with Fibre Channel. To this end, we will first consider the three Fibre Channel topologies point-to-point, fabric and arbitrated loop more closely (Sections 3.4.1. to 3.4.3). We will then introduce some hardware components that are required for the realisation of a Fibre Channel SAN (Section 3.4.4). Building upon this, the networking of small storage network islands to form a large SAN will be discussed (Section 3.4.5). Finally, the question of interoperability in Fibre Channel SANs will be explained (Section 3.4.6).

3.4.1 Point-to-point topology

The point-to-point topology connects just two devices and is not expandable to three or more devices. For storage networks this means that the point-to-point topology connects a server to a storage device. The point-to-point topology may not be very exciting, but it offers two important advantages compared to SCSI cabling. First, significantly greater cable lengths are possible with Fibre Channel than with SCSI because Fibre Channel supports distances up to 10 kilometres without repeaters, whilst SCSI supports only up to 25 metres. Second, Fibre Channel defines various fiber-optic cables in addition to copper cables. Optical transmission via fiber-optic is robust in relation to electromagnetic interference and does not emit electromagnetic signals. This is particularly beneficial in technical environments.

Fibre Channel cables are simpler to lay than SCSI cables. For example, the SCSI SAN shown in Figure 3.7 can very simply be realised using the point-to-point topology. Application servers for the control of production can be set up close to the production machines and the data of the application server can be stored on the shared storage systems, which are located in a room that is protected against unauthorised access and physical influences such as fire, water and extremes of temperature.

3.4.2 Fabric topology

The fabric topology is the most flexible and scalable of the three Fibre Channel topologies. A fabric consists of one or more Fibre Channel switches connected together. Servers and storage devices are connected to the fabric by the Fibre Channel switches. In theory a fabric can connect together up to 15.5 million end devices. Today (2009) medium-sized installations comprise between 500 and 1,000 ports and large installations comprise several thousand ports. However, most installations are in a range below 200 ports.

End devices (servers and storage devices) connected to the various Fibre Channel switches can exchange data by means of switch-to-switch connections (inter switch links, ISLs). Several ISLs can be installed between two switches in order to increase the bandwidth. A transmitting end device only needs to know the Node_ID of the target device; the necessary routing of the Fibre Channel frame is taken care of by the Fibre Channel switches. Fibre Channel switches generally support so-called cut-through routing: cut-through routing means that a Fibre Channel switch forwards an incoming frame before it has been fully received.

The latency describes the period of time that a component requires to transmit a signal or the period of time that a component requires to forward a frame. Figure 3.23 compares the latency of different Fibre Channel SAN components. Light requires approximately 25 microseconds to cover a distance of 10 kilometres. A 10-kilometre long Fibre Channel cable thus significantly increases the latency of an end-to-end connection. For hardware components the rule of thumb is that a Fibre Channel switch can forward a frame in

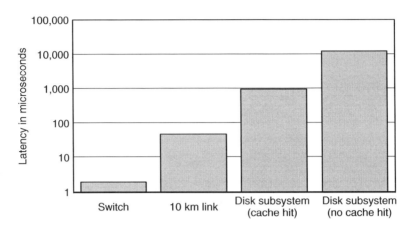

Figure 3.23 The latency of the Fibre Channel switches is low in comparison to the latency of the end devices. The latency of a 10 kilometres link in comparison to the latency of a switch is worth noting (note the logarithmic scale of the y-axis!).

2–4 microseconds; a Fibre Channel HBA requires 2–4 milliseconds to process it. Additional Fibre Channel switches between two end devices therefore only increase the latency of the network to an insignificant degree.

One special feature of the fabric is that several devices can send and receive data simultaneously at the full data rate. All devices thus have the full bandwidth available to them at the same time. Figure 3.24 shows a Fibre Channel SAN with three servers and three storage devices, in which each server works to its own storage device. Each of the three logical connections over the Fibre Channel SAN has the full bandwidth of the link speed available to them.

A prerequisite for the availability of the full bandwidth is good design of the Fibre Channel network. Figure 3.25 shows a similar structure to that in Figure 3.24, the only difference is that the single switch has been replaced by two switches, which are connected via one ISL. It is precisely this ISL that represents the limiting factor because all three logical connections now pass through the same ISL. This means that all three connections have, on average, only a third of the maximum bandwidth available to them. Therefore, despite cut-through routing, switches have a certain number of buffers (frame buffers) available to them, with which they can temporarily bridge such bottlenecks. However, the switch must still reject valid frames if the flow control does not engage quickly enough.

In addition to routing, switches realise the basic services of aliasing, name server and zoning. As described in Section 3.3.6, end devices are differentiated using 64-bit WWNNs or by 64-bit WWPNs and addressed via 24-bit port addresses (N-Port_ID). To make his job easier the administrator can issue alias names to WWNs and ports.

The name server supplies information about all end devices connected to the Fibre Channel SAN (Section 3.3.7). If an end device is connected to a switch, it reports to this and registers itself with the name server. At the same time it can ask the name server

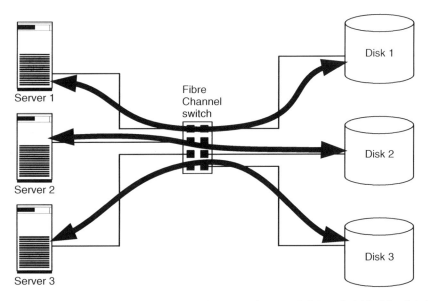

Figure 3.24 The switch can enable several connections at full bandwidth. The fabric thus has a higher total throughput (aggregate bandwidth) than the individual links.

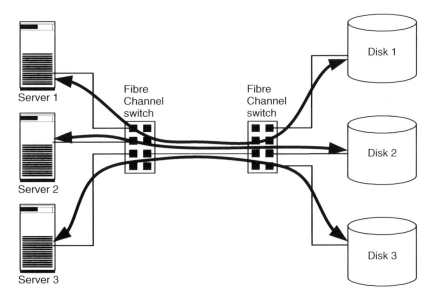

Figure 3.25 Inter-switch links (ISLs) quickly become performance bottlenecks: the total throughput of the three connections is limited by the ISL.

which other devices are still connected to the SAN. The name server administers end devices that are currently active; switched-off end devices are not listed in the name server.

Finally, zoning makes it possible to define subnetworks within the Fibre Channel network to limit the visibility of end devices. With zoning, servers can only see and access storage devices that are located in the same zone. This is particularly important for end devices such as tape drives that do not support LUN masking (Section 2.7.4). Furthermore, incompatible Fibre Channel HBAs can be separated from one another through different zoning.

When the first edition of this book was printed in 2004, we used the following words to introduce the word zoning: "There are many variants of zoning, for which unfortunately no consistent terminology exists. Different manufacturers use the same term for different types of zoning and different terms for the same type of zoning. Therefore, when selecting Fibre Channel switches do not let yourself get fobbed off with statements such as the device supports 'hard zoning'. Rather, it is necessary to ask very precisely what is meant by 'hard zoning'. In the following we introduce various types of zoning". Today (2009) this terminology is not as important because all current switches support the type of hard zoning explained in detail below. However, this confusion with terminology still exists in older white papers and handbooks. As a result, we have decided to keep the following discussion in the new edition, even if it will appear obsolete to some readers.

In zoning, the administrator brings together devices that should see each other in Fibre Channel SAN into a zone, whereby zones can overlap. Zones are described by WWNNs, WWPNs, port addresses or by their alias names. The description on the basis of WWNNs and WWPNs has the advantage that zoning is robust in relation to changes in cabling: it does not need to be changed for a device to be plugged into a different switch port. By contrast, zoning on the basis of port addresses must be altered since every port in the switch has a different port address.

Soft zoning restricts itself to the information of the name server. If an end device asks the name server about other end devices in the Fibre Channel network, it is only informed of the end devices with which it shares at least one common zone. If, however, an end device knows the address (Port_ID) of another device, it can still communicate with it. Soft zoning thus does not protect access to sensitive data. Soft zoning is problematic in relation to operating systems that store the WWNs of Fibre Channel devices that have been found in an internal database or in which WWNs are announced in configuration files because this means that WWNs remain known to the operating system even after a system reboot. Thus, in soft zoning, operating systems continue to have access to all known devices despite changes to the zoning, regardless of whether they lie in a common zone or not.

Hard zoning offers better protection. In hard zoning only devices that share at least one common zone can actually communicate with one another. Both hard zoning and soft zoning can be based upon port addresses or WWNs. Nevertheless, port-based zoning is sometimes known as hard zoning.

Some more modern Fibre Channel switches support LUN masking – described in Section 2.7.3 in relation to disk subsystems – within the switch. To achieve this they read the first bytes of the payload of each Fibre Channel frame. Although reading part of the Fibre Channel payload increases the latency of a Fibre Channel switch, this increase in latency is so minimal that it is insignificant in comparison to the latency of the HBA in the end devices.

Lastly, virtual and logical Fibre Channel SANs also deserve mention. Depending on the supplier, there is a clear difference between the concepts and names for them. Without showing any preference for a particular supplier, we will use virtual Fibre Channel SAN as a common terminology in our discussion. What all forms have in common is that multiple virtual Fibre Channel SANs can be operated over a shared physical Fibre Channel network. In some products separate fabric services such as name server and zoning are implemented for each virtual Fibre Channel SAN. In other products the end devices can be made visible in different virtual Fibre Channel SANs. Beyond pure zoning, virtual Fibre Channels SANs do not only restrict the mutual visibility of end devices but also the mutual visibility of fabric configurations. This is an advantage particularly in installations where the aim is to offer storage services through a consolidated infrastructure to different customers. It is not desirable, especially in this case, if a customer can use the name server to read-out which end devices of other customers are still connected in the storage network – or can even change the configuration of other customers.

3.4.3 Arbitrated loop topology

An arbitrated loop connects servers and storage devices by means of a ring. Data transmission in the ring can only take place in one direction. At any one time only two devices can exchange data with one another – the others have to wait until the arbitrated loop becomes free. Therefore, if six servers are connected to storage devices via an arbitrated loop, each server has on average only one sixth of the maximum bandwidth. Having components with 200 MByte/s components, in such a configuration a server can only send and receive data at 33.3 MByte/s.

In general, hubs are used to simplify the cabling (Figure 3.26). To increase the size of the arbitrated loop several hubs can be cascaded together (Figure 3.27). Hubs are invisible to the connected end devices. The arbitrated loop is less scalable and flexible than the fabric: a maximum of 126 servers and storage devices can be connected in an arbitrated loop. In addition, a switch can connect a loop to a fabric.

Arbitrated loops do not support any additional services such as aliasing, routing, name server and zoning. Therefore, the components for arbitrated loops are significantly cheaper than the components for a fabric. The price advantage of the arbitrated loop has helped the Fibre Channel technology to finally make a breakthrough. The fabric topology is now increasingly displacing the arbitrated loop due to its better scalability in the sense of the number of connected devices and the higher aggregated bandwidth of the storage network as a whole. In new installations arbitrated loop is seldom used for the connection

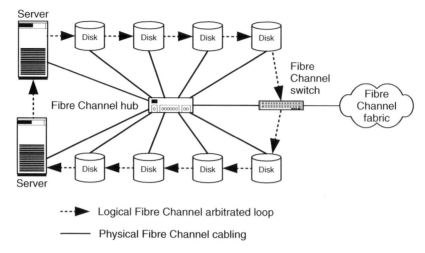

Figure 3.26   Fibre Channel hubs simplify the cabling of Fibre Channel arbitrated loops. A switch can connect a loop to a fabric.

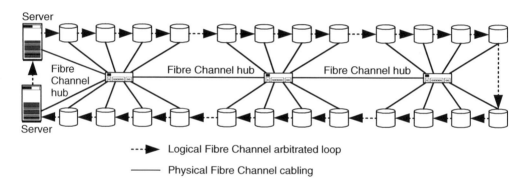

Figure 3.27   An arbitrated loop can span several hubs (cascading).

of servers and storage devices, for example in the connection of individual hard disks or individual tape drives that are fitted with Fibre Channel ports instead of SCSI ports. However, for cost reasons the arbitrated loop still remains important for the realisation of I/O buses within disk subsystems (Figure 3.2).

Arbitrated loops are subdivided into public loops and private loops (Figure 3.28). A private loop is closed in on itself; a public loop is connected to a fabric by a switch. Physically, a public loop can be connected to a fabric via several Fibre Channel switches. However, in an arbitrated loop only one switch can be active at any one time. The other switches serve merely to increase the fault tolerance if one switch fails.

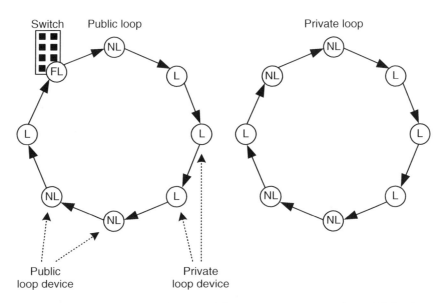

Figure 3.28 In contrast to private loops, public loops are connected to a fabric via a switch. Public loop devices master both the fabric and the loop protocol; private loop devices master only the loop protocol.

The connection of an arbitrated loop to a fabric via a Fibre Channel switch is, however, not enough to permit communication between end devices in the loop and end devices in the fabric. A device in a public arbitrated loop can only communicate with devices in the fabric if it controls both the arbitrated loop protocol and the fabric protocol. This means that the end device must have an NL-port. End devices connected to NL-Ports in arbitrated loops are called public loop devices. Figure 3.29 shows the communication of a public loop device with a device in the fabric.

Unlike public loop devices, private loop devices only have an L-Port. They therefore control only the arbitrated loop protocol. Private loop devices thus cannot be connected to a fabric and cannot communicate with devices in the fabric if they are connected in a public loop.

So-called emulated loops can help here. Emulated loops are vendor specific features produced by the Fibre Channel switch manufacturer that are not compatible with one another. Emulated loops translate between the arbitrated loop protocol and the fabric protocol within the Fibre Channel switch so that private loop devices can nevertheless exchange data with devices in the fabric. Examples of emulated loops are Quickloop from Brocade and Translated Loop from CNT/Inrange. Emulated loops no longer play an important role in new installations because new Fibre Channel devices are generally fitted with NL-Ports. In old devices, however, L-Ports that cannot be exchanged are still encountered.

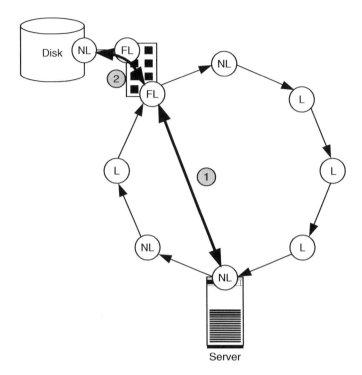

Server

Figure 3.29   For the communication from loop to fabric the public loop device must first arbitrate the arbitrated loop with the loop protocols and build up a connection to the switch (1). From there it can use the fabric protocols to build up a connection to the end device (2). The loop cannot be used by other devices during this time but the fabric can.

3.4.4 Hardware components for Fibre Channel SAN

Within the scope of this book we can only introduce the most important product groups. It is not worth trying to give an overview of specific products or a detailed description of individual products due to the short product cycles. This section mentions once again some product groups that have been discussed previously and introduces some product groups that have not yet been discussed.

It is self-evident that servers and storage devices are connected to a Fibre Channel network. In the server this can be achieved by fitting the host bus adapter cards (HBAs) of different manufacturers, with each manufacturer offering different HBAs with differing performance features. In storage devices the same HBAs are normally used. However, the manufacturers of storage devices restrict the selection of HBAs.

Of course, cables and connectors are required for cabling. In Section 3.3.2 we discussed fiber-optic cables and their properties. Various connector types are currently on offer for all cable types. It may sound banal, but in practice the installation of a Fibre Channel SAN is sometimes delayed because the connectors on the cable do not fit the connectors on the end devices, hubs and switches and a suitable adapter is not to hand.

A further, initially improbable, but important device is the so-called Fibre Channel-to-SCSI bridge. As the name suggests, a Fibre Channel-to-SCSI bridge creates a connection between Fibre Channel and SCSI (Figure 3.30). Old storage devices often cannot be converted from SCSI to Fibre Channel. If the old devices are still functional they can continue to be used in the Fibre Channel SAN by the deployment of a Fibre Channel-to-SCSI bridge. Unfortunately, the manufacturers have not agreed upon a consistent name for this type of device. In addition to Fibre Channel-to-SCSI bridge, terms such as SAN router or storage gateway are also common.

The switch is the control centre of the fabric topology. It provides routing and aliasing, name server and zoning functions. Fibre Channel switches support both cut-through routing and the buffering of frames. The size of switches range from 8 ports to about 250.

Resilient, enterprise-class switches are commonly referred to as 'directors', named after the switching technology used in mainframe ESCON cabling. Like Fibre Channel switches they provide routing, alias names, name server and zoning functions. Fibre Channel directors are designed to avoid any single point of failure, having for instance two backplanes and two controllers. Current directors (2009) have between 64 and 256 ports.

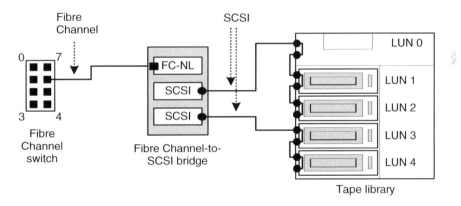

Figure 3.30   Fibre Channel-to-SCSI bridges translate between Fibre Channel FCP and SCSI. This makes it possible to connect old SCSI devices into a Fibre Channel SAN.

Designing a Fibre Channel SAN often raises the question whether several complementary switches or a single director should be preferred. As described, directors are more fault-tolerant than switches, but they are more expensive per port. Therefore, designers of small entry-level SANs commonly choose two complementary Fibre Channel switches, with mutual traffic fail-over in case of a switch or a I/O path failure (Figure 3.31). Designers of larger Fibre Channel SANs often favour directors due to the number of ports currently available per device and the resulting layout simplicity. However, this argument in favour of directors becomes more and more obsolete because today switches with a greater number of ports are available as well.

Fibre Channel SANs running especially critical applications, e.g. stock market banking or flight control, would use complementary directors with mutual traffic fail-over, even though these directors already avoid internal single points of failure. This is similar to wearing trousers with a belt and braces in addition: protecting against double or triple failures. In less critical cases, a single director or a dual complementary switch solution will be considered sufficient.

If we disregard the number of ports and the cost, the decision for a switch or a director in an Open Systems Fibre Channel network primarily comes down to fault tolerance of an individual component. For the sake of simplicity we will use the term 'Fibre Channel switch' throughout this book in place of 'Fibre Channel switch or Fibre Channel director'.

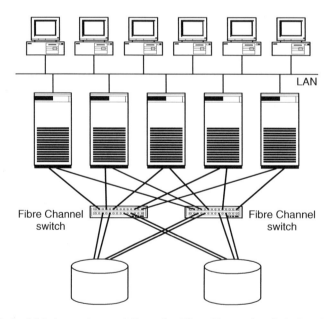

Figure 3.31   A dual fabric, each consisting of a Fibre Channel switch, is a typical entry-level configuration. If the switch, an HBA or a cable fails in a Fibre Channel SAN, then the server can still access the data via the second one. Likewise, zoning errors are isolated to one of the two Fibre Channel SANs.

A hub simplifies the cabling of an arbitrated loop. Hubs are transparent from the point of view of the connected devices. This means that hubs repeat the signals of the connected devices; in contrast to a Fibre Channel switch, however, the connected devices do not communicate with the hub. Hubs change the physical cabling from a ring to a star-shape. Hubs bridge across defective and switched-off devices so that the physical ring is maintained for the other devices. The arbitrated loop protocol is located above this cabling.

Hubs are divided into unmanaged hubs, managed hubs and switched hubs. Unmanaged hubs are the cheap version of hubs: they can only bridge across switched-off devices. However, they can neither intervene in the event of protocol infringements by an end device nor indicate the state of the hub or the arbitrated loop to the outside world. This means that an unmanaged hub cannot itself notify the administrator if one of its components is defective. A very cost-conscious administrator can build up a small Fibre Channel SAN from PC systems, Just a Bunch of Disks (JBODs) and unmanaged hubs. However, the upgrade path to a large Fibre Channel SAN is difficult: in larger Fibre Channel SANs it is questionable whether the economical purchase costs compensate for the higher administration costs.

In contrast to unmanaged hubs, managed hubs have administration and diagnosis functions like those that are a matter of course in switches and directors. Managed hubs monitor the power supply, serviceability of fans, temperature, and the status of the individual ports. In addition, some managed hubs can, whilst remaining invisible to the connected devices, intervene in higher Fibre Channel protocol layers, for example, to deactivate the port of a device that frequently sends invalid Fibre Channel frames. Managed hubs, like switches and directors, can inform the system administrator about events via serial interfaces, Telnet, Hypertext Transfer Protocol (HTTP) and Simple Network Management Protocol (SNMP) (see also Chapter 10).

Finally, the switched hub is mid-way between a hub and a switch. In addition to the properties of a managed hub, with a switched hub several end devices can exchange data at full bandwidth. Fibre Channel switched hubs are cheaper than Fibre Channel switches, so in some cases they represent a cheap alternative to switches. However, it should be noted that only 126 devices can be connected together via hubs and that services such as aliasing and zoning are not available. Furthermore, the protocol cost for the connection or the removal of a device in a loop is somewhat higher than in a fabric (keyword 'Loop Initialisation Primitive Sequence, LIP').

Finally, so-called link extenders should also be mentioned. Fibre Channel supports a maximum cable length of several ten kilometres (Section 3.3.2). A link extender can increase the maximum cable length of Fibre Channel by transmitting Fibre Channel frames using WAN techniques such as ATM, SONET or TCP/IP (Figure 3.32). When using link extenders it should be borne in mind that long distances between end devices significantly increase the latency of a connection. Time-critical applications such as database transactions should therefore not run over a link extender. On the other hand, Fibre Channel SANs with link extenders offer new possibilities for applications such as backup, data sharing and asynchronous data mirroring.

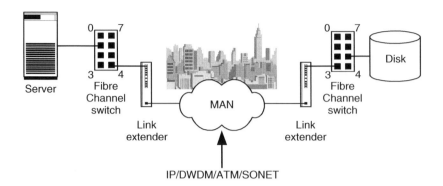

Figure 3.32   A link extender can connect two storage networks over long distances.

3.4.5   InterSANs

Fibre Channel SAN is a comparatively new technology. In many data centres in which Fibre Channel SANs are used, it is currently (2009) more likely that there will be several islands of small Fibre Channel SANs than one large Fibre Channel SAN (Figure 3.33). Over 80% of the installed Fibre Channel SANs consist only of up to four Fibre Channel switches. A server can only indirectly access data stored on a different SAN via the LAN and a second server. The reasons for the islands of small Fibre Channel SANs are that they are simpler to manage than one large Fibre Channel SAN and that it was often unnecessary to install a large one.

Originally, Fibre Channel SAN was used only as an alternative to SCSI cabling. Until now the possibility of flexibly sharing the capacity of a storage device between several servers (storage pooling) and the improved availability of dual SANs have been the main reasons for the use of Fibre Channel SANs. Both can be realised very well with several small Fibre Channel SAN islands. However, more and more applications are now exploiting the possibilities offered by a Fibre Channel SAN. Applications such as backup (Chapter 7), remote data mirroring and data sharing over Fibre Channel SAN (Chapter 6) and storage virtualisation (Chapter 5) require that all servers and storage devices are connected via a single Fibre Channel SAN.

Incidentally, the connection of Fibre Channel SANs to form a large SAN could be one field of application in which a Fibre Channel director is preferable to a Fibre Channel switch (Figure 3.34). As yet these connections are generally not critical. In the future, however, this could change (extreme situation: virtualisation over several data centres). In our opinion these connection points between two storage networks tend to represent a single point of failure, so they should be designed to be particularly fault-tolerant.

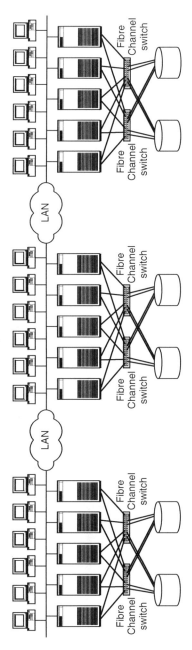

Figure 3.33 Current IT environments (2009) consist almost exclusively of small Fibre Channel SANs that are designed with built-in redundancy. If a server has to access storage devices that are connected to a different SAN, it has to take the indirect route via the LAN and a further server.

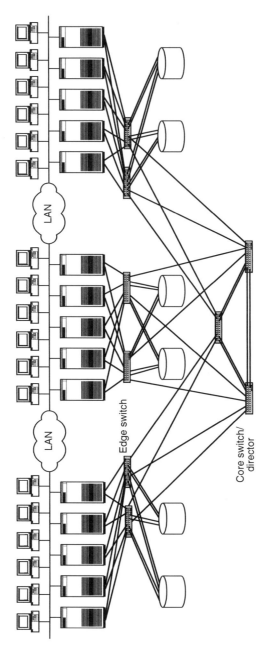

Figure 3.34 More and more applications are exploiting the new possibilities offered by Fibre Channel SAN. In order to make better use of the new possibilities of storage networks it is necessary to connect the individual storage islands.

3.4.6   Interoperability of Fibre Channel SAN

Fibre Channel SANs are currently being successfully used in production environments. Nevertheless, interoperability is an issue with Fibre Channel SAN, as in all new cross-manufacturer technologies. When discussing the interoperability of Fibre Channel SAN we must differentiate between the interoperability of the underlying Fibre Channel network layer, the interoperability of the Fibre Channel application protocols, such as FCP (SCSI over Fibre Channel) and the interoperability of the applications running on the Fibre Channel SAN.

The interoperability of Fibre Channel SAN stands and falls by the interoperability of FCP. FCP is the protocol mapping of the FC-4 layer, which maps the SCSI protocol on a Fibre Channel network (Section 3.3.8).

FCP is a complex piece of software that can only be implemented in the form of a device driver. The implementation of low level device drivers alone is a task that attracts errors as if by magic. The developers of FCP device drivers must therefore test extensively and thoroughly.

Two general conditions make it more difficult to test the FCP device driver. The server initiates the data transfer by means of the SCSI protocol; the storage device only responds to the requests of the server. However, the idea of storage networks is to consolidate storage devices, i.e. for many servers to share a few large storage devices. Therefore, with storage networks a single storage device must be able to serve several parallel requests from different servers simultaneously. For example, it is typical for a server to be exchanging data with a storage device just when another server is scanning the Fibre Channel SAN for available storage devices. This situation requires end devices to be able to multitask. When testing multitasking systems the race conditions of the tasks to be performed come to bear: just a few milliseconds delay can lead to a completely different test result.

The second difficulty encountered during testing is due to the large number of components that come together in a Fibre Channel SAN. Even when a single server is connected to a single storage device via a single switch, there are numerous possibilities that cannot all be tested. If, for example, a Windows server is selected, there was in the year 2004 the choice between NT, 2000 and 2003, each with different service packs. Several manufacturers offer several different models of the Fibre Channel HBA card in the server. If we take into account the various firmware versions for the Fibre Channel HBA cards we find that we already have more than 50 combinations before we even select a switch.

Companies want to use their storage network to connect servers and storage devices from various manufacturers, where some of which are already installed. The manufacturers of Fibre Channel components (servers, switches and storage devices) must therefore perform interoperability tests in order to guarantee that these components work with devices from third-party manufacturers. Right at the top of the priority list are those combinations that are required by most customers, because this is where the expected profit is the highest. The result of the interoperability test is a so-called support matrix. It specifies, for example, which storage device supports which server model with which operating system versions and Fibre Channel cards. Manufacturers of servers and storage devices often limit the Fibre Channel switches that can be used.

Therefore, before building a Fibre Channel SAN you should carefully check whether the manufacturers in question state that they support the planned configuration. If the desired configuration is not listed, you can negotiate with the manufacturer regarding the payment of a surcharge to secure manufacturer support. Although non-supported configurations can work very well, if problems occur, you are left without support in critical situations. If in any doubt you should therefore look for alternatives right at the planning stage.

All this seems absolutely terrifying at first glance. However, manufacturers now support a number of different configurations. If the manufacturers' support matrices are taken into consideration, robust Fibre Channel SANs can now be operated. The operation of up-to-date operating systems such as Windows, AIX, Solaris, HP-UX and Linux is particularly unproblematic.

Fibre Channel SANs are based upon Fibre Channel networks. The incompatibility of the fabric and arbitrated loop topologies and the coupling of fabrics and arbitrated loops have already been discussed in Section 3.4.3. The potential incompatibilities of the Fibre Channel switches within the fabric and from different suppliers should also be mentioned. There was some experience in 2009 with Fibre Channel SANs in which switches of different suppliers were installed. A standard that addresses the interoperability of basic functions such as name servers and zoning was adopted some time ago. However, the assumption is that there will be some teething problems with interoperability with new functions such as SAN security, inter-switch-link-bundling (ISL trunking) and B-ports for linking Fibre Channel SANs over WAN protocols. These problems are likely to surface in heterogeneous networks in particular where restrictions in vendor-specific extensions are expected. When a Fibre Channel SAN is installed using switches of different suppliers, it is therefore recommended that possible restrictions be checked into in advance.

In general, applications can be subdivided into higher applications that model and support the business processes and system-based applications such as file systems, databases and backup systems. The system-based applications are of particular interest from the point of view of storage networks and storage management. The compatibility of network file systems such as NFS and CIFS is now taken for granted and hardly ever queried. As storage networks penetrate into the field of file systems, cross-manufacturer standards are becoming ever more important in this area too. A first offering is Network Data Management Protocol (NDMP, Section 7.9.4) for the backup of NAS servers. Further down the road we expect also a customer demand for cross-vendor standards in the emerging field of storage virtualisation (Chapter 5).

The subject of interoperability will preoccupy manufacturers and customers in the field of storage networks for a long time to come. VIA, InfiniBand and Remote Direct Memory Access (RDMA) represent emerging new technologies that must also work in a cross-manufacturer manner. The same applies for iSCSI and its variants like iSCSI Extensions for RDMA (iSER). iSCSI transmits the SCSI protocol via TCP/IP and, for example, Ethernet. Just like FCP, iSCSI has to serialise the SCSI protocol bit-by-bit and map it onto a complex network topology. Interoperability will therefore also play an important role in iSCSI and FCoE.

3.5 IP STORAGE

Fibre Channel SANs are being successfully implemented in production environments. Nevertheless, a transmission technology is desired that is suitable for both storage traffic (SCSI) and LAN (TCP/IP). IP storage is an approach to build storage networks upon TCP, IP and Ethernet.

This section first introduces various protocols for the transmission of storage data traffic via TCP/IP (Section 3.5.1). Then we explain to what extent TCP/IP and Ethernet are suitable transmission techniques for storage networks at all (Section 3.5.2). Finally, we discuss a migration path from SCSI and Fibre Channel to IP storage (Section 3.5.3).

3.5.1 IP storage standards: iSCSI, iFCP, mFCP, FCIP and iSNS

Three protocols are available for transmitting storage data traffic over TCP/IP: iSCSI, Internet FCP (iFCP) and Fibre Channel over IP (FCIP) – not to be confused with IPFC (Section 3.3.8). They form the family of IP-based storage protocols that are also known as IP storage. These standards have in common that in one form or another they transmit SCSI over IP and thus in practice usually over Ethernet. Of all IP-based storage protocols iSCSI and FCIP are most relevant.

'Storage over IP (SoIP)' is sometimes called a standard in association with IP storage. This is incorrect: SoIP is a discontinued product of the former corporation Nishan Technologies that, according to the manufacturer, is compatible with various IP storage standards.

The basic idea behind iSCSI is to transmit the SCSI protocol over TCP/IP (Figure 3.35). iSCSI thus takes a similar approach to Fibre Channel SAN, the difference being that in iSCSI a TCP/IP/Ethernet connection replaces the SCSI cable. Just like Fibre Channel FCP, iSCSI has to be installed in the operating system as a device driver. Like FCP, this realises the SCSI protocol and maps the SCSI daisy chain onto a TCP/IP network.

The Internet Engineering Task Force (IETF) ratified the iSCSI standard at the beginning of 2003. More and more iSCSI-capable products are now becoming available. Conventional Ethernet network cards, which require the installation of a separate iSCSI device driver that implements the protocol in software to the cost of the server CPU, can be used on servers. Alternatively, iSCSI HBAs, which, compared to TCP/IP offload engines (TOE), realise the iSCSI/TCP/IP Ethernet protocol stack in hardware, are also available. Measurements with iSCSI HBAs (network cards that handle a major part of the

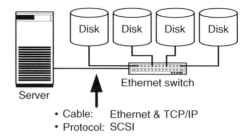

Server

• Cable: Ethernet & TCP/IP
• Protocol: SCSI

Figure 3.35 Like Fibre Channel, iSCSI replaces the SCSI cable by a network over which the SCSI protocol is run. In contrast to Fibre Channel, TCP/IP and Ethernet are used as the transmission technology.

iSCSI/TCP/IP/Ethernet protocol stack on the network) prove that these ease the burden on the server CPU considerably. Therefore, iSCSI HBAs can be used for higher-performance requirements, whereas conventional and thus clearly less expensive Ethernet cards are adequate for low or perhaps even average performance requirements. It is conceivable that future developments of the iSCSI standard, as an alternative to TCP/IP, will also be run on UDP/IP, IP or in the form of SDP and iSER on RDMA, thereby improving performance even more.

Fibre Channel still dominates in production environments. One is seeing more and more references to the successful use of iSCSI, as well as descriptions of iSCSI SANs, on the web pages of relevant vendors. A frequent field of application is the area of Windows servers in small and medium companies or at the department level. Another one is with virtualised Windows servers running on products such as VMware or XEN. iSCSI-capable storage devices or iSCSI-to-Fibre-Channel gateways are deployed to make the storage capacity in a Fibre Channel SAN available to the servers over iSCSI. Another field of application is the booting of the diskless nodes of cluster nodes by means of iSCSI (Figure 3.36). These take their hard disks, including boot image and operating system, over iSCSI. This kind of transfer of storage capacity from internal hard disks to external storage systems comes with the normal cost benefits associated with storage networks. The IETF ratified booting over iSCSI as a standard in 2005.

In contrast to iSCSI, which defines a new protocol mapping of SCSI on TCP/IP, iFCP describes the mapping of Fibre Channel FCP on TCP/IP. The idea is to protect the investment in a large number of Fibre Channel devices that have already been installed and merely replace the Fibre Channel network infrastructure by an IP/Ethernet network infrastructure. The developers of iFCP expect that this will provide cost benefits in relation to a pure Fibre Channel network. For the realisation of iFCP, LAN switches must either provide a Fibre Channel F-Port or an FL-Port. Alternatively, Fibre Channel FCP-to-iFCP gateways could also be used (Figure 3.37).

The difference between Metro FCP (mFCP) and iFCP is that mFCP is not based upon TCP/IP but on UDP/IP. This means that mFCP gains performance at the expense of the reliability of the underlying network connection. The approach of replacing TCP/IP

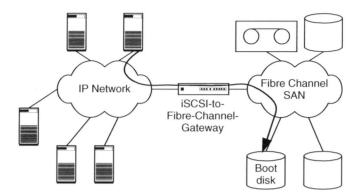

Figure 3.36   When booting over iSCSI the internal hard disk and the Fibre Channel host bus adapter can be dispensed with.

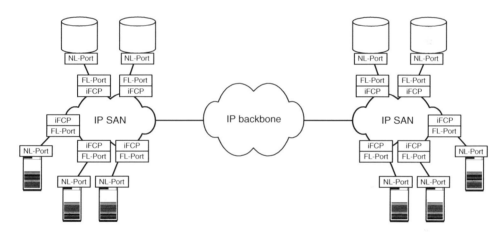

Figure 3.37   iFCP is a gateway protocol that connects Fibre Channel devices via a TCP/IP network.

by UDP/IP has proved itself many times. For example, NFS was originally based upon TCP/IP, but today it can be based upon TCP/IP or UDP/IP. Error correction mechanisms in the application protocol (in this case NFS or mFCP) ensure that no data is lost. This is only worthwhile in low-error networks such as LANs.

In order to provide fabric services, iFCP/mFCP must evaluate the Fibre Channel frames received from the end devices and further process these accordingly. It forwards data packets for a different end device to the appropriate gateway or switch via TCP/IP. Likewise, it also has to map infrastructure services of the fabric such as zoning and name service on TCP/IP.

The IETF ratified iFCP as a standard in 2005. However, in our opinion, the benefits of iFCP and mFCP remain to be proven. Both protocols make a very elegant attempt at protecting investments by connecting existing Fibre Channel storage devices into IP-based storage networks. In addition, the use of iFCP/mFCP makes it possible to take advantage of the fully developed techniques, services and management tools of an IP network. However, iFCP and mFCP are complex protocols that require intensive testing before cross-manufacturer compatibility can be ensured. They offer few new benefits for the transmission of Fibre Channel FCP over IP. Today Fibre Channel-to-iSCSI gateways and the FCIP mentioned below offer alternative methods for connecting existing Fibre Channel devices over IP. Therefore, the benefits, and, hence, the future of iFCP/mFCP remain in doubt in view of the implementation and testing effort required by manufacturers of iFCP/mFCP components.

The third protocol for IP storage, FCIP, was designed as a supplement to Fibre Channel, in order to remove the distance limitations of Fibre Channel. Companies are increasingly requiring longer distances to be spanned, for example for data mirroring or to backup data to backup media that is a long way from the production data in order to prevent data loss in the event of large-scale catastrophes. Until now, such requirements meant that either the tapes had to be sent to the backup data centre by courier or comparatively expensive and difficult to manage WAN techniques such as Dark Fiber, DWDM or SONET/SDH had to be used.

FCIP represents an alternative to the conventional WAN techniques: it is a tunnelling protocol that connects two Fibre Channel islands together over a TCP/IP route (Figure 3.38). FCIP thus creates a point-to-point connection between two Fibre Channel SANs and simply encapsulates all Fibre Channel frames into TCP/IP packets. The use of FCIP remains completely hidden from the Fibre Channel switches, so both of the Fibre Channel SANs connected using FCIP merge into a large storage network. Additional services and drivers are unnecessary.

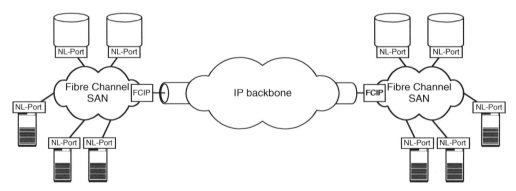

Figure 3.38   FCIP is a tunnelling protocol that connects two Fibre Channel SANs by means of TCP/IP.

A further advantage of FCIP compared to the connection of Fibre Channel networks using conventional WAN techniques lies in the encryption of the data to be transmitted. Whereas encryption techniques are still in their infancy in Dark Fibre, DWDM and SONET/SDH, the encryption of the data traffic between two IP routers by means of IPSec has now become a standard technique.

FCIP components have been on the market for a number of years and are being used successfully in production environments. The IETF ratified the FCIP standard in 2004. Nevertheless, as with all new network techniques, it will be necessary to conduct a very careful study of the extent of interoperability of the FCIP components of different manufacturers. This particularly applies when enhancements that have not yet been standardised are to be used. When any doubt exists, one should use FCIP-to-Fibre-Channel gateways from the same manufacturer at both ends of an FCIP route.

A common feature of all the protocols introduced here is that they transmit SCSI data traffic over IP in one form or another. In addition to the data transmission, a service is required to help scan for devices or communication partners in the IP network and to query device properties. Internet Storage Name Service (iSNS) is a standard that defines precisely such a service. iSNS is a client-server application, in which the clients register their attributes with the server, which for its part informs clients about changes to the topology. Both iSCSI and iFCP integrate iSNS. FCIP does not need to do this because it only provides a transmission route between two Fibre Channel SANs and thus has the same function as a Fibre Channel cable.

iSCSI, iFCP and FCIP are similar protocols that can easily be mistaken for one another. Therefore it makes sense to contrast these protocols once again from different points of view. Figure 3.39 compares the protocol stacks of the different approaches: FCP is realised completely by Fibre Channel protocols. FCIP creates a point-to-point connection between two Fibre Channel SANs, with all Fibre Channel frames simply being packetised in TCP/IP packets. iFCP represents an expansion of FCIP, since it not only tunnels Fibre Channel frames but also realises fabric services such as routing, name server and zoning over TCP/IP. Finally, iSCSI is based upon TCP/IP without any reference to Fibre Channel. Table 3.4 summarises which parts of the protocol stack in question are realised by Fibre Channel and which of those by TCP/IP/Ethernet. Finally, Figure 3.40 compares the frame formats.

3.5.2 TCP/IP and Ethernet as an I/O technology

From a technical point of view, Fibre Channel has some advantages in relation to IP storage: the Fibre Channel protocol stack is integrated and thus very efficient. In comparison to Fibre Channel, TCP/IP has a significantly higher protocol overhead. Whilst Fibre Channel assumes an underlying reliable network, TCP includes protective mechanisms for unreliable transmission techniques such as the detection and re-transmission of lost messages. Furthermore, Fibre Channel has for some years been successfully used in production environments. IP storage entered only a few niche markets. By contrast to Fibre

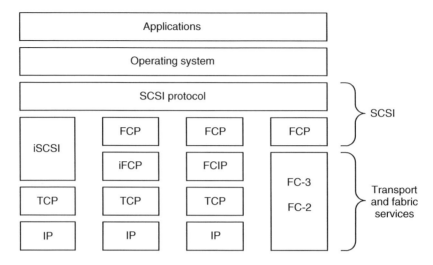

Figure 3.39 All protocols are addressed from the operating system via SCSI.

Table 3.4 In contrast to FCIP, iFCP manages without Fibre Channel networks.

	End device	Fabric services	Transport
FCP	Fibre Channel	Fibre Channel	Fibre Channel
FCIP	Fibre Channel	Fibre Channel	IP/Ethernet
iFCP	Fibre Channel	IP/Ethernet	IP/Ethernet
iSCSI	IP/Ethernet	IP/Ethernet	IP/Ethernet

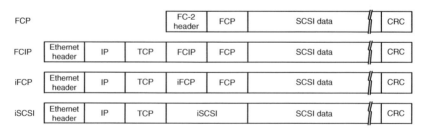

Figure 3.40 Fibre Channel FCP has the lowest protocol overhead in comparison to the IP-based SCSI protocols.

Channel, only a few production environments exist which use alternatives such as iSCSI and iFCP.

In what follows we will describe the reasons why we nevertheless believe that IP storage must be considered as an appropriate technique for storage networks. To this end, we will first explain the advantages and disadvantages of IP storage and then show in Section 3.5.3 a migration path from Fibre Channel to IP storage.

Proponents of IP storage cite the following advantages in relation to Fibre Channel:

- Common network for LAN, MAN, WAN, SAN, voice and probably video;
- Standardisation and maturity of technology since TCP/IP and Ethernet have been in use for decades;
- More personnel are available with TCP/IP knowledge than with knowledge of Fibre Channel;
- TCP/IP have no distance limits;
- Cheaper hardware, since competition is greater in the field of TCP/IP than Fibre Channel due to the higher market volume;
- Availability of comprehensive administration tools for TCP/IP networks

In the following we will discuss how these supposed advantages of IP storage are not as clear-cut as they might appear. However, let us first also mention the supposed disadvantages of IP storage:

- Lack of interoperability of IP storage;
- High CPU use for storage data traffic via TCP/IP;
- Greater TCP/IP overhead, since the protocol is not designed for bulk data;
- High latency of TCP/IP/Ethernet switches;
- Low exploitation of the bandwidth of Ethernet (20–30%) due to the typical collisions for Ethernet

In what follows we will also investigate the listed disadvantages, some of which contradict the previously listed advantages of IP storage that are often put forward.

It is correct that when using IP storage LAN, MAN, WAN and SAN can be operated via common physical IP networks (Figure 3.41). However, it should be borne in mind that in many environments the LAN-MAN-WAN network is already utilised at its limit. This means that when using IP storage, just as when using Fibre Channel SAN, additional network capacity must be installed. It is questionable whether it is organisationally possible for the IP network for LAN to be managed by the same people who manage the IP network for IP storage; with LAN, access to data is restricted by the applications so that the LAN administrator cannot simply access confidential data. In IP storage, on the other hand, the administrator can access significantly more data.

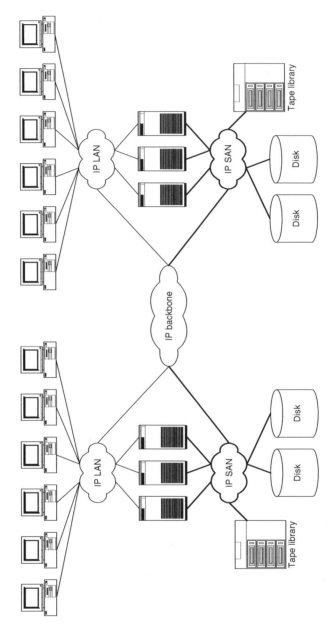

Figure 3.41   With IP storage LAN, MAN, WAN and SAN can be based upon a common network infrastructure. In practice, the LAN is usually heavily loaded for the data traffic between clients and servers, which means that a further LAN must be installed for SAN data traffic. LANs and SANs can be connected via the same IP backbone.

Nevertheless, there is increasingly a trend towards handling all data traffic over IP and Ethernet. Conventional data networks use almost traditional TCP/IP and its application protocols such as HTTP, FTP, NFS, CIFS or SMTP. In pilot projects Gigabit Ethernet is already being used for the networking of schools, authorities and households in MANs. It is therefore easily possible that Gigabit Ethernet will at some point supersede Digital Subscriber Line (DSL) for connecting companies, authorities and households to the broadband Internet (the Internet of the future). In addition, telephoning over IP (Voice over IP, VoIP) has been in use in new office buildings for some time. If locations a long distance apart frequently have to be in telephone contact, telephoning over the Internet can save an immense amount of money.

The standardisation of all data traffic – from telephony through LAN to storage networks – to IP networks would have certain advantages. If only IP networks were used in office buildings, the available bandwidth could be provided to different types of data traffic as required. In an extreme case, the capacity could be used for different purposes depending upon the time of day, for example, for telephone calls during the day and for network backup during the night.

In addition, many companies rent dedicated IP connections for the office data traffic, which are idle during the night. FCIP allows the network capacity to be used to copy data without renting additional lines. A throughput of between 30 and 40 MBytes/s was measured already in 2002 for tapes written over a Gigabit-Ethernet route of 10 kilometres in length using FCIP. A higher throughput was limited by the tape drive used. Considerable cost savings are thus possible with FCIP because the WAN connections that are already available and occasionally not utilited can also be used.

Furthermore, the standardisation of all communications to TCP/IP/Ethernet ensures further cost savings because the market volume of TCP/IP/Ethernet components is significantly greater than that of any other network technology segment. For example, the development and testing cost for new components is distributed over a much larger number of units. This gives rise to greater competition and ultimately to lower prices for Ethernet components than for Fibre Channel components. However, high-end LAN switches and high-end LAN routers also come at a price, so we will have to wait and see how great the price advantage is.

The availability of personnel with knowledge of the necessary network technology is a point in favour of IP and Gigabit Ethernet. IP and Ethernet have been in use for LANs for many years. Knowledge regarding these technologies is therefore widespread. Fibre Channel, on the other hand, is a young technology that is mastered by few people in comparison to IP and Ethernet. There is nothing magical about learning to use Fibre Channel technology. However, it costs money and time for the training of staff, which is usually not necessary for IP and Ethernet. However, training is also necessary for IP SANs, for example for iSCSI and iSCSI SAN concepts.

It is correct to say that, compared to TCP/IP/Ethernet, there are very few tools in the market today that can help in the management of heterogeneous Fibre Channel SANs. Some tools now available display the topology of Fibre Channel SANs, even in cases where network components and end devices from different manufacturers are used. Administration tools for TCP/IP networks are more advanced. For example, they offer more options for controlling data traffic over specific links. However, extensions for TCP/IP/Ethernet are also still needed, for example, for storage management it is necessary to know which servers are using how much storage on which storage devices and how great the load is on the storage devices in question as a result of write and read access. Although this information can be found for individual servers, there are no tools available today that help to determine the storage resource usage of all the servers in a heterogeneous environment (compare with Chapter 10).

In connection with IP storage, the vision is sometimes put forward that servers will store their data on storage systems that export virtual hard disks on the Internet – TCP/IP makes this possible. However, we have to keep in mind the fact that the Internet today has a high latency and the transmission rates achieved sometimes fluctuate sharply. This means that storage servers on the Internet are completely unsuitable for time-critical I/O accesses such as database transactions. Even if the performance of the Internet infrastructure increases, the transmission of signals over long distances costs time. For this reason, a database server in London will never access virtual hard disks in New York. This scenario is therefore only of interest for services that tolerate a higher network latency, such as the copying, backup, replication or asynchronous mirroring of data.

The assertion that IP storage will have no interoperability problems because the underlying TCP/IP technology has been in use for decades is nonsense. The protocols based upon TCP/IP such as iSCSI or iFCP have to work together in a cross-manufacturer manner just like Fibre Channel SAN. In addition, there is generally room for interpretation in the implementation of a standard. Experiences with Fibre Channel show that, despite standardisation, comprehensive interoperability testing is indispensable (Section 3.4.6). Interoperability problems should therefore be expected in the first supposedly standard-compliant products from different manufacturers.

It is correct that TCP/IP data traffic is very CPU-intensive. Figure 3.42 compares the CPU load of TCP/IP and Fibre Channel data traffic. The reason for the low CPU load of Fibre Channel is that a large part of the Fibre Channel protocol stack is realised on the Fibre Channel HBA. By contrast, in current network cards a large part of the TCP/IP protocol stack is processed on the server CPU. The communication between the Ethernet network card and the CPU takes place via interrupts. This costs additional computing power, because every interrupt triggers an expensive process change in the operating system. However, more and more manufacturers are now offering so-called TCP/IP offload engines (TOEs). These are network cards that handle most of the TCP/IP protocol stack and thus greatly free up the CPU. Now even iSCSI HBAs are available, which in addition to TCP/IP also realise the iSCSI protocol in hardware. Measurements have shown that the CPU load can be significantly reduced.

The Fibre Channel protocol stack is a integrated whole. As a result, cut-through routing is comparatively simple to realise for Fibre Channel switches. By contrast, TCP/IP and Ethernet were developed independently and not harmonised to one another. In the TCP/IP/Ethernet protocol stack the IP layer is responsible for the routing. So-called Level-3 routers permit the use of cut-through routing by analysing the IP data traffic

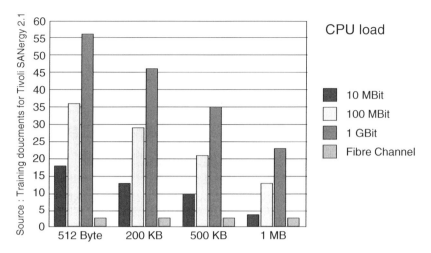

Figure 3.42 TCP/IP data traffic places a load on the server CPU. The CPU load of Fibre Channel is low because its protocol is mainly processed on the HBA. (The y-axis shows the CPU load of a 400 MHz CPU).

and then realising the cut-through routing a layer below on the Ethernet layer. It is therefore highly probable that the latency of an Ethernet/IP switch will always be poorer than the latency of a Fibre Channel switch. How relevant this is to the performance of IP storage is currently unknown: Figure 3.23 shows that in today's Fibre Channel SANs the latency of the switches is insignificant in comparison to the latency of the end devices. The economic advantages of IP storage discussed above would presumably be negated if IP storage required different IP/Ethernet switches than the switches for LAN/MAN/WAN data traffic.

Proponents of Fibre Channel sometimes assert that TCP/IP and Ethernet is inefficient where there are several simultaneous transmitters because in this situation the collisions that occur in Ethernet lead to the medium only being able to be utilised at 20–30%. This statement is simply incorrect. Today's Ethernet switches are full duplex just like Fibre Channel switches. Full duplex means that several devices can exchange data in pairs using the full bandwidth, without interfering with each other.

To summarise the discussion above, IP storage will at least fulfil the performance requirements of many average applications. The question 'To what degree IP storage is also suitable for business critical applications with extremely high performance requirements?' is yet to be answered. An I/O technique for high performance applications must guarantee a high throughput at a low CPU load and a low latency (delay) of data transmission. Since no practical experience is available regarding this question, only theoretical considerations are possible.

As discussed, even now the CPU load is under control with iSCSI HBAs. 10-Gigabit Ethernet components provide sufficient bandwidth between servers and storage devices, the more so as trunking brings together several 10-Gigabit Ethernet connections into one virtual connection that provides an even greater bandwidth.

It is more difficult to make predictions regarding the effect of the latency of IP storage, which will probably be higher, on the performance of applications. We will have to wait for relevant experiences in production environments. The current reference installations are not yet sufficient to make a comprehensive judgement that is proven by practical experience. Furthermore, in the more distant future we can hope for improvements if techniques such as RDMA, VI and InfiniBand are drawn into storage networks, and protocols such as iSCSI, iSER and SDP are based directly upon these new techniques (Sections 3.6).

It will be a long time before IP storage can displace Fibre Channel techniques in business-critical areas of operation. Current experience (2009) shows that IP storage works well with applications with low and medium performance requirements. A growing number of known manufacturers are offering more and more IP storage products. During the beginnings of IP Storage, at least one of the authors was convinced that IP storage would gain a large share of the market for storage networks because of the economic advantages over Fibre Channel discussed above. Thus in earlier editions of this book we acknowledged IP storage the potential of displacing Fibre Channel in the long term, much like ATM and FDDI that once had similar ambitions. Meanwhile FCoE, which is a further alternative to Fibre Channel and IP storage, is entering the data centers (Section 3.7). It may be the case that FCoE displaces IP storage before IP storage displaces Fibre Channel.

3.5.3 Migration from Fibre Channel to IP storage

Anyone investing in a storage network today wants the use of that investment to be long term. This means investing in technologies that solve real problems. It also means investing in techniques that have a long life cycle ahead of them. Ultimately, it is not only a matter of procuring the necessary hardware and software but also training staff and gathering experience in production environments.

Anyone who wants to use storage networks today (2009) can choose between IP storage and Fibre Channel for small and medium environments with average performance and availability requirements. NAS (Section 4.2.2) offers even a third alternative in this area.

In large environments it is almost impossible to avoid Fibre Channel as the basis for storage infrastructures. There is simply a lack of products and experience for IP storage in large environments with high requirements for performance and availability. However, IP storage is offered as a supplemental option here – for example, to supply storage capacity already available in Fibre Channel SANs to less critical servers via iSCSI-to-Fibre-Channel gateways. Another use of IP storage is for copying data to backup data centres over FCIP. Equally, NAS can be integrated as a third technique.

The performance capabilities of IP storage will not become obvious until there is also an increase in the use of IP storage for business-critical applications in large environments. The current reference installations do not provide enough of a basis for forming

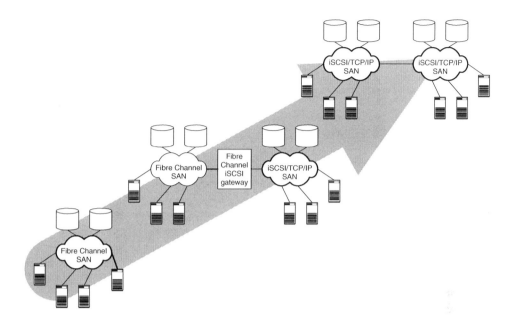

Figure 3.43   Possible migration path from Fibre Channel to iSCSI.

a comprehensive verdict underpinned by practical experience. Therefore, one can only philosophise theoretically about whether IP storage is also suitable for this area. IP storage can only prove itself as an alternative for large installations after it has been used for a long time in actual practice.

A migration path to IP storage exists for current investments in Fibre Channel components, thereby guaranteeing a long life cycle (Figure 3.43). In the first stage, pure Fibre Channel SANs are used for high requirements. These Fibre Channel SANs are gradually transferred into IP storage over iSCSI-to-Fibre Channel gateways. For some customers, the performance of iSCSI is sufficient, thus they change over completely to iSCSI. In the mid-term we expect a co-existence of Fibre Channel, iSCSI and FCoE.

3.6   INFINIBAND-BASED STORAGE NETWORKS

At the end of the 1990s one could tell, that the parallel PCI bus would not meet future performance requirements. Two industry consortia started to work on the initially competing technologies Future I/O and Next Generation I/O (NGIO) which have been merged later on as InfiniBand though, the claims of InfiniBand were not limited to be a successor of PCI. Within a data center a so-called 'System Area Network' should consolidate all

I/O including Fibre Channel and Ethernet. Today (2009) InfiniBand is mostly used as fast interconnect for the interprocess communication within parallel cluster computers. Against InfiniBand, the serial PCI Express has established as the successor of the parallel PCI bus.

In this section we start with an overview about InfiniBand (Section 3.6.1) and then we dig into some details of the Virtual Interface Architecture (VIA) which is integral part of InfiniBand (Section 3.6.2). Finally we present several approaches for transferring SCSI data traffic via InfiniBand (Section 3.6.3).

3.6.1 InfiniBand

InfiniBand replaces the PCI bus with a serial network (Figure 3.44). In InfiniBand the devices communicate by means of messages, with an InfiniBand switch forwarding the

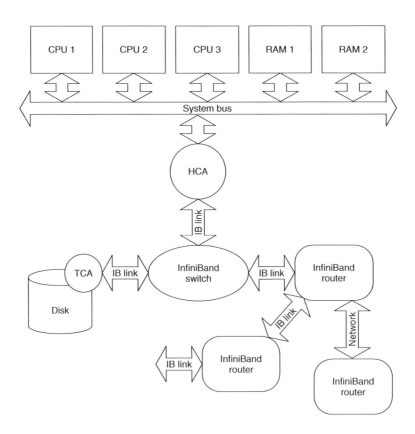

Figure 3.44 InfiniBand replaces the PCI bus by a serial network.

data packets to the receiver in question. The communication is full duplex and a transmission rate of 2.5 Gbit/s in each direction is supported. If we take into account the fact that, like Fibre Channel, InfiniBand uses 8b/10b encoding, this yields a net data rate of 250 MByte/s per link and direction. InfiniBand makes it possible to bundle four or twelve links so that a transmission rate of 10 Gbit/s (1 GByte/s net) or 30 Gbit/s (3 GByte/s net) is achieved in each direction.

As a medium, InfiniBand defines various copper and fiber-optic cables. A maximum length of 17 metres is specified for copper cable and up to 10,000 metres for fiber-optic cable. There are also plans to realise InfiniBand directly upon the circuit board using conductor tracks.

The end points in an InfiniBand network are called channel adapters. InfiniBand differentiates between Host Channel Adapters (HCAs) and Target Channel Adapters (TCAs). HCAs bridge between the InfiniBand network and the system bus to which the CPUs and the main memory (RAM) are connected. TCAs make a connection between InfiniBand networks and peripheral devices that are connected via SCSI, Fibre Channel or Ethernet. In comparison to PCI, HCAs correspond with the PCI bridge chips and TCAs correspond with the Fibre Channel HBA cards or the Ethernet network cards.

The originators of InfiniBand had in mind to completely change the architecture of servers and storage devices. We have to consider this: network cards and HBA cards can be located 100 metres apart. This means that mainboards with CPU and memory, network cards, HBA cards and storage devices are all installed individually as physically separate, decoupled devices. These components are connected together over a network. Today it is still unclear which of the three transmission technologies will prevail in which area.

Figure 3.45 shows what such an interconnection of CPU, memory, I/O cards and storage devices might look like. The computing power of the interconnection is provided by two CPU & RAM modules that are connected via a direct InfiniBand link for the benefit of lightweight interprocess communication. Peripheral devices are connected via the InfiniBand network. In the example a tape library is connected via Fibre Channel and the disk subsystem is connected directly via InfiniBand. If the computing power of the interconnection is no longer sufficient a further CPU & RAM module can be added.

Intelligent disk subsystems are becoming more and more powerful and InfiniBand facilitates fast communication between servers and storage devices that reduces the load on the CPU. It is therefore at least theoretically feasible for subfunctions such as the caching of file systems or the lock synchronisation of shared disk file systems to be implemented directly on the disk subsystem or on special processors (Chapter 4).

Right from the start, the InfiniBand protocol stack was designed so that it could be realised efficiently. A conscious decision was made only to specify performance features that could be implemented in the hardware. Nevertheless, the InfiniBand standard incorporates performance features such as flow control, zoning and various service classes. However, we assume that in InfiniBand – as in Fibre Channel – not all parts of the standard will be realised in the products.

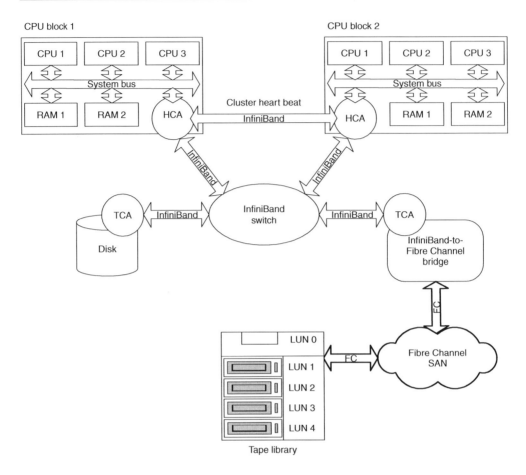

Figure 3.45   InfiniBand could radically change the architecture of server clusters.

3.6.2   Virtual interface architecture (VIA)

The Virtual Interface Architecture (VIA) significantly influenced the design of InfiniBand. VIA facilitates fast and efficient data exchange between applications that run on different computers. This requires an underlying network with a low latency and a low error rate. This means that the VIA can only be used over short distances, perhaps within a data centre or within a building. Originally, VIA was launched in 1997 by Compaq, Intel and Microsoft. Today it is an integral part of InfiniBand. Furthermore, protocol mappings exist for Fibre Channel and Ethernet.

Today communication between applications is still relatively complicated. Incoming data is accepted by the network card, processed in the kernel of the operating system and finally delivered to the application. As part of this process, data is copied repeatedly from one buffer to the next. Furthermore, several process changes are necessary in the operating system. All in all this costs CPU power and places a load upon the system bus. As a result the communication throughput is reduced and its latency increased.

The idea of VIA is to reduce this complexity by making the application and the network card exchange data directly with one another, bypassing the operating system. To this end, two applications initially set up a connection, the so-called Virtual Interface (VI): a common memory area is defined on both computers by means of which application and local network card exchange data (Figure 3.46). To send data the application fills the common memory area in the first computer with data. After the buffer has been filled with all data, the application announces by means of the send queue of the VI and the so-called doorbell of the VI hardware that there is data to send. The VI hardware reads the data directly from the common memory area and transmits it to the VI hardware on the second computer. This does not inform the application until all data is available in the common memory area. The operating system on the second computer is therefore bypassed too.

The VI is the mechanism that makes it possible for the application (VI consumer) and the network card (VI Network Interface Controller, VI NIC) to communicate directly with each other via common memory areas, bypassing the operating system. At a given point

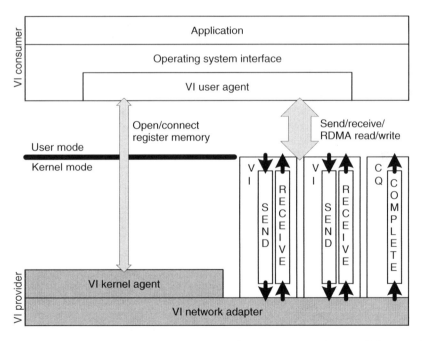

Figure 3.46   The Virtual Interface Architecture (VIA) allows applications and VI network cards to exchange data, bypassing the operating system.

in time a VI is connected with a maximum of one other VI. VIs therefore only ever allow point-to-point communication with precisely one remote VI.

A VI provider consists of the underlying physical hardware (VI NIC) and a device driver (kernel agent). Examples of VI NICs are VI-capable Fibre Channel HBAs, VI-capable Ethernet network cards and InfiniBand HCAs. The VI NIC realises the VIs and completion queues and transmits the data to other VI NICs. The kernel agent is the device driver of a VI NIC that is responsible for the management of VIs. Its duties include the generation and removal of VIs, the opening and closing of VI connections to remote VIs, memory management and error handling. In contrast to communication over the VI and the completion queue, communication with the kernel agent is associated with the normal overhead such as process switching. This extra cost can, however, be disregarded because after the VI has been set up by the kernel agent all data is exchanged over the VI.

Applications and operating systems that communicate with each other via VIs are called VI consumers. Applications generally use a VI via an intermediate layer such as sockets or Message Passing Interface (MPI). Access to a VI is provided by the user agent. The user agent first of all contacts the kernel agent in order to generate a VI and to connect this to a VI on a remote device (server, storage device). The actual data transfer then takes place via the VI as described above, bypassing the operating system, meaning that the transfer is quick and the load on the CPU is lessened.

The VI consists of a so-called work queue pair and the doorbells. The work queue pair consists of a send queue and a receive queue. The VI consumer can charge the VI NIC with the sending or receiving of data via the work queues, with the data itself being stored in a common memory area. The requests are processed asynchronously. VI consumer and VI NIC let each other know when new requests are pending or when the processing of requests has been concluded by means of a doorbell. A VI consumer can bundle the receipt of messages about completed requests from several VIs in a common completion queue. Work queues, doorbell and completion queue can be used bypassing the operating system.

Building upon VIs, the VIA defines two different communication models. It supports the old familiar model of sending and receiving messages, with the messages in this case being asynchronous sent and asynchronously received. An alternative communication model is the so-called Remote Direct Memory Access (RDMA). Using RDMA, distributed applications can read and write memory areas of processes running on a different computer. As a result of VI, access to the remote memory takes place with low latency and a low CPU load.

3.6.3   SCSI via InfiniBand and RDMA

It has already proved that RDMA can improve the performance of commercial applications, but none of these RDMA-enabled applications is commercially successful. This is mostly due to the fact that RDMA-capable network cards had not been interoperable and thus added costs for owning and managing RDMA-enabled applications. Therefore in

May 2002 several companies founded the RDMA Consortium (www.rdmaconsortium.org) to standardise the RDMA protocol suite. The consortium has standardised all interfaces required to implement the software and the hardware for RDMA over TCP. In addition to that, it has been defined two upper layer protocols – the Socket Direct Protocol (SDP) and the iSCSI Extension for RDMA iSER – which exploit RDMA for fast and CPU light communication – and handed over the further standardisation to other organisations.

RDMA over TCP offloads much of TCP protocol processing overhead from the CPU to the Ethernet network card. Furthermore, each incoming network packet has enough information, thus its payload can be placed directly to the proper destination memory location, even when packets arrive out of order. That means RDMA over TCP gains the benefits of the VIA whilst it uses the existing TCP/IP/Ethernet network infrastructure. RDMA over TCP is layered on top of TCP, needs no modification of the TCP/IP protocol suite and thus can benefit from underlying protocols like IPsec. The IETF Remote Direct Data Placement (RDDP) Working Group refined RDMA over TCP as Internet Wide Area RDMA Protocol (iWARP). iWARP comprises the sub-protocols MPA, Direct Data Placement (DDP) and RDMAP.

RDMA over TCP has some advantages in comparison to TCP/IP Offload Engines TOEs. TOEs move the load for TCP protocol processing from the CPU to the network card, but the zero copy of incoming data streams is very proprietary in the TOE design, the operating systems interfaces, and the applications communication model. Thus in many cases, TOEs do not support a zero copy model for incoming data thus the data must be copied multiple times until it is available for the application. In contrast, RDMA over TCP benefits from its superior specification, thus a combination of TOEs and RDMA provides the optimal architecture for 10-Gigabit Ethernet by reducing the CPU load and avoiding the need for copying data from buffer to buffer. Meanwhile RDMA-enabled network interface controllers (RNIC) are available on the market.

RDMA over TCP, iWARP, VIA and InfiniBand each specify a form of RDMA, but these are not exactly the same. The aim of VIA is to specify a form of RDMA without specifying the underlying transport protocol. On the other hand, InfiniBand specifies an underlying transmission technique which is optimised to support RDMA semantics. Finally, the RDMA over TCP and iWARP each specify a layer which will interoperate with the standard TCP/IP protocol stack. As a result, the protocol verbs of each RDMA variant are slightly different, thus these RDMA variants are not interoperable.

However, the RDMA Consortium specified two upper layer protocols which utilise RDMA over TCP. The Socket Direct Protocol (SDP) represents an approach to accelerate TCP/IP communication. SDP maps the socket API of TCP/IP onto RDMA over TCP so that protocols based upon TCP/IP such as NFS and CIFS can benefit from RDMA without being modified. SDP benefits from offloading much of the TCP/IP protocol processing burden from CPU and its ability to avoid copying packets from buffer to buffer. It is very interesting to observe that applications using SDP think that they are using native TCP when the real transport of the data is performed by an integration of RDMA and TCP. The Software Working Group (SWG) of the InfiniBand Trade Association ratified SDP in 2002 as part of the InfiniBand Architecture Specification.

iSCSI Extension for RDMA iSER is the second upper layer protocol specified by the RDMA Consortium. It is an extension of the iSCSI protocol (Section 3.5.1) which enables iSCSI to benefit from RDMA eliminating TCP/IP processing overhead on generic RNICs. This is important as Ethernet and therefore iSCSI approach 10 Gbit/s. iSER is not a replacement for iSCSI, it is complementary. iSER requires iSCSI components such as login negotiation, discovery, security and boot. It only changes the data mover model of iSCSI. It is expected the iSCSI end nodes and iSCSI/iSER end nodes will be interoperable. During iSCSI login both end nodes will exchange characteristics, thus each node is clearly aware of the other's node transport capabilities. The IETF ratified iSER in 2007.

Finally, the SCSI RDMA Protocol (SRP) describes a mapping of SCSI to RDMA, thus RDMA acts as transmission technology between an SCSI initiator and an SCSI target. The Technical Committee T10 of the International Committee for Information Technology Standards (INCITS) published SRP in 2002. In 2003, the work on the follow-on version SRP-2 was discontinued due to a lack of interest.

In 2009 RDMA and its related protocols are not yet widespread in current applications; however, it opens up new possibilities for the implementation of distributed synchronisation mechanisms for caching and locking in databases and file systems. All distributed applications will benefit from RDMA-enabled transport via SDP and iSER whilst the applications themselves remain unchanged. Furthermore, communication intensive applications will be adapted to utilise the native RDMA communication, for instance, file systems, databases, and applications in parallel computing. Section 4.2.5 shows in the example of the Direct Access File System (DAFS) how RDMA changes the design of network file systems.

3.7   FIBRE CHANNEL OVER ETHERNET (FCoE)

So far neither Fibre Channel nor IP storage, InfiniBand or Ethernet has succeeded in establishing a transmission technology which is widely used for storage networks and LAN networks. Fibre Channel and InfiniBand tried to migrate LAN traffic from Ethernet to a new transmission technology. Fibre Channel over Ethernet (FCoE) tries the opposite approach using Ethernet as transmission technology for storage traffic.

This section discusses the general advantages of I/O consolidation (Section 3.7.1) and then introduces FCoE (Section 3.7.2). We illustrate the use of FCoE on the basis of some examples (Section 3.7.3) and discuss necessary enhancements of Ethernet which are required for FCoE (Section 3.7.4). Finally we give an outlook about how the market for FCoE could develop in the future (Section 3.7.5).

3.7.1   I/O Consolidation based on Ethernet

I/O consolidation refers to the consolidation of all I/O traffic types onto the same transmission technology. Today (2009) different transmission technologies are used for traditional LAN traffic (Ethernet), storage traffic (Fibre Channel), and where required cluster traffic (InfiniBand). Hence, inside a data center three different network types are installed and the servers are equipped with a bunch of I/O cards for all these transmission technologies. With I/O consolidation these different networks are replaced by a single one which is shared for all traffic types (Figure 3.47). As a side effect, servers only need two I/O ports (for redundancy) instead of multiple ports for all the different transmission technologies.

Fibre Channel and InfiniBand originally intended to offer a transmission technology for I/O consolidation. But both provided only rudimentary support for multicast and broadcast, thus neither has ever been seriously considered for LAN traffic. The originators of FCoE look at it from the other way round, trying to migrate storage traffic from Fibre Channel to the omnipresent Ethernet. This seems to be a reasonable approach, considering the successful consolidation of Internet, voice and traditional LAN onto TCP/IP and thus the huge investments in Ethernet and considering the failed consolidation efforts based on Fibre Channel, IP storage and InfiniBand.

According to the product roadmaps of relevant vendors, 10-Gigabit Ethernet is going to replace Gigabit Ethernet within the next few years. Thus, servers which comprise two 10-Gigabit Ethernet ports will have sufficient bandwidth to consolidate all traffic types on

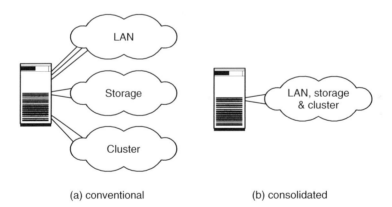

(a) conventional   (b) consolidated

Figure 3.47   I/O consolidation uses the same transmission technology for all traffic types. This saves adapter cards, cables, and switches.

Ethernet. This is also true for setups where a single physical server hosts multiple virtual servers. PCI Express provides a fast connection between both consolidated 10-Gigabit Ethernet ports and the memory (RAM).

I/O consolidation based on two Ethernet ports is especially handsome for server farms built upon huge amounts of blade servers or 1U high rack mounted servers, which have a very limited capacity of PCI slots. The announcements about on-board components for 10-Gigabit Ethernet promise additional savings. Furthermore, the replacement of the three different networks, Ethernet, Fibre Channel and InfiniBand, by a single, consolidated Ethernet network results in a reduced consumption of ports, switches and cables in contrast to the three previous networks.

In general, Ethernet-based I/O consolidation can leverage the already proven iSCSI protocol for storage traffic. However, the originators of FCoE have identified two drawbacks which induced them to develop an alternative to iSCSI. On the one hand, TCP has a certain protocol overhead. Mechanisms like the detection and the re-transmission of lost TCP packages and the re-assembly of fragmented TCP packages require a stateful protocol engine. Therefore, the implementation of TCP in hardware is complex and costly and the implementation in software consumes too much CPU for providing high bandwidths. Hence, an alternative is needed which transmits storage traffic (SCSI) directly via Ethernet bypassing TCP to achieve a better performance.

On the other hand, iSCSI is not compatible with Fibre Channel. Both have different management mechanisms (such as discovery, addressing and zoning), thus a gateway is needed to connect iSCSI and Fibre Channel. Considering the significant investments in Fibre Channel SANs, the management mechanisms of the new alternative to iSCSI should be compatible with the management of Fibre Channel SANs.

3.7.2 FCoE Details

Therefore, the originators of FCoE intend, as already indicated by its name, to establish Ethernet as the transmission technology for Fibre Channel whilst avoiding using TCP and IP (Figure 3.48). From the operating systems point of view there is no difference between FCoE and Fibre Channel. A server still acts as an SCSI initiator, which uses the SCSI protocol to establish a connection to a storage device which acts as an SCSI target. The SCSI protocol is still mapped to Fibre Channel FCP. Only the transmission of the SCSI data no longer uses Fibre Channel (FC-0 and FC-1) as transmission technology but FCoE and Ethernet. The upper layers of Fibre Channel (FC-3 and FC-4) remain unchanged. Later in this section we will see that FCoE needs some enhancements of traditional Ethernet, which are referred as Data Center Bridging (DCB, Section 3.7.4).

FCoE encapsulates each Fibre Channel frame into exactly one Ethernet frame. Beginning with a frame size of 2,112 Byte for Fibre Channel, an FCoE frame has a size of 2,180 Bytes (Figure 3.49). Reserving some space for future enhancements the underlying Ethernet network must support a frame size of 2.5 KByte (Baby Jumbo Frames). The net transfer rate of 96.2% is only 1.1% worse than the net transfer rate of pure Fibre Channel.

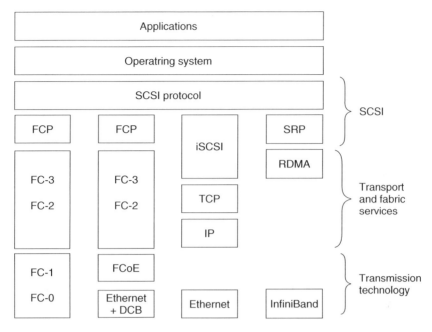

Figure 3.48   FCoE remains the upper layers of Fibre Channel unchanged. Only the transmission technology is replaced by FCoE, where FCoE requires some enhancements of traditional Ethernet.

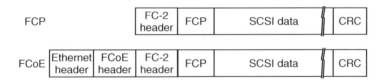

Figure 3.49   FCoE encapsulates Fibre Channel Frames 1:1 into Ethernet Frames.

The 1:1 mapping of Fibre Channel frames onto Ethernet frames has the advantage that frames are not fragmented. The encapsulation and decapsulation can be performed by a stateless and thus simple protocol. This reduces the complexity of FCoE enabling an implementation in hardware.

An end device (server, storage) which comprises an FCoE-capable Ethernet adapter is also called FCoE Node or just ENode (Figure 3.50). Each such Ethernet card comprises a so-called FCoE Controller for each of its Ethernet ports. The FCoE Controller connects via its Ethernet port to an FCoE switch and instantiates by the means of the Fibre Channel Entity (FC Entity) a respective Virtual N_Port (VN_Port). A VN_Port emulates a Fibre

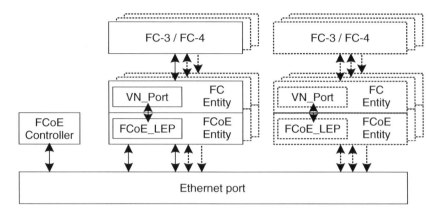

Figure 3.50 FCoE end devices (server, storage) are also called FCoE Node or just ENode. The FCoE Controller initialises the Ethernet port and instantiates, depending on the configuration, one or more VN_Ports.

Channel N_Port which is used by the upper layers of the Fibre Channel protocol stack. The FC Entity implements the Fibre Channel FC-2 protocol. Each FC Entity is coupled with an FCoE Entity, which comprises an FCoE Link Endpoint (FCoE_LEP) for the encapsulation and decapsulation of Fibre Channel Frames to FCoE frames and its transmission via Ethernet. A FCoE Controller can instantiate multiple VN_Ports when it connects via an Ethernet layer-2 switch (bridge) to multiple FCoE Switches (see also Figure 3.52) or when different virtual machines use different VN_Ports.

FCoE switches have a similar design to ENodes (Figure 3.51). Though the FCoE Controller instantiates a Virtual F_Port (VF_Port) for a connection to an FCoE end device (ENode) and a Virtual E_Port (VE_Port) for a connection to another FCoE Switch. VF_Ports and VE_Ports are connected to the so-called Fibre Channel Switching Element which provides the services of a Fibre Channel switch. In addition to that, FCoE-capable Fibre Channel switches also comprise usual Fibre Channel F_Ports and Fibre Channel E_Ports.

FCoE switches provide the same management mechanism like traditional Fibre Channel switches. Zoning works in the same manner as well as Name Service, Registered State Change Notification (RSCN) and Fibre Channel Shortest Path First (FSPF). Therefore, it should be possible to connect Fibre Channel switches and FCoE switches easily and build an integrated storage network. It is expected that existing monitoring and management tools for Fibre Channel SANs can be used for FCoE SANs as well.

There is a significant difference between Fibre Channel and FCoE with regard to links. A Fibre Channel link is always a point-to-point connection between two ports of switches or end devices. In contrast with FCoE, each Ethernet port can connect via layer-2 switches (Ethernet bridge) to multiple other Ethernet ports, thus Ethernet ports of FCoE ENodes and FCoE switches can establish multiple virtual links (Figure 3.52). Exactly for these reasons,

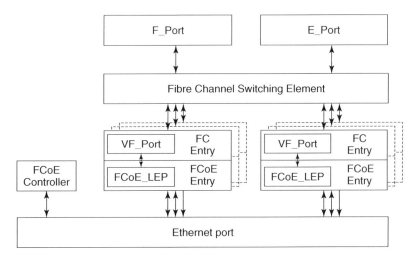

Figure 3.51   The structure of an FCoE switch is similar to ENodes. In addition to that, FCoE-capable Fibre Channel switches comprise F_Ports and E_Ports as well.

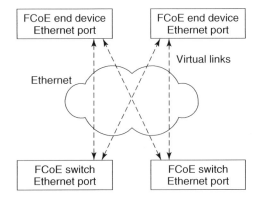

Figure 3.52   In contrast to Fibre Channel, an FCoE Ethernet port can connect to multiple other FCoE Ethernet ports at the same time. Therefore, at a single point in time a physical Ethernet port may establish multiple virtual links.

FCoE ENodes and FCoE switches must be capable to instantiate multiple VN_Ports or multiple VF_Ports and VE_Ports respectively for each Ethernet port.

FCoE sends Fibre Channel frames along virtual links. Though, additional means are required to establish these virtual links which are provided by the FCoE Initialization protocol (FIP, Figure 3.53). FCoE ENodes and FCoE switches find each other during the FIP discovery phase and exchange their properties. Once matching capabilities are found, the FIP login phase follows to instantiate respective virtual links and virtual ports which

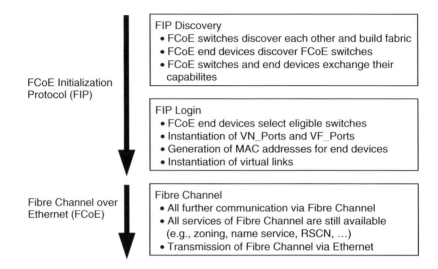

FCoE Initialization Protocol (FIP)

FIP Discovery
• FCoE switches discover each other and build fabric
• FCoE end devices discover FCoE switches
• FCoE switches and end devices exchange their capabilites

FIP Login
• FCoE end devices select eligible switches
• Instantiation of VN_Ports and VF_Ports
• Generation of MAC addresses for end devices
• Instantiation of virtual links

Fibre Channel over Ethernet (FCoE)

Fibre Channel
• All further communication via Fibre Channel
• All services of Fibre Channel are still available (e.g., zoning, name service, RSCN, ...)
• Transmission of Fibre Channel via Ethernet

Figure 3.53   The FCoE Initialization Protocol (FIP) discovers matching FCoE devices and instantiates virtual links between them. All further communication uses Fibre Channel commands which are transferred via FCoE.

results in the arise of a Fibre Channel fabric. To this purpose Fibre Channel commands such as Fabric Login (FLOGI) and Fibre Channel Discovery (FDISC) have been adopted to cope with the specifics of FCoE. Once the virtual links are established, all further communication uses unmodified Fibre Channel commands which are transferred via FCoE. For instance, during the establishment phase of FCP the Fibre Channel commands Port Login (PLOGI) and Process Login (PRLI) are used.

FCoE stipulates a two-staged mechanism for addressing and routing. FCoE Controllers and FCoE_LEPs use Ethernet MAC addresses to send FCoE frames between FCoE switches and FCoE ENodes. The FCoE Controller assumes the MAC address of the physical Ethernet port. The MAC addresses of the FCoE_LEPs are generated. Furthermore, Fibre Channel Node IDs are still used for routing Fibre Channel frames in an FCoE SAN, for instance, from an FCoE server via one or more FCoE switches to an FCoE storage device. Since FCoE is immediately mapped onto Ethernet, the routing mechanisms of IPv4 or IPv6 are not available.

FCoE defines two different methods for the generation of MAC addresses which are assigned to the FCoE_LEPs of end devices. With Fabric Provided MAC Address (FPMA), the FCoE switch generates the MAC address during the initialisation of a virtual link. The FCoE switch has a 24-bit identifier, which is unique within each FCoE SAN, and appends the 24-bit Fibre Channel Port ID, which results in a 48-bit MAC address.

With Server Provided MAC Address (SPMA), the end device generates the MAC address. In virtualised environments, dedicated VN_Ports and thus dedicated SPMA MAC addresses can be assigned to each virtual server. In this way, the context of an I/O connection can be moved together with a virtual server, when it is moved from one physical server to another physical server. This is advantageous for a continuous IT operation. The management software for the virtual servers must assure the uniqueness of the SPMAs. SPMA only enables the migration of the context of current storage traffic. Nevertheless the management software for virtual servers must be enhanced to provide the migration of current storage traffic.

The Technical Committee T11 of the INCITS standardises FCoE as a subset of 'Fibre Channel–Fibre Channel Backbone–5' (FC-BB-5). The ratification of FC-BB-5 as American National Standards Institute (ANSI) standard is expected for 2009.

In the year 2008, customers were running first test setups with prototypes of FCoE adapters and traditional Ethernet switches. FCoE-capable switches will enter the market in 2009. Shortly afterwards FCoE could make the first step into production environments.

3.7.3 Case studies

At first glance, a homogeneous FCoE SAN has the same architecture like a Fibre Channel SAN (Figure 3.54). The Fibre Channel Components (HBAs, Switches, cables) are just replaced by respective FCoE components.

Figure 3.55 depicts the integration of an FCoE SAN into an already existing Fibre Channel SAN and thus illustrates a migration path from Fibre Channel to FCoE. New FCoE servers and storage devices are connected to FCoE switches where the FCoE switches are connected to the existing Fibre Channel switches. The result is an integrated storage network comprising FCoE and Fibre Channel components. Little by little the old Fibre Channel components could be replaced by new FCoE components thus current investments in Fibre Channel components are well protected and FCoE could be introduced step by step. The compatibility of FCoE and Fibre Channel enables a smooth migration to FCoE without needing gateways or other components which typically cause performance bottlenecks and interoperability issues.

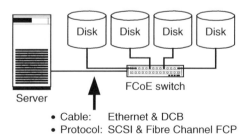

Figure 3.54   FCoE still uses Fibre Channel FCP for sending data via the SCSI protocol. However, FCoE uses Ethernet as transmission technology and not Fibre Channel.

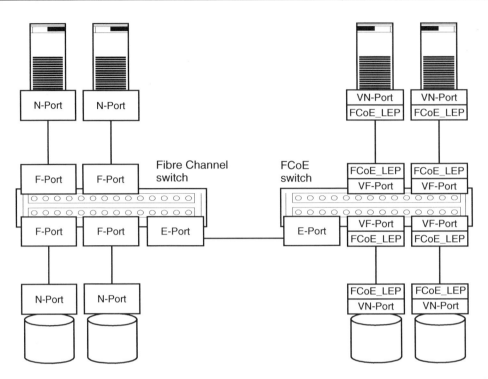

Figure 3.55   A FCoE switch can comprise a Fibre Channel E_Port to hook into an already existing Fibre Channel SAN. This enables the integration of new FCoE end devices into an already existing Fibre Channel SAN. More and more FCoE components can be integrated step by step.

Blade servers which have only very limited capacity for I/O adapters are a very interesting domain for FCoE (Figure 3.56). The internal layer-2 Ethernet switches of the blade center housing connect the Ethernet ports of the internal blade server modules and the external FCoE switches. FCoE not only intends to replace Fibre Channel as transmission technology but also intends to consolidate I/O. Thus the FCoE switches are connected to FCoE storage devices as well as to the LAN backbone. The FCoE switches inspect incoming Ethernet frames and forward them accordingly to the respective network. An Ethernet adapter which is capable to run traditional TCP as well as FCoE and cluster traffic is also called Converged Network Adapter (CNA).

3.7.4 Data Center Bridging (DCB)

As already mentioned, certain enhancements of Ethernet are required to make Ethernet a suitable transmission technology for I/O consolidation. Ethernet started as shared medium

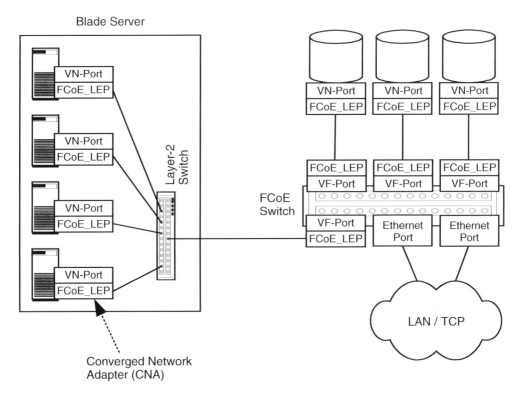

Figure 3.56   The internal layer-2 Ethernet switch connects the FCoE ports of the internal blade server modules and the ports of the external FCoE switch. The FCoE switch forwards incoming FCoE frames to the storage devices. The blade server modules use the same path to connect to the LAN.

and evolved to a network of full-duplex point-to-point links. The loss of Ethernet frames due to collisions is a thing of past times and its bit error rate matches Fibre Channel. However, it is still common praxis that Ethernet switches drop frames during congestion.

The loss of frames is not a big deal for LAN traffic because TCP is designed to detect lost messages quickly and to re-transmit them. Though for storage traffic a loss of Ethernet frames causes more harm. The SCSI protocol and its time-outs are designed for a very reliable transmission technology, thus in an FCoE SAN it could take minutes until a dropped Ethernet frame is noticed and re-transmitted.

The Data Center Bridging Task Group (DCB TG) of IEEE is working on several enhancements of Ethernet. It is the goal to make Ethernet a suitable transmission technology for traditional LAN traffic as well as for storage traffic and for cluster traffic, where the latter two typically remain within a data center or two collocated data centers.

Enhanced Transmission Selection (ETS, IEEE 802.1Qaz) defines multiple priority groups which can be used to separate LAN, storage and cluster communication and guarantees a configurable minimal bandwidth for each priority group (Figure 3.57). Each priority group can consume more bandwidth if a port is not fully utilised. If a port is overloaded, ETS cuts the traffic of each priority group thus its respective guaranteed bandwidth is not exceeded. Each priority group can support multiple traffic types such as VoIP, peer-to-peer networks (P2P), web browsing and similar traffic types for LAN. This two-staged prioritisation of data helps to control the bandwidth and to enforce other Quality-of-Service (QoS) parameters like latency and jitter.

Priority-based Flow Control (PFC, IEEE 801.1Qbb) refines the Ethernet PAUSE mechanism. An Ethernet port uses the conventional PAUSE frame to signal its peer port to suspend the transmission of further Ethernet frames for a certain period of time and thus disables the whole port. PFC adopts the concept of PAUSE frames to priority groups, thus the transmission can be suspended for each priority group without disabling the whole port.

The effect of PAUSE and PFC is – especially in smaller networks within a data center – similar to the buffer credit mechanisms of Fibre Channel. However PAUSE and PFC increase the consumption of the blocked switches buffer memory, thus the congestion of the network grows. Congestion Notification (CN, IEEE 802.1Qau) propagates such a situation to the end devices, thus these throttle their data rate to relieve the switches inside the network.

Finally, Data Center Bridging Exchange (DCBX) refers to an enhancement of the Link Layer Discovery Protocol (LLDP, IEEE 802.1AB-2005) to manage components which support these new protocols. DCBX helps to discover these new capabilities, to configure the components and to detect configuration mismatches. In early 2009 it is not yet determined whether DCBX will become a separate standard or whether it will be integrated into ETS.

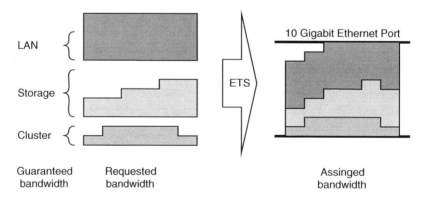

Figure 3.57 Enhanced Transmission Selection (ETS) guarantees for each priority group a configurable minimal bandwidth. A priority group can exceed its actual bandwidth when other priority groups actually consume less bandwidth than guaranteed.

Beyond DCB, additional protocols are discussed. For instance, storage networks typically comprise multiport end devices and redundant inter switch links (ISLs) to increase bandwidth and fault tolerance. The Spanning Tree Protocol (SPT, IEEE 802.1D) and enhancements such as Multiple Spanning Tree Protocol (MSPT, IEEE 802.1Q-2003) would use only one of the redundant links actively and only use one of the other redundant links if the active link fails.

One approach to close this gap is called TRILL (Transparent Interconnection of Lots of Links) which is developed by the IETF. The TRILL Working Group aims for a liaison with the IEEE 802.1 Working Group to assure TRILL's compatibility with the IEEE 802.1 protocol family. An alternate approach is the Shortest Path Bridging (SPB, IEEE 801.1aq). SPB specifies routing mechanisms for unicast and for multicast frames which support redundant links and parallel VLAN (virtual LAN) configurations.

Unfortunately, the industry has not yet agreed upon a consistent name to refer to these enhancements of Ethernet. The IEEE DCB TG defines the scope of the DCB protocol family exactly. However some vendors use the terms Converged Enhanced Ethernet (CEE) and Data Center Ethernet (DCE). Typically both include the DCB protocol family and FCoE. But you still have to check very carefully what else is included in CEE or DCE. For instance, some vendors include SPB but others do not. This opens the door for interoperability issues widely which probably makes the operation of FCoE SANs with switches of different vendors more difficult or even impossible.

3.7.5   Outlook

FCoE is a promising approach. However, the future will show how the market share of FCoE, iSCSI and Fibre Channel will evolve. One potential scenario would be the increasing adoption of I/O consolidation based on Ethernet. In the coming years, Fibre Channel could lose market share to FCoE or disappear completely. iSCSI may have a niche market in entry-level configurations, for instance for the attachment of branch offices and smaller virtual server landscapes. Many migrations paths from Fibre Channel to FCoE exists, thus actual investments in Fibre Channel have a long-term investment protection.

All design decisions with regard to FCoE must take into account that I/O consolidation has impact to conventional LAN traffic as well as to cluster communication. Connecting servers to a single switch which provides LAN and storage network raises the question 'Who controls this switch: the storage administrator or the LAN administrator?' It may be the case that at least larger corporations have to reorganise its IT organisation.

3.8   SUMMARY

I/O techniques connect CPUs to primary storage and to peripheral devices. In the past I/O techniques were based on the bus architecture. Buses are subdivided into system buses, host I/O buses and I/O buses. The most important I/O buses for servers are SCSI, Fibre Channel and the family of IP storage protocols as well as FCoE. SCSI enables storage devices to be addressed in a block-oriented way via targets and LUNs. The SCSI protocol also surfaces again in Fibre Channel and IP storage. These two new transmission techniques replace the parallel SCSI cable with a serial network and continue to use the SCSI protocol over this network. Fibre Channel is a new transmission technique that is particularly well suited to storage networks. Using point-to-point, arbitrated loop and fabric, it defines three different network topologies that – in the case of fabric – can together connect up to 15.5 million servers and storage devices. IP storage follows a similar approach to Fibre Channel. However, in contrast to Fibre Channel, it is based on the tried and tested TCP/IP and therefore usually on Ethernet. In contrast to IP storage, FCoE maps Fibre Channel directly to Ethernet and thus is a further approach to combine the advantages of Fibre Channel and Ethernet. Today (2009) Fibre Channel and IP storage can be used for small- and medium-sized environments. In contrast, larger setups cannot manage without Fibre Channel. The next few years will show what market share each of these three – Fibre Channel, IP storage and FCoE – will gain. For many years, InfiniBand and its upper layer protocols such as VIA, RDMA, SDP, SRP and iSER, have also been discussed as transmission technology for storage networks. However, InfiniBand was only rarely used for storage networks, although it has evolved into the dominant transmission technology for communication inside parallel cluster computers.

With the disk subsystems discussed in the previous chapter and the Fibre Channel and IP storage I/O techniques discussed in this chapter we have introduced the technologies that are required to build storage-centric IT systems. However, intelligent disk subsystems and storage networks represent only the physical basis for storage-centric IT systems. Ultimately, software that exploits the new storage-centric infrastructure and thus fully develops its possibilities will also be required. Therefore, in the next chapter we show how intelligent disk subsystems and storage networks can change the architecture of file systems.

4

File Systems and Network Attached Storage (NAS)

Disk subsystems provide block-oriented storage. For end users and for higher applications the handling of blocks addressed via cylinders, tracks and sectors is very cumbersome. File systems therefore represent an intermediate layer in the operating system that provides users with the familiar directories or folders and files and stores these on the block-oriented storage media so that they are hidden to the end users. This chapter introduces the basics of files systems and shows the role that they play in connection with storage networks.

This chapter first of all describes the fundamental requirements that are imposed upon file systems (Section 4.1). Then network file systems, file servers and the Network Attached Storage (NAS) product category are introduced (Section 4.2). We will then show how shared disk file systems can achieve a significantly higher performance than classical network file systems (Section 4.3). The chapter concludes with a comparison with block-oriented storage networks (Fibre Channel SAN, FCoE SAN, iSCSI SAN) and Network Attached Storage (NAS) (Section 4.4).

4.1 LOCAL FILE SYSTEMS

File systems form an intermediate layer between block-oriented hard disks and applications, with a volume manager often being used between the file system and the hard disk

Storage Networks Explained: Basics and Application of Fibre Channel SAN, NAS, iSCSI, InfiniBand and FCoE, Second Edition
U. Troppens R. Erkens W. Müller-Friedt N. Haustein R. Wolafka © 2009 John Wiley & Sons, Ltd

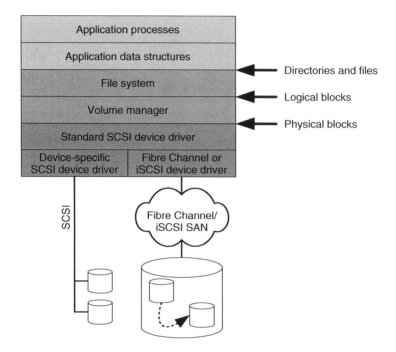

Figure 4.1 File system and volume manager manage the blocks of the block-oriented hard disks. Applications and users thus use the storage capacity of the disks via directories and files.

(Figure 4.1). Together, these manage the blocks of the disk and make these available to users and applications via the familiar directories and files.

4.1.1 File systems and databases

File systems and volume manager provide their services to numerous applications with various load profiles. This means that they are generic applications; their performance is not generally optimised for a specific application.

Database systems such as DB2 or Oracle can get around the file system and manage the blocks of the hard disk themselves (Figure 4.2). As a result, although the performance of the database can be increased, the management of the database is more difficult. In practice, therefore, database systems are usually configured to store their data in files that are managed by a file system. If more performance is required for a specific database, database administrators generally prefer to pay for higher performance hardware than to reconfigure the database to store its data directly upon the block-oriented hard disks.

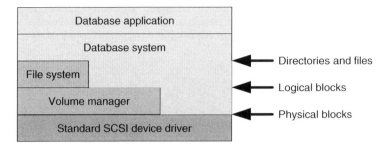

Figure 4.2 To increase performance, databases can get around the file system and manage the blocks themselves.

4.1.2 Journaling

In addition to the basic services, modern file systems provide three functions – journaling, snapshots and dynamic file system expansion. Journaling is a mechanism that guarantees the consistency of the file system even after a system crash. To this end, the file system first of all writes every change to a log file that is invisible to applications and end users, before making the change in the file system itself. After a system crash the file system only has to run through the end of the log file in order to recreate the consistency of the file system.

In file systems without journaling, typically older file systems like Microsoft's FAT32 file system or the Unix File System (UFS) that is widespread in Unix systems, the consistency of the entire file system has to be checked after a system crash (file system check); in large file systems this can take several hours. In file systems without journaling it can therefore take several hours after a system crash – depending upon the size of the file system – before the data and thus the applications are back in operation.

4.1.3 Snapshots

Snapshots represent the same function as the instant copies function that is familiar from disk subsystems (Section 2.7.1). Snapshots freeze the state of a file system at a given point in time. Applications and end users can access the frozen copy via a special path. As is the case for instant copies, the creation of the copy only takes a few seconds. Likewise, when creating a snapshot, care should be taken to ensure that the state of the frozen data is consistent.

Table 4.1 compares instant copies and snapshots. An important advantage of snapshots is that they can be realised with any hardware. On the other hand, instant copies within a disk subsystem place less load on the CPU and the buses of the server, thus leaving more system resources for the actual applications.

Table 4.1 Snapshots are hardware-independent, however, they load the server's CPU.

	Instant copy	Snapshot
Place of realisation	Disk subsystem	File system
Resource consumption	Loads disk subsystem's controller and its buses	Loads server's CPU and all buses
Availability	Depends upon disk subsystem (hardware-dependent)	Depends upon file system (hardware-independent)

4.1.4 Volume manager

The volume manager is an intermediate layer within the operating system between the file system or database and the actual hard disks. The most important basic function of the volume manager is to aggregate several hard disks to form a large virtual hard disk and make just this virtual hard disk visible to higher layers. Most volume managers provide the option of breaking this virtual disk back down into several smaller virtual hard disks and enlarging or reducing these (Figure 4.3). This virtualisation within the volume manager makes it possible for system administrators to quickly react to changed storage requirements of applications such as databases and file systems.

The volume manager can, depending upon its implementation, provide the same functions as a RAID controller (Section 2.4) or an intelligent disk subsystem (Section 2.7). As in snapshots, here too functions such as RAID, instant copies and remote mirroring are realised in a hardware-independent manner in the volume manager. Likewise, a RAID controller or an intelligent disk subsystem can take the pressure off the resources of the server if the corresponding functions are moved to the storage devices. The realisation of RAID in the volume manager loads not only on the server's CPU, but also on its buses (Figure 4.4).

4.2 NETWORK FILE SYSTEMS AND FILE SERVERS

Network file systems are the natural extension of local file systems. End users and applications can access directories and files that are physically located on a different computer – the file server – over a network file system (Section 4.2.1). File servers are so important in modern IT environments that preconfigured file servers, called Network Attached Storage (NAS), have emerged as a separate product category (Section 4.2.2). We highlight the performance bottlenecks of file servers (Section 4.2.3) and discuss the possibilities for the acceleration of network file systems. Finally, we introduce the Direct Access File System (DAFS), a new network file system that relies upon RDMA and VI instead of TCP/IP.

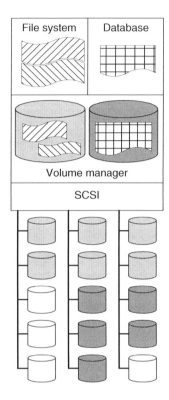

Figure 4.3   The volume manager aggregates physical hard disks into virtual hard disks, which it can break back down into smaller virtual hard disks. In the illustration one virtual hard disk is used directly from a database, the others are shared between two file systems.

4.2.1   Basic principle

The metaphor of directories and files for the management of data is so easy to understand that it was for a long time the prevailing model for the access of data over networks. So-called network file systems give end users and applications access to data stored on a different computer (Figure 4.5).

The first widespread network file system was the Network File System (NFS) developed by Sun Microsystems, which is now *the* standard network file system on all Unix systems. Microsoft developed its own network file system – the Common Internet File System (CIFS) – for its Windows operating system and this is incompatible with NFS. Today, various software solutions exist that permit the exchange of data between Unix and Windows over a network file system.

With the aid of network file systems, end users and applications can work on a common data set from various computers. In order to do this on Unix computers the system

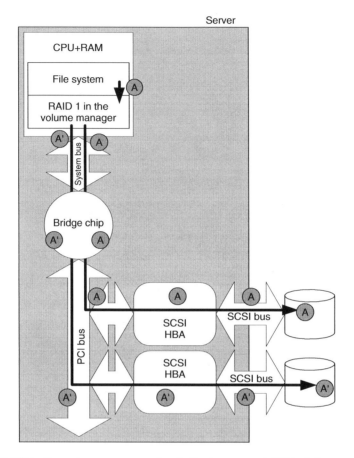

Figure 4.4 RAID in the volume manager loads the buses and CPU of the server. In RAID 1, for example, each block written by the file system must be passed through all the buses twice.

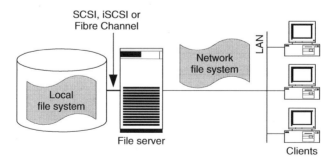

Figure 4.5 Network file systems make local files and directories available over the LAN. Several end users can thus work on common files (for example, project data, source code).

administrator must link a file system exported from an NFS server into the local directory structure using the `mount` command. On Windows computers, any end user can do this himself using the Map Network Drive command. Then, both in Unix and in Windows, the fact that data is being accessed from a network file system, rather than a local file system, is completely hidden apart from performance differences.

Long before the World Wide Web (WWW), the File Transfer Protocol (FTP) provided a mechanism by means of which users could exchange files over the Internet. Even today, FTP servers remain an important means of distributing freely available software and freely available documents. Unlike network file systems, access to FTP servers is clearly visible to the end user. Users require a special FTP client with which they can copy back and forwards between the FTP server and their local computer.

The Hyper Text Markup Language (HTML) and the Hyper Text Transfer Protocol (HTTP) radically changed the usage model of the Internet. In contrast to FTP, the data on the Internet is linked together by means of HTML documents. The user on the Internet no longer accesses individual files, instead he 'surfs' the WWW. He views HTML documents on his browser that are sometimes statically available on a HTTP server in the form of files or today are increasingly dynamically generated. Currently, graphical HTTP clients – the browsers – without exception have an integrated FTP client, with which they can easily 'download' files.

4.2.2 Network Attached Storage (NAS)

File servers are so important in current IT environments that they have developed into an independent product group in recent years. Network Attached Storage (NAS) is the name for preconfigured file servers. They consist of one or more internal servers, preconfigured disk capacity and usually a stripped-down or special operating system (Figure 4.6).

NAS servers are usually connected via Ethernet to the LAN, where they provide their disk space as file servers. Web servers represent a further important field of application for NAS servers. By definition, the clients are located at the other end of the WAN so there is no alternative to communication over IP. Large NAS servers offer additional functions such as snapshots, remote mirroring and backup over Fibre Channel SAN.

NAS servers were specially developed for file sharing. This has two advantages: since, by definition, the purpose of NAS servers is known, NAS operating systems can be significantly better optimised than generic operating systems. This means that NAS servers can operate more quickly than file servers on comparable hardware that are based upon a generic operating system.

The second advantage of NAS is that NAS servers provide Plug & Play file systems, i.e. connect – power up – use. In contrast to a generic operating system all functions can be removed that are not necessary for the file serving. NAS storage can therefore excel due to low installation and maintenance costs, which takes the pressure off system administrators.

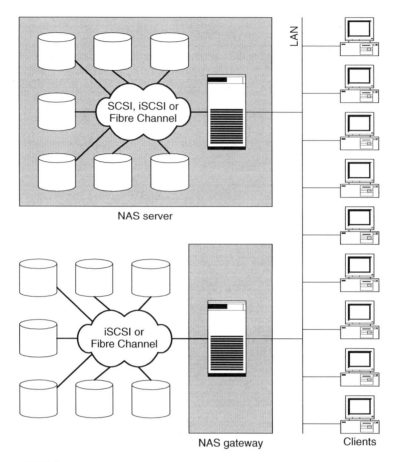

Figure 4.6   A NAS server is a preconfigured file server with internal hard disks, which makes its storage capacity available via LAN. A NAS gateway is a preconfigured file server that provides the storage capacity available in the storage network via the LAN.

NAS servers are very scalable. For example the system administrator can attach a dedicated NAS server for every project or for every department. In this manner it is simple to expand large websites. E-mail file system full? No problem, I simply provide another NAS server for the next 10,000 users in my Ethernet. However, this approach can become a management nightmare if the storage requirement is very large, thus tens of NAS servers are required.

One disadvantage of NAS servers is the unclear upgrade path. For example, the internal server cannot simply be replaced by a more powerful server because this goes against the principle of the preconfigured file server. The upgrade options available in this situation are those offered by the manufacturer of the NAS server in question.

Performance bottlenecks for more I/O-intensive applications such as databases, backup, batch processes or multimedia applications represent a further important disadvantage of NAS servers. These are described in the following subsection.

4.2.3 Performance bottlenecks in file servers

Current NAS servers and NAS gateways, as well as classical file servers, provide their storage capacity via conventional network file systems such as NFS and CIFS or Internet protocols such as FTP and HTTP. Although these may be suitable for classical file sharing, such protocols are not powerful enough for I/O-intensive applications such as databases or video processing. Nowadays, therefore, I/O-intensive databases draw their storage from disk subsystems rather than file servers.

Let us assume for a moment that a user wishes to read a file on an NFS client, which is stored on a NAS server with internal SCSI disks. The NAS server's operating system first of all loads the file into the main memory from the hard disk via the SCSI bus, the PCI bus and the system bus, only to forward it from there to the network card via the system bus and the PCI bus. The data is thus shovelled through the system bus and the PCI bus on the file server twice (Figure 4.7). If the load on a file server is high enough, its buses can thus become a performance bottleneck.

When using classical network file systems the data to be transported is additionally copied from the private storage area of the application into the buffer cache of the kernel

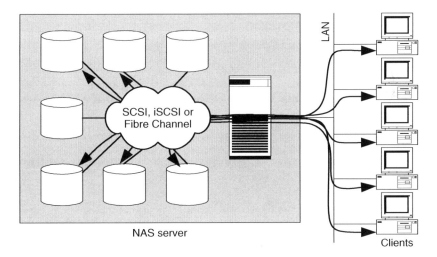

Figure 4.7   The file server becomes like the eye of the needle: en route between hard disk and client all data passes through the internal buses of the file server twice.

on the transmitting computer before this copies the data via the PCI bus into the packet buffer of the network card. Every single copying operation increases the latency of the communication, the load on the CPU due to costly process changes between application processes and kernel processes, and the load on the system bus between CPU and main memory.

The file is then transferred from the network card to the NFS client via IP and Gigabit Ethernet. At the current state of technology most Ethernet cards can only handle a small part of the TCP/IP protocol independently, which means that the CPU itself has to handle the rest of the protocol. The communication from the Ethernet card to the CPU is initiated by means of interrupts. Taken together, this can cost a great deal of CPU time (Section 3.5.2, 'TCP/IP and Ethernet as an I/O technology').

4.2.4 Acceleration of network file systems

If we look at the I/O path from the application to the hard disks connected to a NAS server (Figure 4.15 on page 157), there are two places to start from to accelerate file sharing: (1) the underlying communication protocol (TCP/IP); and (2) the network file system (NFS, CIFS) itself.

TCP/IP was originally developed to achieve reliable data exchange via unreliable transport routes. The TCP/IP protocol stack is correspondingly complex and CPU-intensive. This can be improved first of all by so-called TCP/IP offload engines (TOEs), which in contrast to conventional network cards process a large part of the TCP/IP protocol stack on their own processor and thus significantly reduce the load on the server CPU (Section 3.5.2).

It would be even better to get rid of TCP/IP all together. This is where communication techniques such as VIs and RDMA come into play (Section 3.6.2). Today there are various approaches for accelerating network file systems with VI and RDMA. The Socket Direct Protocol (SDP) represents an approach which combines the benefits of TOEs and RDMA-enabled transport (Section 3.6.3). Hence, protocols based on TCP/IP such as NFS and CIFS can – without modification – benefit via SDP from RDMA-enabled transport.

Other approaches map existing network file systems directly onto RDMA. For example, a subgroup of the Storage Networking Industry Association (SNIA) is working on the protocol mapping of NFS on RDMA. Likewise, it would also be feasible for Microsoft to develop a CIFS implementation that uses RDMA instead of TCP/IP as the communication protocol. The advantage of this approach is that the network file systems NFS or CIFS that have matured over the years merely have a new communication mechanism put underneath them. This makes it possible to shorten the development and testing cycle so that the quality requirements of production environments can be fulfilled comparatively quickly.

A greater step is represented by newly developed network file systems, which from the start require a reliable network connection, such as the DAFS (Section 4.2.5), the family of the so-called shared disk file systems (Section 4.3) and the virtualisation on the file-level (Section 5.5).

4.2.5 Case study: The Direct Access File System (DAFS)

The Direct Access File System (DAFS) is a newly developed network file system that is tailored to the use of RDMA. It is based upon NFS version 4, requires VI and can fully utilise its new possibilities. DAFS makes it possible for several DAFS servers together to provide the storage space for a large file system (Figure 4.8). It remains hidden from the application server – as the DAFS client – which of these DAFS servers the actual data is located in (Figure 4.9).

The communication between DAFS client and DAFS server generally takes place by means of RDMA. The use of RDMA means that access to data that lies upon a DAFS server is nearly as quick as access to local data. In addition, typical file system operations such as the address conversion of files to SCSI block addresses, which naturally also require the CPU, are offloaded from the application server to the DAFS server.

An important function of file sharing in general is the synchronisation of concurrent accesses to file entries – i.e. metadata such as file names, access rights, etc. – and file contents, in order to protect the consistency of the data and metadata. DAFS makes it possible to cache the locks at the client side so that a subsequent access to the same

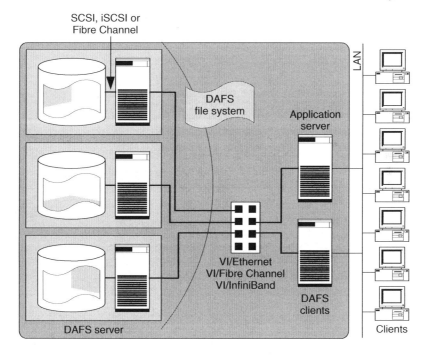

Figure 4.8 A DAFS file system can extend over several DAFS servers. All DAFS servers and DAFS clients are connected via a VI-capable network such as InfiniBand, Fibre Channel or Ethernet.

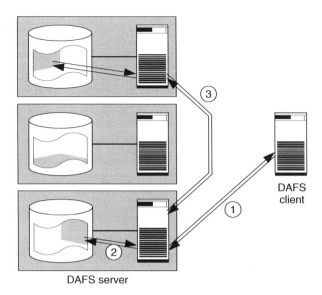

Figure 4.9   The DAFS client communicates with just one DAFS server (1). This processes file access, the blocks of which it manages itself (2). In the case of data that lies on a different DAFS server, the DAFS server forwards the storage access to the corresponding DAFS server, with this remaining hidden from the DAFS client (3).

data requires no interaction with the file server. If a node requires the lock entry of a different node, then this transmits the entry without time-out. DAFS uses lease-based locking in order to avoid the permanent blocking of a file due to the failure of a client. Furthermore, it possesses recovery mechanisms in case the connection between DAFS client and DAFS server is briefly interrupted or a different server from the cluster has to step in. Similarly, DAFS takes over the authentication of client and server and furthermore can also authenticate individual users in relation to a client-server session.

Two approaches prevail in the discussion about the client implementation. It can either be implemented as a shared library (Unix) or Dynamic Link Library (DLL) (Windows) in the user space or as a kernel module (Figure 4.10). In the user space variant – known as uDAFS – the DAFS library instructs the kernel to set up an exclusive end-to-end connection with the DAFS server for each system call (or for each API call under Windows) by means of a VI provider layer (VIPL), which is also realised as a library in user space. The VI-capable NIC (VI-NIC) guarantees the necessary protection against accesses or faults caused by other processes. The user space implementation can utilise the full potential of DAFS to increase the I/O performance because it completely circumvents the kernel. It offers the application explicit control over the access of the NIC to its private storage area. Although control communication takes place between the VIPL in the user space and the VI-NIC driver in the kernel, the CPU cost that this entails can be disregarded due to the low data quantities.

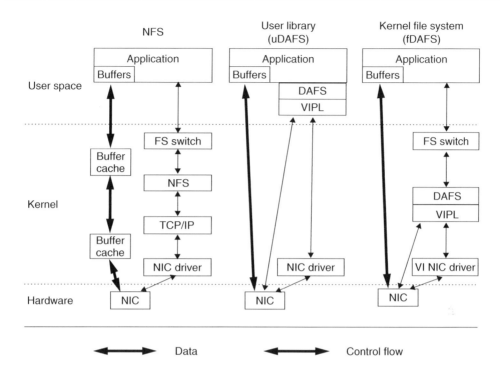

Figure 4.10   A comparison between NFS, uDAFS and fDAFS.

The disadvantage of the methods is the lack of compatibility with already existing applications, which without exception require an upgrade in order to use DAFS. Such a cost is only justified for applications for which a high I/O performance is critical.

In the second, kernel-based variant the implementation is in the form of a loadable file system module (fDAFS) underneath the Virtual File System (VFS layer) for Unix or as an Installable File System (IFS) for Windows. Each application can address the file system driver as normal by means of the standard system calls and VFS or API calls and the I/O manager; DAFS then directs a query to the DAFS server. The I/O performance is slowed due to the fact that all file system accesses run via the VFS or the I/O manager because this requires additional process changes between user and kernel processes. On the other hand, there is compatibility with all applications.

Some proponents of DAFS claim to have taken measurements in prototypes showing that data access over DAFS is quicker than data access on local hard disks. In our opinion this comparison is dubious. The DAFS server also has to store data to hard disks. We find it barely conceivable that disk access can be quicker on a DAFS server than on a conventional file server.

Nevertheless, DAFS-capable NAS servers could potentially support I/O-intensive applications such as databases, batch processes or multi-media applications. Integration with

RDMA makes it irrelevant whether the file accesses take place via a network. The separation between databases and DAFS servers even has the advantage that the address conversion of files to SCSI block addresses is offloaded from the database server to the DAFS server, thus reducing the load on the database server's CPU.

However, file servers and database servers will profit equally from InfiniBand, VI and RDMA. There is therefore the danger that a DAFS server will only be able to operate very few databases from the point of view of I/O, meaning that numerous DAFS servers may have to be installed. A corresponding number of DAFS-capable NAS servers could be installed comparatively quickly. However, the subsequent administrative effort could be considerably greater.

The development of DAFS was mainly driven by the company Network Appliance, a major NAS manufacturer. The standardisation of the DAFS protocol for communication between server and client and of the DAFS API for the use of file systems by applications took place under the umbrella of the DAFS Collaborative. Its website www.dafscollaborative can no longer be reached. Since the adoption of Version 1.0 in September 2001 (protocol) and November 2001 (API) the standardisation of DAFS has come to a standstill. Originally, DAFS was submitted as an Internet standard to the Internet Engineering Task Force (IETF) in September 2001; however, it has found very little support in the storage industry. Instead, widespread attention is being given to an alternative, namely an extension of NFS with RDMA as a transport layer and the addition of DAFS-like lock semantics.

DAFS is an interesting approach to the use of NAS servers as storage for I/O-intensive applications. Due to a lack of standardisation and widespread support across the industry, current DAFS offerings (2009) should be considered as a temporary solution for specific environments until alternatives like NFS over RDMA and CIFS over RDMA emerge. Furthermore, iSCSI is of interest to those who see DAFS as a way of avoiding an investment in Fibre Channel, the more so because iSCSI – just like NFS and CIFS – can benefit from TOEs, the SDP and a direct mapping of iSCSI on RDMA (iSER) (Section 3.6.3). Shared-disk file systems (Section 4.3) also offer a solution for high-speed file sharing and in addition to that, storage virtualisation (Chapter 5) provides high-speed file sharing while it addresses the increasingly expensive management of storage and storage networks as well.

4.3 SHARED DISK FILE SYSTEMS

The greatest performance limitation of NAS servers and self-configured file servers is that each file must pass through the internal buses of the file servers twice before the files arrive at the computer where they are required (Figure 4.7). Even DAFS and its alternatives like NFS over RDMA cannot get around this 'eye of the needle'.

With storage networks it is possible for several computers to access a storage device simultaneously. The I/O bottleneck in the file server can be circumvented if all clients fetch the files from the disk directly via the storage network (Figure 4.11).

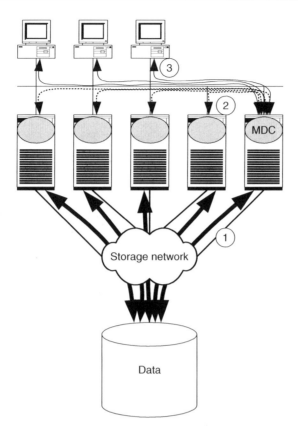

Figure 4.11   In a shared disk file system all clients can access the disks directly via the storage network (1). LAN data traffic is now only necessary for the synchronisation of the write accesses (2). The data of a shared disk file system can additionally be exported over the LAN in the form of a network file system with NFS or CIFS (3).

   The difficulty here: today's file systems consider their storage devices as local. They concentrate upon the caching and the aggregation of I/O operations; they increase performance by reducing the number of disk accesses needed.

   So-called shared disk file systems can deal with this problem. Integrated into them are special algorithms that synchronise the simultaneous accesses of several computers to common disks. As a result, shared disk file systems make it possible for several computers to access files simultaneously without causing version conflict.

   To achieve this, shared disk file systems must synchronise write accesses in addition to the functions of local file systems. It should be ensured locally that new files are written to different areas of the hard disk. It must also be ensured that cache entries are marked as invalid. Let us assume that two computers each have a file in their local cache and one of the computers changes the file. If the second computer subsequently reads the file again it may not take the now invalid copy from the cache.

The great advantage of shared disk file systems is that the computers accessing files and the storage devices in question now communicate with each other directly. The diversion via a central file server, which represents the bottleneck in conventional network file systems and also in DAFS and RDMA-enabled NFS, is no longer necessary.

In addition, the load on the CPU in the accessing machine is reduced because communication via Fibre Channel places less of a load on the processor than communication via IP and Ethernet. The sequential access to large files can thus more than make up for the extra cost for access synchronisation. On the other hand, in applications with many small files or in the case of many random accesses within the same file, we should check whether the use of a shared disk file system is really worthwhile.

One side-effect of file sharing over the storage network is that the availability of the shared disk file system can be better than that of conventional network file systems. This is because a central file server is no longer needed. If a machine in the shared disk file system cluster fails, then the other machines can carry on working. This means that the availability of the underlying storage devices largely determines the availability of shared disk file systems.

4.3.1 Case study: The General Parallel File System (GPFS)

We have decided at this point to introduce a product of our employer, IBM, for once. The General Parallel File System (GPFS) is a shared disk file system that has for many years been used on cluster computers of type RS/6000 SP (currently IBM eServer Cluster 1600). We believe that this section on GPFS illustrates the requirements of a shared disk file system very nicely. The reason for introducing GPFS at this point is quite simply that it is the shared disk file system that we know best.

The RS/6000 SP is a cluster computer. It was, for example, used for Deep Blue, the computer that beat the chess champion Gary Kasparov. An RS/6000 SP consists of up to 512 conventional AIX computers that can also be connected together via a so-called high performance switch (HPS). The individual computers of an RS/6000 SP are also called nodes.

Originally GPFS is based upon so-called Virtual Shared Disks (VSDs) (Figure 4.12). The VSD subsystem makes hard disks that are physically connected to a computer visible to other nodes of the SP. This means that several nodes can access the same physical hard disk. The VSD subsystem ensures that there is consistency at block level, which means that a block is either written completely or not written at all. From today's perspective we could say that VSDs emulate the function of a storage network. In more recent versions of GPFS the VSD layer can be replaced by an Serial Storage Architecture (SSA) SAN or a Fibre Channel SAN.

GPFS uses the VSDs to ensure the consistency of the file system, i.e. to ensure that the metadata structure of the file system is maintained. For example, no file names are

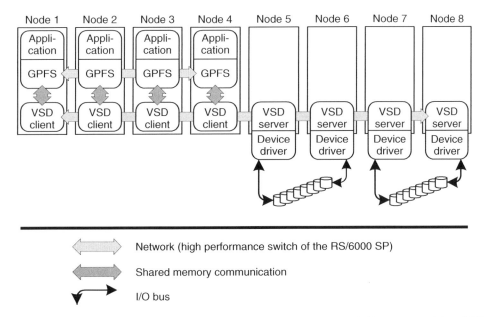

Network (high performance switch of the RS/6000 SP)

Shared memory communication

I/O bus

Figure 4.12 Applications see the GPFS file system like a local file system. The GPFS file system itself is a distributed application that synchronises parallel accesses. The VSD subsystem permits access to hard disks regardless of where they are physically connected.

allocated twice. Furthermore, GPFS realises some RAID functions such as the striping and mirroring of data and metadata.

Figure 4.12 illustrates two benefits of shared disk file systems. First, they can use RAID 0 to stripe the data over several hard disks, host bus adapters and even disk subsystems, which means that shared disk file systems can achieve a very high throughput. All applications that have at least a partially sequential access pattern profit from this.

Second, the location of the application becomes independent of the location of the data. In Figure 4.12 the system administrator can start applications on the four GPFS nodes that have the most resources (CPU, main memory, buses) available at the time. A so-called workload manager can move applications from one node to the other depending upon load. In conventional file systems this is not possible. Instead, applications have to run on the nodes on which the file system is mounted since access via a network file system such as NFS or CIFS is generally too slow.

The unusual thing about GPFS is that there is no individual file server. Each node in the GPFS cluster can mount a GPFS file system. For end users and applications the GPFS file system behaves – apart from its significantly better performance – like a conventional local file system.

GPFS introduces the so-called node set as an additional management unit. Several node sets can exist within a GPFS cluster, with a single node only ever being able to belong to

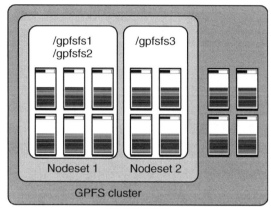

/gpfsfs1
/gpfsfs2

/gpfsfs3

Nodeset 1   Nodeset 2

GPFS cluster

RS/6000 SP

Figure 4.13 GPFS introduces the node sets as an additional management unit. GPFS file systems are visible to all nodes of a node set.

a maximum of one node set (Figure 4.13). GPFS file systems are only ever visible within a node set. Several GPFS file systems can be active in every node set.

The GPFS Daemon must run on every node in the GPFS cluster. GPFS is realised as distributed application, with all nodes in a GPFS cluster having the same rights and duties. In addition, depending upon the configuration of the GPFS cluster, the GPFS Daemon must take on further administrative functions over and above the normal tasks of a file system.

In the terminology of GPFS the GPFS Daemon can assume the following roles:

- *Configuration Manager*
  In every node set one GPFS Daemon takes on the role of the Configuration Manager. The Configuration Manager determines the File System Manager for every file system and monitors the so-called quorum. The quorum is a common procedure in distributed systems that maintains the consistency of the distributed application in the event of a network split. For GPFS more than half of the nodes of a node set must be active. If the quorum is lost in a node set, the GPFS file system is automatically deactivated (unmount) on all nodes of the node set.

- *File System Manager*
  Every file system has its own File System Manager. Its tasks include the following:

  – Configuration changes of the file system

  – Management of the blocks of the hard disk

  – Token administration

– Management and monitoring of the quota and

– Security services

Token administration is particularly worth highlighting. One of the design objectives of GPFS is the support of parallel applications that read and modify common files from different nodes. Like every file system, GPFS buffers files or file fragments in order to increase performance. GPFS uses a token mechanism in order to synchronise the cache entries on various computers in the event of parallel write and read accesses (Figure 4.14). However, this synchronisation only ensures that GPFS behaves precisely in the same way as a local file system that can only be mounted on one computer. This means that in GPFS – as in every file system – parallel applications still have to synchronise the accesses to common files, for example, by means of locks.

- *Metadata Manager*
  Finally, one GPFS Daemon takes on the role of the Metadata Manager for every open file. GPFS guarantees the consistency of the metadata of a file because only the Metadata Manager may change a file's metadata. Generally, the GPFS Daemon of the node on which the file has been open for the longest is the Metadata Manager for the file. The assignment of the Metadata Manager of a file to a node can change in relation to the access behaviour of the applications.

The example of GPFS shows that a shared disk file system has to achieve a great deal more than a conventional local file system, which is only managed on one computer. GPFS has been used successfully on the RS/6000 SP for several years. The complexity of shared disk file systems is illustrated by the fact that IBM only gradually transferred the GPFS file system to other operating systems such as Linux, which is strategically supported by IBM, and to new I/O technologies such as Fibre Channel and iSCSI.

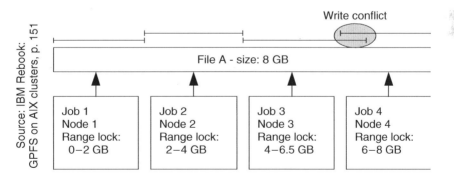

Figure 4.14   GPFS synchronises write accesses for file areas. If several nodes request the token for the same area, GPFS knows that it has to synchronise cache entries.

4.4 COMPARISON: FIBRE CHANNEL SAN, FCoE SAN, iSCSI SAN AND NAS

Fibre Channel SAN, FCoE SAN, iSCSI SAN and NAS are four techniques with which storage networks can be realised. Figure 4.15 compares the I/O paths of the four techniques and Table 4.2 summarises the most important differences.

In contrast to NAS, in Fibre Channel, FCoE and iSCSI the data exchange between servers and storage devices takes place in a block-based fashion. Storage networks are more difficult to configure. On the other hand, Fibre Channel at least supplies optimal performance for the data exchange between server and storage device.

NAS servers, on the other hand, are turnkey file servers. They can only be used as file servers, but they do this very well. NAS servers have only limited suitability as data storage for databases due to lack of performance. Storage networks can be realised with NAS servers by installing an additional LAN between NAS server and the application servers (Figure 4.16). In contrast to Fibre Channel, FCoE and iSCSI this storage network transfers files or file fragments.

One supposed advantage of NAS is that NAS servers at first glance have a higher pre-fetch hit rate than disk subsystems connected via Fibre Channel, FCoE or iSCSI (or just SCSI). However, it should be borne in mind that NAS servers work at file system level and disk subsystems only at block level. A file server can move the blocks of an opened file from the hard disk into the main memory and thus operate subsequent file accesses more quickly from the main memory.

Disk subsystems, on the other hand, have a pre-fetch hit rate of around 40% because they only know blocks; they do not know how the data (for example, a file system or database) is organised in the blocks. A self-configured file server or a NAS server that uses hard disks in the storage network can naturally implement its own pre-fetch strategy in addition to the pre-fetch strategy of the disk subsystem and, just like a NAS server, achieve a pre-fetch hit rate of 100%.

Today (2009) Fibre Channel, iSCSI and NAS have been successfully implemented in production environments. It is expected the FCoE will enter the market in 2009. Fibre Channel satisfies the highest performance requirements – it is currently (2009) the only transmission technique for storage networks that is suitable for I/O intensive databases. Since 2002, iSCSI has slowly been moving into production environments. It is said that iSCSI is initially being used for applications with low or medium performance requirements. It remains to be seen in practice whether iSCSI also satisfies high performance requirements (Section 3.5.2) or whether FCoE establishes as *the* technology for storage networks based on Ethernet. NAS is excellently suited to web servers and for the file sharing of work groups. With RDMA-enabled NFS and CIFS, NAS could also establish itself as a more convenient data store for databases.

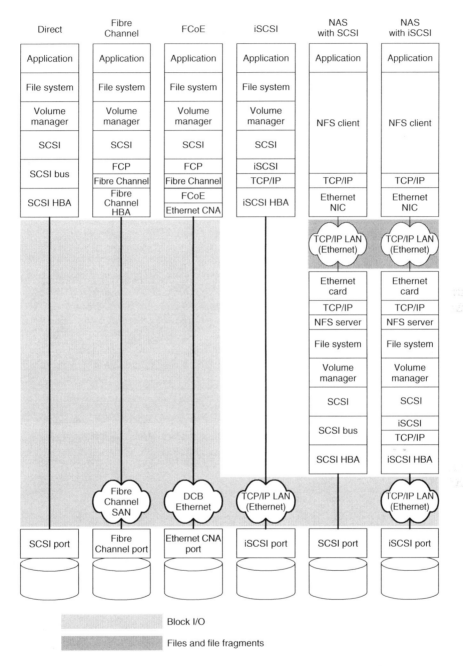

Figure 4.15   Comparison of the different I/O paths of SCSI, iSCSI, Fibre Channel, FCoE and NAS.

Table 4.2 Comparison of Fibre Channel, iSCSI and NAS.

	Fibre channel	FCoE	iSCSI	NAS
Protocol	FCP (SCSI)	FCP (SCSI)	iSCSI (SCSI)	NFS, CIFS, HTTP
Network	Fibre Channel	DCB Ethernet	TCP/IP	TCP/IP
Source/target	Server/storage device	Server/storage device	Server/storage device	Client/NAS server, application server/NAS server
Transfer objects	Device blocks	Device blocks	Device blocks	Files, file fragments
Access via the storage device	Directly via Fibre Channel	Directly via Fibre Channel	Directly via iSCSI	Indirectly via the NAS-internal computer
Embedded file system	No	No	No	Yes
Pre-fetch hit rate	40%	40%	40%	100%
Configuration	By end user (flexible)	By end user (flexible)	By end user (flexible)	Preconfigured by NAS manufacturers (Plug&Play)
Suitability for databases	Yes	Yes	To a limited degree (2009)	To a limited degree
Production-readiness	Yes	Market entry in 2009	Yes	Yes

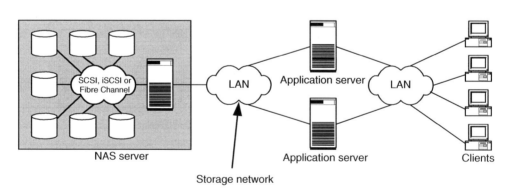

Figure 4.16 For performance reasons a separate LAN, which serves as a storage network, is installed here between the NAS server and the application servers.

4.5 SUMMARY

Hard disks provide their storage in the form of blocks that are addressed via cylinders, sectors and tracks. File systems manage the blocks of the hard disks and make their storage capacity available to users in the form of directories and files. Network file systems and shared disk file systems make it possible to access to the common data set from various computers. Modern file systems have additional functions, which can increase the availability of data in various situations. Journaling ensures that file systems become available again quickly after a system crash; snapshots allows data sets to be virtually copied within a few seconds; the volume manager makes it possible to react to changed storage requirements without interrupting operation. Network file systems export local file systems over the LAN so that it is possible to work on a common data set from different computers. The performance of network file systems is limited by two factors: (1) all data accesses to network file systems have to pass through a single file server; and (2) current network file systems such as NFS and CIFS and the underlying network protocols are not suitable for a high throughput. We introduced two approaches to circumventing these performance bottlenecks: RDMA-enabled file systems such as DAFS and shared disk file systems such as GPFS. NAS represents a new product category. NAS servers are preconfigured file servers, the operating systems of which have been optimised for the tasks of file servers. Storage networks can also be realised with NAS servers. However, current NAS servers are not suitable for providing storage space for I/O intensive databases.

In the previous chapters we have introduced all the necessary techniques for storage networks. Many of these techniques deal with the virtualisation of storage resources like RAID, instant copy, remote mirroring, volume manager, file systems, and file system snapshots. All these virtualisation techniques have in common that they present given storage resources to the upper layers as logical resources which are easier to use and administrate and which very often are also more performant and more failure tolerant. The next chapter will discuss how the concept of storage virtualisation changes in a storage-centric IT architecture.

5

Storage Virtualisation

Although the cost of storage has fallen considerably in recent years, at the same time the need for storage has risen immensely, so that we can observe of a real data explosion. The administrative costs associated with these quantities of data should not, however, increase to the same degree. The introduction of storage networks is a first step towards remedying the disadvantages of the server-centric IT architecture (Section 1.1). Whereas in smaller environments the use of storage networks is completely adequate for the mastery of data, practical experience has shown that, in large environments, a storage network alone is not sufficient to efficiently manage the ever-increasing volumes of data.

In this chapter we will introduce the storage virtualisation in the storage network, an approach that simplifies the management of large quantities of data. The basic idea behind storage virtualisation is to move the storage virtualisation functions from the servers (volume manager, file systems) and disk subsystems (caching, RAID, instant copy, remote mirroring, LUN masking) into the storage network (Figure 5.1). This creates a new virtualisation entity which, as a result of its positioning in the storage network, spans all servers and storage systems. This new virtualisation in the storage network permits the full utilisation of the potential of a storage network with regard to the efficient use of resources and data, the improvement of performance and protection against failures.

As an introduction into storage virtualisation we repeat the I/O path from the disk to the main memory: Section 5.1 contrasts the virtualisation variants discussed so far once again. Then we describe the difficulties relating to storage administration and the requirements of data and data users that occur in a storage network, for which storage virtualisation aims to provide a solution (Section 5.2). We will then define the term 'storage virtualisation' and consider the concept of storage virtualisation in more detail (Section 5.3). We will see that for storage virtualisation a virtualisation entity is required. The requirements for

Storage Networks Explained: Basics and Application of Fibre Channel SAN, NAS, iSCSI, InfiniBand and FCoE, Second Edition
U. Troppens R. Erkens W. Müller-Friedt N. Haustein R. Wolafka © 2009 John Wiley & Sons, Ltd

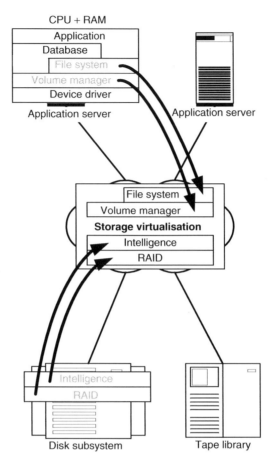

Figure 5.1 Storage virtualisation in the storage network moves virtualisation functions from servers and storage devices into the storage network. This creates a new virtualisation entity which, as a result of its central position in the storage network, spans all servers and storage systems and can thus centrally manage all available storage resources.

this virtualisation entity are defined and some implementation considerations investigated (Section 5.4). Then we will consider the two different forms of virtualisation (on block and file level) (Section 5.5), before going on to consider on which different levels a virtualisation entity can be positioned in the storage network (server, storage device or network) and the advantages and disadvantages of each (Section 5.6). We will underpin this by reintroducing some examples of virtualisation methods that have already been discussed. Finally, we will introduce two new virtualisation approaches – symmetric and asymmetric storage virtualisation – in which the virtualisation entity is positioned in the storage network (Section 5.7).

5.1 ONCE AGAIN: VIRTUALISATION IN THE I/O PATH

The structure of Chapters 2, 3 and 4 was based upon the I/O path from the hard disk to the main memory (Figure 1.7). Consequently, several sections of these chapters discuss different aspects of virtualisation. This section consolidates the various realisation locations for storage virtualisation which we have presented so far. After that we will move on to the virtualisation inside of the storage network.

Virtualisation is the name given to functions such as RAID, caching, instant copies and remote mirroring. The objectives of virtualisation are:

- Improvement of availability (fault-tolerance)
- Improvement of performance
- Improvement of scalability
- Improvement of maintainability

At various points of the previous chapters we encountered virtualisation functions. Figure 5.2 illustrates the I/O path from the CPU to the storage system and shows at what points of the I/O path virtualisation is realised. We have already discussed in detail virtualisation within a disk subsystem (Chapter 2) and, based upon the example of volume manager and file system, in the main memory and CPU (Chapter 4). The host bus adapter and the storage network itself should be mentioned as further possible realisation locations for virtualisation functions.

Virtualisation in the disk subsystem has the advantage that tasks are moved from the computer to the disk subsystem, thus freeing up the computer. The functions are realised at the point where the data is stored: at the hard disks. Measures such as mirroring (RAID 1) and instant copies only load the disk subsystem itself (Figure 5.3). This additional cost is not even visible on the I/O channel between computer and disk subsystem. The communication between servers and other devices on the same I/O bus is thus not impaired.

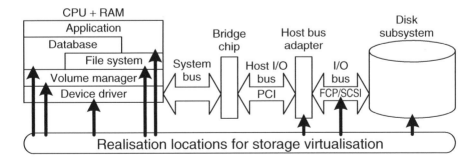

Figure 5.2 Virtualisation functions can be realised at various points in the I/O path.

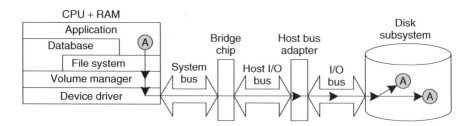

Figure 5.3   Virtualisation in the disk subsystem frees up server and storage network.

Virtualisation in the storage network has the advantage that the capacity of all available storage resources (e.g. disks and tapes) is centrally managed (Figure 5.4). This reduces the costs for the management of and access to storage resources and permits a more efficient utilisation of the available hardware. For example, a cache server installed in the storage network can serve various disk subsystems. Depending upon the load on the individual disk subsystems, sometimes one and sometimes the other requires more cache. If virtualisation is realised only within the disk subsystems, the cache of one disk subsystem that currently has a lower load cannot be used to support a different disk subsystem operating at a higher load. A further advantage of virtualisation within the storage network is that functions such as caching, instant copy and remote mirroring can be used even with cheaper disk subsystems (JBODs, RAID arrays).

The I/O card provides the option of realising RAID between the server and the disk subsystem (Figure 5.5). As a result, virtualisation takes place between the I/O bus and the host I/O bus. This frees up the computer just like virtualisation in the disk subsystem, however, for many operations the I/O buses between computer and disk subsystem are more heavily loaded.

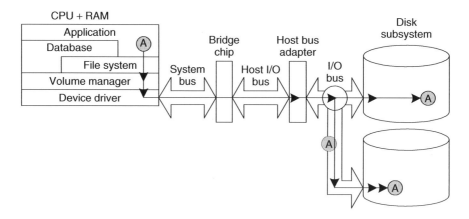

Figure 5.4   Virtualisation in the storage network is not visible to the server.

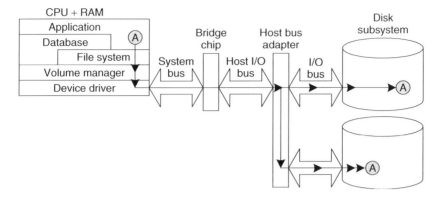

Figure 5.5   Virtualisation in the host bus adapter also remains hidden to the server.

Virtualisation in the main memory can either take place within the operating system in the volume manager or in low-level applications such as file systems or databases (Figure 5.6). Like all virtualisation locations described previously, virtualisation in the volume manager takes place at block level; the structure of the data is not known. However, in virtualisation in the volume manager the cost for the virtualisation is fully passed on to the computer: internal and external buses in particular are now more heavily loaded. On the other hand, the CPU load for volume manager mirroring can generally be disregarded.

The alternative approach of realising copying functions such as remote mirroring and instant copy in system-near applications such as file systems and databases is of interest. The applications know the structure of the data and can therefore sometimes perform the copying functions significantly more efficiently than when the structure is not known.

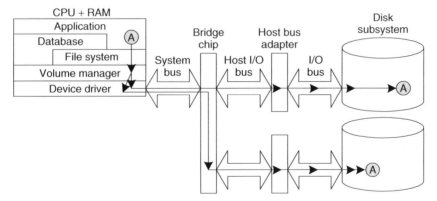

Figure 5.6   Virtualisation in the volume manager facilitates load distribution over several buses and disk subsystems.

It is not possible to give any general recommendation regarding the best location for the realisation of virtualisation. Instead it is necessary to consider the requirements of resource consumption, fault-tolerance, performance, scalability and ease of administration for the specific individual case when selecting the realisation location. From the point of view of the resource load on the server it is beneficial to realise the virtualisation as close as possible to the disk subsystem. However, to increase performance when performance requirements are high, it is necessary to virtualise within the main memory: only thus can the load be divided amongst several host bus adapters and host I/O buses (Figure 5.6). Virtualisation in the volume manager is also beneficial from the point of view of fault-tolerance (Section 6.3.3).

As is the case of many design decisions, the requirements of simple maintainability and high performance are in conflict in the selection of the virtualisation location. In Section 2.5.3 we discussed how both the performance and also the fault-tolerance of disk subsystems can be increased by the combination of RAID 0 (striping) and RAID 1 (mirroring), with the striping of mirrored disks being more beneficial in terms of fault-tolerance and maintainability than the mirroring of striped disks.

However, things are different if data is distributed over several disk subsystems. In terms of high fault-tolerance and simple maintainability it is often better to mirror the blocks in the volume manager and then stripe them within the disk subsystems by means of RAID 0 or RAID 5 (Figure 5.7). This has the advantage that the application can be kept in operation if an entire disk subsystem fails.

In applications with very high performance requirements for write throughput, it is not always feasible to mirror in the volume manager because the blocks have to be transferred through the host I/O bus twice, as shown in Figure 5.7. If the host I/O bus is the performance bottleneck, then it is better to stripe the blocks in the volume manager and only mirror them within the disk subsystems (Figure 5.8). The number of blocks to be written can thus be halved at the expense of fault-tolerance.

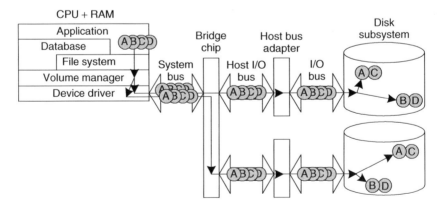

Figure 5.7   Mirroring in the volume manager protects against the failure of a disk subsystem. The blocks are striped within the disk subsystems to increase performance.

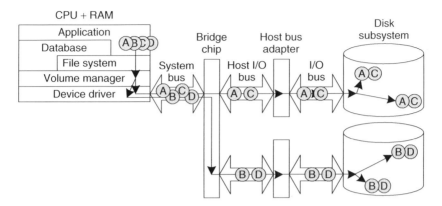

Figure 5.8   Striping in the volume manager for very high write performance: the load is distributed over several disk subsystems and I/O buses. The blocks are not mirrored until they reach the disk subsystems.

5.2   LIMITATIONS AND REQUIREMENTS

In this section we will first discuss architecture-related (Section 5.2.1) and implementation-related limitations (Section 5.2.2) of storage networks. We will then go on to describe further requirements for an efficient management of storage in large storage networks (Section 5.2.3). Finally, in Section 5.2.4, we will summarise the requirements of storage virtualisation.

5.2.1   Architecture-related limitations of non-virtualised storage networks

The storage resources brought together in a storage network are easier to manage than those in a server-centric IT architecture due to the separation of servers and storage devices. In many cases, the number of storage devices to be administered can also be drastically reduced by the introduction of a storage network. As a result, resource management becomes easier and more flexible (Section 1.3).

The flexible assignment of free disk capacity (Figure 2.3) alone is not, however, enough to fully exploit the savings potential of the storage-centric IT architecture such as resource and data sharing and the simplification of the management of storage resources. Without storage virtualisation in the storage network there remains a direct connection between the storage devices that provide the storage and the servers that use it. If, for example, changes are made to the configuration of the storage devices, then these remain visible on the servers, meaning that appropriate modifications are necessary there. In smaller environments with moderate data quantities this may be acceptable. In large environments the examples mentioned below represent a greater challenge to system administrators.

The replacement of storage devices in the event of a defect or an upgrade to newer, more powerful devices can only be performed at significant extra cost because the data from the old media must be transferred to the new. Additionally, administrators often are faced with an impossible task if this is to take place in a 24×7 environment without application failures. The modifications that have to be carried out on servers and applications to use the new storage devices after a replacement give rise to additional costs.

Without storage virtualisation, every change to the storage resources requires changes to the operating system and to the applications on the servers that use this. In many configurations a volume manager shields changes to the storage hardware from the applications. We will see later that a volume manager represents a form of storage virtualisation. In addition to storage virtualisation on the server by a volume manager we will also discover further possibilities for storage virtualisation, which solve the problem of the tight coupling of storage devices and servers.

Often there is also an inefficient use of the storage resources. For example, the following situation can occur (Figure 5.9): a storage network has two servers, A and B, which own disk space allocated on a disk subsystem. If the disks on server A have been filled, this server cannot simply share the free storage space on the drives of server B.

This is because the volumes on a disk subsystem are still statically assigned to a server and no suitable mechanisms have been implemented in the disk subsystems for the sharing of block-level resources. So, although on most disk subsystems it is possible to assign the same volume to several servers, the possible data inconsistencies that could arise as a result of such disk sharing are not rectified by the disk subsystem itself. The consequence of this is that additional new storage must be purchased or at least further storage resources allocated in order to solve the space problems of server A, even though the capacity allocated to server B has not yet been used up. Storage virtualisation software can realise the sharing of resources in a storage network. The shared disk file systems introduced in Section 4.3 represent one approach to this. However, the complexity of the software to be used for this increases the administrative cost.

5.2.2 Implementation-related limitations of storage networks

In previous sections we discussed only the conceptual boundaries of storage networks. In real environments matters are even worse. Incompatibilities between the storage systems of different manufacturers represent the main evil. For example, the manufacturers of disk subsystems are supplying special device drivers for their storage devices which provide advanced functions like handling multiple paths between a server and the storage device (Section 6.3.1). Unfortunately these disk subsystem specific device drivers only work with

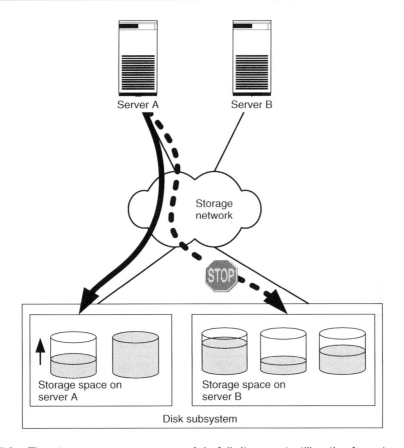

Figure 5.9 The storage space on server A is full. It cannot utilise the free storage space which is available on the volumes for server B.

one disk subsystem type. To make matters worse, the installation of the device drivers for different disk subsystems onto one computer at the same time is usually not supported.

This incompatibility between disk subsystem specific device drivers means that each server is linked to a very specific disk subsystem model or a very specific range of disk subsystem models, depending upon the respective device driver (Figure 5.10). As a result, one advantage of storage networks that has been mentioned several times in this book – namely the flexible allocation of free storage – only works on paper and in homogeneous storage landscapes. In real systems, a server can only use free storage capacity if this exists on a storage device with a compatible device driver. However, it looks as though the functionality of device drivers will, to an increasing degree, be integrated into operating systems, so that this limitation will very probably be rectified in the future. Furthermore, some disk subsystem vendors are working on the interoperability of their device drivers and they can thus be installed in parallel on the same server.

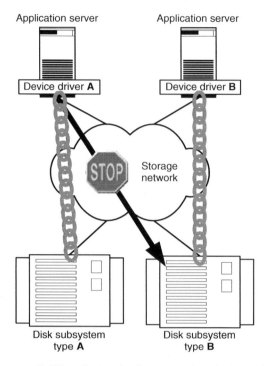

Figure 5.10 The incompatibilities that exist between the device drivers for different disk subsystems mean that a server can only use the storage capacity of the corresponding disk subsystems.

In practice this leads directly to economic disadvantages: let us assume that high performance requirements exist for 20% of the data of an application, whilst only average performance requirements exist for the remaining 80%. In this case it would be desirable to put just the 20% of the data on a high-end disk subsystem and store the rest of the data on a cheaper mid-range disk subsystem. Due to the incompatibility of the disk subsystem device drivers, however, the application server can only be connected to one type of disk subsystem, which means that all data must be stored on the high-end disk subsystem. Ultimately, this means that 80% of the data is taking up storage capacity that is more expensive than necessary.

A further interoperability problem in the practical use of storage systems is the lack of standardisation of the interfaces for the disk subsystem-based remote mirroring (Section 2.7.2). As a result of this, the data cannot be mirrored between any two storage systems. Usually, remote mirroring is only possible between two disk subsystems of the same type, only very exceptionally is it possible between different models or between two devices from different manufacturers. Here too, economic disadvantages result because this restricts the choice of disk subsystems that can be combined, meaning that it is not always possible to use the most economical storage system.

The inadequate interoperability of disk subsystem device drivers and remote mirroring gives rise to very real difficulties when replacing an old disk subsystem with a new one: this operation requires that the data from the old device is copied to the new device. However, it is precisely this copying process that can be very difficult in practice. Data mirroring by the volume manager would be preferable, so that the data could be copied to the new disk subsystem largely without interruptions. However, it is precisely this that is not possible if the device drivers of the two participating disk subsystems are not compatible with each other. In this situation, remote mirroring does not offer an alternative due to the lack of interoperability. Ultimately, economic disadvantages arise here, too, because it is not always possible to choose the most economical new storage system.

5.2.3 Requirements of the data

The introduction of storage networks is a first step towards the efficient management of storage resources, but it alone cannot solve the problems associated with the data explosion. In large environments with several ten of servers, each working on several disks, a storage administrator has to deal with hundreds of virtual and even more physical disks. In the following we wish to describe some difficulties associated with the administration of storage resources and the requirements of data and data users in such large environments.

Storage networks offer the possibility of realising requirements such as scalability, availability, performance, data protection and migration. Often, however, this calls for complex configurations, the administration of which again requires additional time and expense. Furthermore, requirements regarding availability, performance, data protection or migration are different for different data. For example, certain data has to be available even in the event of the failure of its storage resources, whereas it is possible to temporarily manage without other data. Some data has to be available to applications quickly, for example in order to keep the response times of a database low. For other data the response time is not so critical, but it may have to be backed up frequently. Still other data tends to change so seldom that frequent backups are not required. All this has to be taken into consideration when distributing the data on the resources. Data that requires a high throughput should therefore be stored on faster – and thus more expensive – media such as disks, whereas other data can be moved to slower – and thus cheaper – tapes.

A storage administrator is not able to match the requirements for the data to the physical storage devices. In large environments it would be too much to deal with the needs of every single virtual disk which is assigned to a server. Therefore, corresponding data profiles have to be created that specify the requirements of the data. An entity is required here that creates such data profiles automatically based particularly upon the data usage, and realises these accordingly. In this manner load peaks in the use of individual data can be recognised so that this data can be moved to fast media or can be distributed over several resources in order to achieve a balanced resource utilisation.

Furthermore, users of storage do not want to think about the size, type and location of the storage media when using their applications. They demand an intuitive handling of the storage media. The actual processes that are necessary for the administration of storage should remain hidden and run so that they are invisible to users. Users of storage space are, however, very interested in response times, data throughput and the availability of their applications. In short: users only think about what is related to their applications and not to the physical aspects of their data. The creation of this user-friendliness and the respective service level agreements for storage resources imposes additional requirements on storage administrators.

5.2.4 Proposed solution: storage virtualisation

To sum up, we can say that the implementation of a storage network alone does not meet the requirements for the management of large quantities of data. This requires additional mechanisms that simplify administration and at the same time make it possible to make full use of the storage resources. The use of storage virtualisation software offers the appropriate possibilities for, on the one hand, simplifying the administration of data and storage resources and, on the other, making their use by the users easier.

The objectives of storage virtualisation can be summed up by the following three points:

- *Simplification of the administration and access of storage resources*
- *Full utilisation of the possibilities of a storage network*
  The possibilities of a storage network should be fully utilised with regard to the efficient use of resources and data, the improvement of performance and protection in the event of failures by a high level of data availability.

- *Realisation of advanced storage functions*
  Storage functions such as data backups and archiving, data migration, data integrity, access controls and data sharing should be oriented towards data profiles and run automatically.

5.3 DEFINITION OF STORAGE VIRTUALISATION

The term 'storage virtualisation' is generally used to mean the separation of the storage into the physical implementation level of the storage devices and the logical representation level of the storage for use by operating systems, applications and users.

In the following we will also use the term 'virtualisation', i.e. dropping the word 'storage'. This is always used in the sense of the above definition of storage virtualisation. Various uses of the term 'storage virtualisation' and 'virtualisation' are found in the literature depending upon which level of the storage network the storage virtualisation takes place on. The various levels of the storage network here are the server, the storage devices and the network. Some authors only speak of storage virtualisation if they explicitly mean storage virtualisation within the network. They use the term virtualisation, on the other hand, to mean the storage virtualisation in the storage devices (for example, in the disk subsystems) or on servers (such as in a volume manager). However, these different types of storage virtualisation are not fundamentally different. Therefore, we do not differentiate between the two terms and always use 'storage virtualisation' and 'virtualisation' in the sense of the above definition.

In Section 5.1 we revised various types of storage virtualisation on the various levels of the storage network and we will pick these up again later. First of all, however, we want to deal in detail with the conceptual realisation of storage virtualisation.

Storage virtualisation inserts – metaphorically speaking – an additional layer between storage devices and storage users (Figure 5.11). This forms the interface between virtual and physical storage, by mapping the physical storage onto the virtual and conversely the virtual storage onto the physical. The separation of storage into the physical implementation level and the logical representation level is achieved by abstracting the

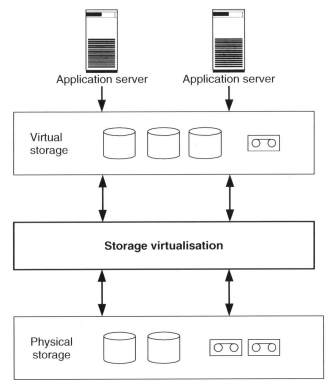

Figure 5.11   In storage virtualisation an additional layer is inserted between the storage devices and servers. This forms the interface between virtual and physical storage.

physical storage to the logical storage by aggregating several physical storage units to form one or more logical, so-called virtual, storage units. The operating system or applications no longer have direct access to the physical storage devices, they use exclusively the virtual storage. Storage accesses to the physical storage resources take place independently and separately from the storage accesses to the virtual storage resources.

For example, the physical hard disks available on a disk stack (JBOD) are brought together by the volume manager of a server to form a large logical volume. In this manner the volume manager thus forms an additional layer between the physical disks of the disk stack and the logical and thus virtual volume with which the applications (e.g. file systems and databases) of the server work. Within this layer, the mapping of physical hard disks onto logical volumes and vice versa is performed.

This means that storage virtualisation always calls for a virtualisation entity that maps from virtual to physical storage and vice versa. On the one hand it has to make the virtual storage available to the operating system, the applications and the users in usable form and, on the other, it has to realise data accesses to the physical storage medium. This entity can be implemented both as hardware and software on the various levels in a storage network.

It is also possible for several virtualisation entities to be used concurrently. For example, an application can use the virtualised volume of a volume manager on server level, which for its part is formed from a set of virtualised volumes which are exported by one or more disk subsystems (Section 5.1).

5.4 IMPLEMENTATION CONSIDERATIONS

In the following we want to draw up general requirements and considerations for the implementation of the virtualisation entity and illustrate how the difficulties described in Section 5.2 can be solved with the aid of storage virtualisation. For example, storage virtualisation also facilitates the integration of higher storage functions that previously had to be realised by means of other software products.

5.4.1 Realisation of the virtualisation entity

First of all, it is important that a storage virtualisation entity can be administered from a central console regardless of whether it is implemented as hardware or software and where it is positioned in the storage network. It is desirable for all tools that are required for the administration of the storage device to run via this console.

All operations performed by the virtualisation entity should take place in a rule-based manner and orientate themselves to the applicable data profiles. Policy-based operation allows the storage administrator to configure and control the operations of the virtualisation entity. Profile-orientation makes it possible for the data to be automated according to its specific properties and requirements.

Because virtualisation always intervenes in the data stream, correct implementation is indispensable if data corruption is to be avoided. The virtualisation entity itself should therefore also be backed up so that access to the virtualised storage resources is still possible in the event of a failure. The concepts for server-clustering introduced in Section 6.3.2 are suitable here.

In order to achieve the greatest possible degree of compatibility to servers and applications and also to win acceptance amongst users it is necessary for a virtualisation entity to remain hidden from its users. Servers, applications and users must always have the impression that they are working with physical storage media and must not notice the existence of a virtualisation entity. Furthermore, for reasons of compatibility, access on both file and block level, including the required protocols, must be supported (Section 5.5).

In addition to virtualised storage, the classical non-virtualised storage access options should continue to exist. This facilitates first of all an incremental introduction of the virtualisation technique into the storage network and second, allows applications, servers and storage devices that are incompatible with virtualisation to continue to be operated in the same storage network. For example, in our practical work during the testing of virtualisation software we found that the connection of a Windows server functioned perfectly, whilst the connection of a Solaris server failed. This was because of minor deviations from the Fibre Channel standard in the realisation of the Fibre Channel protocol in the virtualisation software used.

5.4.2   Replacement of storage devices

When using storage virtualisation the replacement of storage devices is relatively easy to perform, since the servers no longer access the physical devices directly, instead only working with virtual storage media. The replacement of a storage device in this case involves the following steps:

1. Connection of the new storage device to the storage network.
2. Configuration and connection of the new storage device to the virtualisation entity.
3. Migration of the data from the old to the new device by the virtualisation entity whilst the applications are running.
4. Removal of the old storage device from the configuration of the virtualisation entity.
5. Removal of the old storage device from the storage network.

The process requires no configuration changes to the applications. These continue to work on their virtual hard disks throughout the entire process.

5.4.3 Efficient use of resources by dynamic storage allocation

Certain mechanisms, such as the insertion of a volume manager within the virtualisation entity, permit the implementation of various approaches for the efficient use of resources. First, all storage resources can be shared. Furthermore, the virtualisation entity can react dynamically to the capacity requirements of virtual storage by making more physical capacity available to a growing data set on virtual storage and, in the converse case, freeing up the storage once again if the data set shrinks. Such concepts can be more easily developed on the file level than on the block level since on the file level a file system holds the information on unoccupied blocks, whereas on the block level this information is lacking. Even if such concepts have not previously been realised, they can be realised with storage virtualisation.

In this manner it is possible to practically imitate a significantly larger storage, of which only part is actually physically present. By the dynamic allocation of the physical storage, additional physical storage can be assigned to the virtual storage when needed. Finally by dynamic, data-oriented storage allocation it is possible to achieve a more efficient utilisation of resources.

5.4.4 Efficient use of resources by data migration

If a virtualisation entity is oriented towards the profiles of the data that it administers, it can determine which data is required and how often. In this manner it is possible to control the distribution of the data on fast and slow storage devices in order to achieve a high data throughput for frequently required data. Such data migration is also useful if it is based upon the data type. In the case of video data, for example, it can be worthwhile to store only the start of the file on fast storage in order to provide users with a short insight into the video file. If the user then accesses further parts of the video file that are not on the fast storage, this must first be played back from the slower to the fast storage.

5.4.5 Performance increase

Performance can be increased in several ways with the aid of storage virtualisation. First of all, caching within the virtualisation entity always presents a good opportunity for reducing the number of slow physical accesses (Section 2.6).

Techniques such as striping or mirroring within the virtualisation entity for distributing the data over several resources can also be used to increase performance (Section 2.5).

Further options for increasing performance are presented by the distribution of the I/O load amongst several virtualisation entities working together and amongst several data paths between server and virtual storage or virtual storage and physical storage devices (Section 5.1).

5.4.6 Availability due to the introduction of redundancy

The virtualisation entity can ensure the redundancy of the data by itself since it has complete control over the resources. The appropriate RAID techniques are suitable here (Section 2.5). For example, in the event of the failure of a storage device, operation can nevertheless be continued. The virtualisation entity can then immediately start to mirror the data once again in order to restore the redundancy of the data. As a result, a device failure is completely hidden from the servers – apart from possible temporary reductions in performance.

It is even more important that information about a device failure is reported to a central console so that the device is replaced immediately. The message arriving at the console can also be forwarded by e-mail or pager to the responsible person (Section 10.3).

Multiple access paths, both between servers and virtual storage and also between virtual storage and physical storage devices can also contribute to the improvement of fault-tolerance in storage virtualisation (Section 6.3.1).

5.4.7 Backup and archiving

A virtualisation entity is also a suitable data protection tool. By the use of appropriate rules the administrator can, for example, define different backup intervals for different data. Since the virtualisation entity is responsible for the full administration of the physical storage it can perform the backup processes in question independently. All network backup methods (Chapter 7) can be integrated into storage virtualisation.

5.4.8 Data sharing

Data sharing can be achieved if the virtualisation entity permits access to the virtual storage on file level. In this case, the virtualisation entity manages the file system centrally. By means of appropriate protocols, the servers can access the files in this file system in parallel. Currently this is permitted primarily by classical network file systems such as NFS and CIFS (Section 4.2) and fast shared disk file systems (Section 4.3). Some manufacturers are working on cross-platform shared disk file systems with the corresponding protocol mechanisms that permit the fast file sharing even in heterogeneous environments.

5.4.9   Privacy protection

The allocation of user rights and access configurations can also be integrated into a virtualisation entity, since it forms the interface between virtual and physical storage and thus prevents direct access to the storage by the user. In this manner, the access rights of the data can be managed from a central point.

5.5   STORAGE VIRTUALISATION ON BLOCK OR FILE LEVEL

In Section 5.3 we saw that the virtualisation of storage requires an entity that maps between virtual and physical storage and vice versa. The virtualisation entity can be located on the servers (for example, in the form of a volume manager), on the storage devices (for example, in a disk subsystem) or in the network (for example, as a special device). Regardless of which level of the storage network (server, network or storage device) the virtualisation entity is located on, we can differentiate between two basic types of virtualisation: virtualisation on block level and virtualisation on file level.

Virtualisation on block level means that storage capacity is made available to the operating system or the applications in the form of virtual disks (Figure 5.12). Operating system and applications on the server then work to the blocks of this virtual disk. To this end, the blocks are managed as usual – like the blocks of a physical disk – by a file system or by a database on the server. The task of the virtualisation entity is to map these virtual blocks to the physical blocks of the real storage devices. It can come about as part of this process that the physical blocks that belong to the virtual blocks of a file in the file system of the operating system are stored on different physical storage devices or that they are virtualised once more by a further virtualisation entity within a storage device.

By contrast, virtualisation on file level means that the virtualisation entity provides virtual storage to the operating systems or applications in the form of files and directories (Figure 5.13). In this case, the applications work with files instead of blocks and the conversion of the files to virtual blocks is performed by the virtualisation entity itself. The physical blocks are presented in the form of a virtual file system and not in the form of virtual blocks. The management of the file system is shifted from the server to the virtualisation entity.

To sum up, virtualisation on block or file level can be differentiated as follows: in virtualisation on block level, access to the virtual storage takes place by means of blocks, in virtualisation on file level it takes place by means of files. In virtualisation on block level the task of file system management is the responsibility of the operating system or the applications, whereas in virtualisation on file level this task is performed by the virtualisation entity.

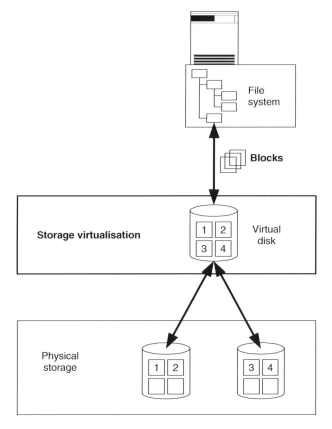

Figure 5.12   In virtualisation on block level the virtualisation entity provides the virtual storage to the servers in the form of a virtual disk.

Virtualisation on block level is suitable if the storage is to be virtualised for as many different operating systems and applications as possible. Virtualisation on block level is actually necessary when dealing with applications that handle their storage access on block level and cannot work on file level. Classic representatives of this category are, for example, databases that can only work with raw devices.

Virtualisation on file level, on the other hand, is indispensable for those who want to establish data sharing between several servers. To achieve this, the virtualisation entity must allow several servers access to the same files. This can only be achieved if the file system is implemented in the form of a shared resource as in a network file system (Section 4.2) or a shared disk file system (Section 4.3) or, just like virtualisation on file level, is held centrally by the virtualisation entity.

In this chapter we constrain the virtualisation on block level to the virtualisation of disks. Later on in Chapter 11 we will expand this approach and discuss the virtualisation of removeable media like tapes and opticals.

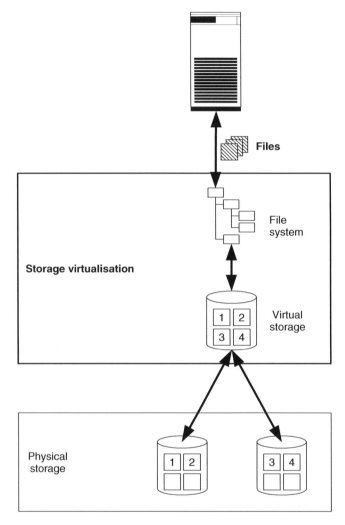

Figure 5.13   In virtualisation on file level the virtualisation entity provides the virtual storage to the servers in the form of files and directories.

5.6   STORAGE VIRTUALISATION ON VARIOUS LEVELS OF THE STORAGE NETWORK

In the following we will concern ourselves with the locations at which a virtualisation entity can be positioned in the storage network. The following three levels can be defined

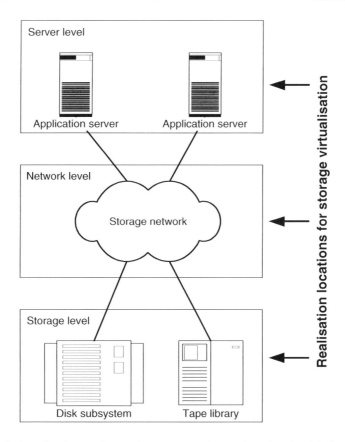

Figure 5.14   A virtualisation entity can be positioned on various levels of the storage network.

here (Figure 5.14): the server (Section 5.6.1), the storage devices (Section 5.6.2) and the network (Section 5.6.3). This will be explained in what follows.

5.6.1   Storage virtualisation in the server

A classic representative of virtualisation in the server is the combination of file system and volume manager (Section 4.1.4). A volume manager undertakes the separation of the storage into logical view and physical implementation by encapsulating the physical hard disk into logical disk groups and logical volumes. These are then made available to the applications via file systems. File systems and databases positioned on the server now work with these logical volumes and cease to work directly with the physical hard disks. Some volume managers additionally have further storage functions such as RAID, snapshots or dynamic reconfiguration options, which permit the addition and removal of storage during operation. With shared disk file systems (Section 4.3) storage virtualisation can be expanded to several servers, in order to allow fast file sharing among several servers. These cannot, however, be used in a straightforward manner in heterogeneous environments due to the incompatibilities that prevail.

Virtualisation on block level can be performed on a server by the host bus adapter itself. Virtualisation on block level is found, for example, in the use of a RAID controller. This performs the mapping of the logical blocks that are used by the file system or the volume manager of the operating system to the physical blocks of the various drives.

The benefits of virtualisation on server level are:

- Tried and tested virtualisation techniques are generally used.

- The virtualisation functions can integrate multiple storage systems.

- No additional hardware is required in the storage network to perform the virtualisation. Thus additional error sources can be ruled out. The approach remains cost-effective.

The disadvantages of a virtualisation on server level are:

- The administration of the storage virtualisation must take place on every single server. To achieve this, the appropriate software must be installed and maintained upon the computers.

- The storage virtualisation software running on the server can cost system resources and thus have a negative impact upon the server performance.

- Incompatibilities may occur between the virtualisation software and certain applications.

- The virtualisation extends only to those areas of a storage network that are accessible or assigned to those servers running a virtualisation entity.

- The virtualisation only ever takes place on individual servers. This disadvantage can be remedied by complex cluster approaches, which, however, come at an additional administration cost.

5.6.2 Storage virtualisation in storage devices

Virtualisation on block level in storage devices is, for example, found within intelligent disk subsystems (Section 2.7). These storage systems make their storage available to several servers via various I/O channels by means of LUN masking and RAID. The physical hard disks are brought together by the storage devices to form virtual disks, which the servers access using protocols such as SCSI, Fibre Channel FCP, FCoE and iSCSI. In this manner, the mapping of virtual to physical blocks is achieved.

Virtualisation on file level in storage devices is, for example, achieved by NAS servers (Section 4.2.2). The file system management is the responsibility of the NAS server. Access by the server to the storage resources takes place on file level by means of protocols such as NFS and CIFS.

The advantages of virtualisation on storage device level are:

- The majority of the administration takes place directly upon the storage device, which is currently perceived as easier and more reliable since it takes place very close to the physical devices.
- Advanced storage functions such as RAID and instant copies are realised directly at the physical storage resources, meaning that servers and I/O buses are not loaded.
- The uncoupling of the servers additionally eases the work in heterogeneous environments since a storage device is able to make storage available to various platforms.
- The servers are not placed under additional load by virtualisation operations.

The disadvantages of virtualisation on storage device level are:

- Configuration and implementation of virtualisation are manufacturer-specific and may thus become a proprietary solution in the event of certain incompatibilities with other storage devices.
- It is very difficult – and sometimes even impossible – to get storage devices from different manufacturers to work together.
- Here too, virtualisation takes place only within a storage system and cannot effectively be expanded to include several such storage devices without additional server software.

5.6.3   Storage virtualisation in the network

Storage virtualisation by a virtualisation entity in the storage network is realised by symmetric or asymmetric storage virtualisation (Section 5.7). First, however, we want to discuss the general advantages and disadvantages of storage virtualisation in the network.

The advantages of virtualisation in the storage network are:

- The virtualisation can extend over the storage devices of various manufacturers.
- The virtualisation is available to servers with different operating systems that are connected to the storage network.
- Advanced storage functions, such as mirroring or snapshots can be used on storage devices that do not themselves support these techniques (for example, JBODs and low cost RAID arrays).

- The administration of storage virtualisation can be performed from a central point.
- The virtualisation operations load neither the server nor the storage device.

The disadvantages are:

- Additional hardware and software are required in the storage network.
- A virtualisation entity in the storage network can become a performance bottleneck.
- Storage virtualisation in the storage network is in contrast to other storage technologies currently (2009) still a new product category. Whilst storage virtualisation on the block level has been successfully established in production environments, there is still very limited experience with file level storage virtualisation which is located in the storage network.

5.7 SYMMETRIC AND ASYMMETRIC STORAGE VIRTUALISATION IN THE NETWORK

The symmetric and asymmetric virtualisation models are representatives of storage virtualisation in the network. In both approaches it is possible to perform virtualisation both on block and on file level. In both models the virtualisation entity that undertakes the separation between physical and logical storage is placed in the storage network in the form of a specialised server or a device. This holds all the meta-information needed for the virtualisation. The virtualisation entity is therefore also called the metadata controller. Its duties also include the management of storage resources and the control of all storage functions that are offered in addition to virtualisation.

Symmetric and asymmetric virtualisation differ primarily with regard to their distribution of data and control flow. Data flow is the transfer of the application data between the servers and storage devices. The control flow consists of all metadata and control information necessary for virtualisation between virtualisation entity and storage devices and servers. In symmetric storage virtualisation the data flow and the control flow travel down the same path. By contrast, in asymmetric virtualisation the data flow is separated from the control flow.

5.7.1 Symmetric storage virtualisation

In symmetric storage virtualisation the data and control flow go down the same path (Figure 5.15). This means that the abstraction from physical to logical storage necessary for virtualisation must take place within the data flow. As a result, the metadata controller is positioned precisely in the data flow between server and storage devices, which is why symmetric virtualisation is also called in-band virtualisation.

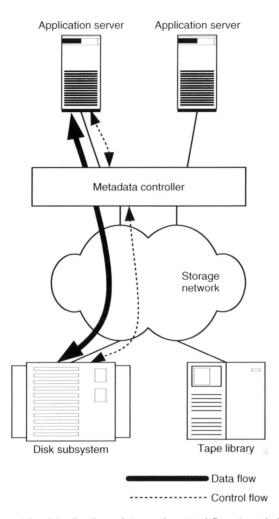

Figure 5.15   In symmetric virtualisation, data and control flow travel down the same path. The abstraction from physical to logical storage takes place within the data stream.

In addition to the control of the virtualisation, all data between servers and storage devices now flow through the metadata controller. To this end virtualisation is logically structured in two layers: the layer for the management of the logical volumes and the data access layer (Figure 5.16):

1. The volume management layer is responsible for the management and configuration of the storage devices that can be accessed directly or via a storage network and it provides the aggregation of these resources into logical disks.

2. The data access layer makes the logical drives available for access either on block or file level, depending upon what degree of abstraction is required. These logical drives can thus be made available to the application servers by means of appropriate protocols. In the case of virtualisation on block level, this occurs in the form of a virtual disk and in the case of virtualisation on file level it takes place in the form of a file system.

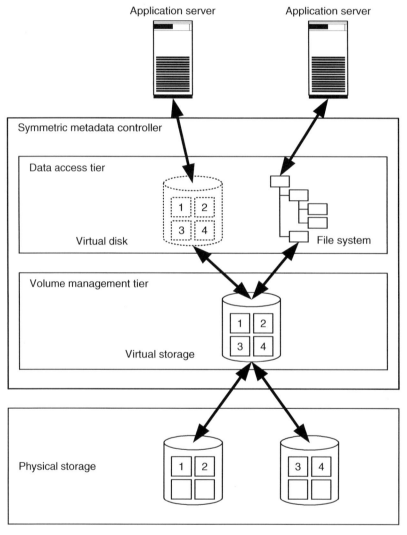

Figure 5.16 In symmetric virtualisation the metadata controller consists of a data access layer and a volume management layer.

In symmetric virtualisation all data flow through the metadata controller, which means that this represents a potential bottleneck. To increase performance, therefore, the metadata controller is upgraded by the addition of a cache. With the use of caching and symmetric virtualisation it is even possible to improve the performance of an existing storage network as long as exclusively write-intensive applications are not used.

A further issue is fault-tolerance. A single metadata controller represents a single point of failure. The use of cluster technology (Section 6.3.2) makes it possible to remove the single point of failure by using several metadata controllers in parallel. In addition, a corresponding load distribution provides a performance increase. However, a configuration failure or a software failure of that cluster can lead to data loss on all virtualised resources. In the case of a network-based virtualisation spanning several servers and storage devices, this can halt the activity of a complete data centre (Section 6.3.4).

Thus the advantages of symmetric virtualisation are evident:

- The application servers can easily be provided with data access both on block and file level, regardless of the underlying physical storage devices.

- The administrator has complete control over which storage resources are available to which servers at a central point. This increases security and eases the administration.

- Assuming that the appropriate protocols are supported, symmetric virtualisation does not place any limit on specific operating system platforms. It can thus also be used in heterogeneous environments.

- The performance of existing storage networks can be improved by the use of caching and clustering in the metadata controllers.

- The use of a metadata controller means that techniques such as snapshots or mirroring can be implemented in a simple manner, since they control the storage access directly. They can also be used on storage devices such as JBODs or simple RAID arrays that do not provide to these techniques themselves.

The disadvantages of a symmetric virtualisation are:

- Each individual metadata controller must be administered. If several metadata controllers are used in a cluster arrangement, then the administration is relatively complex and time-consuming particularly due to the cross-computer data access layer. This disadvantage can, however, be reduced by the use of a central administration console for the metadata controller.

- Several controllers plus cluster technology are indispensable to guarantee the fault-tolerance of data access.

- As an additional element in the data path, the controller can lead to performance problems, which makes the use of caching or load distribution over several controllers indispensable.

- It can sometimes be difficult to move the data between storage devices if this is managed by different metadata controllers.

5.7.2   Asymmetric storage virtualisation

In contrast to symmetric virtualisation, in asymmetric virtualisation the data flow is separated from the control flow. This is achieved by moving all mapping operations from logical to physical drives to a metadata controller outside the data path (Figure 5.17). The metadata controller now only has to look after the administrative and control tasks of virtualisation, the flow of data takes place directly from the application servers to the storage devices. As a result, this approach is also called out-band virtualisation.

The communication between metadata controller and agents generally takes place via the LAN (out-band) but can also be realised in-band via the storage network. Hence, in our opinion the terms 'in-band virtualisation' and 'out-band virtualisation' are a little misleading. Therefore, we use instead the terms 'symmetric virtualisation' and 'asymmetric virtualisation' to refer to the two network-based virtualisation approaches.

Like the symmetric approach, the metadata controller is logically structured in two layers (Figure 5.18). The volume management layer has the same duties as in the symmetric approach. The second layer is the control layer, which is responsible for the communication with an agent software that runs on the servers.

The agent is required in order to enable direct access to the physical storage resources. It is made up of a data access layer with the same tasks as in symmetric virtualisation and a control layer (Figure 5.18). Via the latter it loads the appropriate location and access information about the physical storage from the metadata controller when the virtual storage is accessed by the operating system or an application. In this manner, access control to the physical resources is still centrally managed by the metadata controller.

An agent need not necessarily run in the memory of the server. It can also be integrated into a host bus adapter. This has the advantage that the server can be freed from the processes necessary for virtualisation.

In asymmetric storage virtualisation – as is also the case for symmetric storage virtualisation – advanced storage functions such as snapshots, mirroring or data migration can be realised. The asymmetric model is, however, not so easy to realise as the symmetric one, but performance bottlenecks as a result of an additional device in the data path do not occur here. If we want to increase performance by the use of caching for both application as well as metadata, this caching must be implemented locally on every application server. The caching algorithm to be used becomes very complex since it is a distributed environment, in which every agent holds its own cache (Section 4.3).

Data inconsistencies as a result of different cache contents for the same underlying physical storage contents must be avoided and error situations prevented in which an application crashes, that still has data in the cache. Therefore, additional mechanisms are necessary to guarantee the consistency of the distributed cache. Alternatively, the installation of a dedicated cache server in the storage network that devotes itself exclusively to the caching of the data flow would also be possible. Unfortunately, such products are not currently (2009) available on the market.

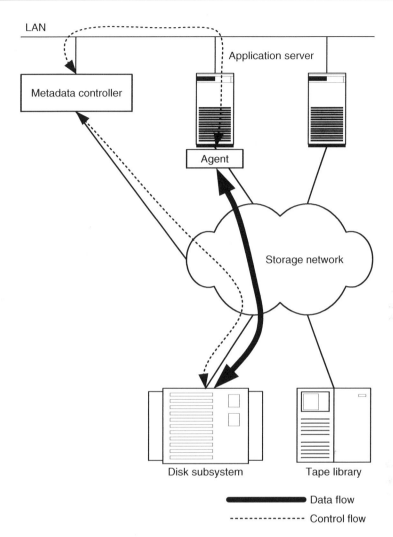

Figure 5.17 In contrast to symmetric virtualisation, in asymmetric virtualisation the data flow is separated from the control flow. The abstraction of physical to logical storage thus takes place outside the data flow.

Metadata controllers can also be constructed as clusters for the load distribution of the control flow and to increase fault-tolerance. The implementation is, however, easier with the asymmetric approach than it is with the symmetric since only the control flow has to be divided over several computers. In contrast to the symmetric approach, the splitting of the data flow is dispensed with.

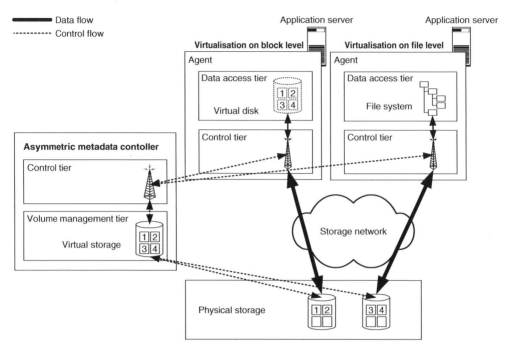

Figure 5.18   In asymmetric virtualisation the metadata controller takes on only the administrative control tasks for the virtualisation. Access to the physical storage is realised by means of an agent software.

The following advantages of asymmetric virtualisation can be established:

- Complete control of storage resources by an absolutely centralised management on the metadata controller.

- Maximum throughput between servers and storage devices by the separation of the control flow from the data flow, thus avoiding additional devices in the data path.

- In comparison to the development and administration of a fully functional volume manager on every server, the porting of the agent software is associated with a low cost.

- As in the symmetric approach, advanced storage functions such as snapshots or mirroring can be used on storage devices that do not themselves support these functions.

- To improve fault-tolerance, several metadata controllers can be brought together to form a cluster. This is easier than in the symmetric approach, since no physical connection from the servers to the metadata controllers is necessary for the data flow.

The disadvantages of asymmetric virtualisation are:

- A special agent software is required on the servers or the host bus adapters. This can make it more difficult to use this approach in heterogeneous environments, since such software or a suitable host bus adapter must be present for every platform. Incompatibilities between the agent software and existing applications may sometimes make the use of asymmetric virtualisation impossible.

- The agent software must be absolutely stable in order to avoid errors in storage accesses. In situations where there are many different platforms to be supported, this is a very complex development and testing task.

- The development cost increases further if the agent software and the metadata controller are also to permit access on file level in addition to access on block level.

- A performance bottleneck can arise as a result of the frequent communication between agent software and metadata controller. These performance bottlenecks can be remedied by the caching of the physical storage information.

- Caching to increase performance requires an ingenious distributed caching algorithm to avoid data inconsistencies. A further option would be the installation of a dedicated cache server in the storage network.

- In asymmetric virtualisation there is always the risk of a server with no agent software being connected to the storage network. In certain cases it may be possible for this server to access resources that are already being used by a different server and to accidentally destroy these. Such a situation is called a rogue host condition.

5.8 SUMMARY

By the separation of storage into physical implementation and logical view, storage virtualisation opens up a number of possibilities for simplifying administration. In addition, it makes it possible to fully utilise the capabilities of a storage network. In this manner, resources can be used more efficiently in order to increase performance and improve fault-tolerance due to higher availability of the data. In addition, more extensive functions such as data protection, data migration, privacy protection and data sharing can be integrated into the storage virtualisation, making the use of additional software for these functions unnecessary. The storage virtualisation itself is performed by a virtualisation entity, which forms the interface between virtual and physical storage. Virtualisation takes place in two different forms: on block or on file level. The virtualisation entity can be positioned on different levels of the storage network such as at the servers, storage devices or network. The advantages and disadvantages of the individual levels come to bear here. In particular, symmetric and asymmetric virtualisation have great potential, since due to the positioning in the storage network they can provide their services to different servers and at the same time are also able to integrate the various storage devices into the virtualisation.

In the previous chapters we introduced all the necessary techniques for storage networks. In the chapters that follow we will discuss how the previously discussed techniques can be used in order to realise requirements such as storage pooling and clustering (Chapter 6) backup (Chapter 7) or digital archiving (Chapter 8). Furthermore, in (Chapter 11) we will broaden the concept of storage virtualisation by including the virtualisation of tapes and other removeable media.

Part II

Application and Management of Storage Networks

6

Application of Storage Networks

In the first part of the book we introduced the fundamental building blocks of storage networks such as disk subsystems, file systems, virtualisation and transmission techniques. In the second part of the book our objective is to show how these building blocks can be combined in order to fulfil the requirements of IT systems such as flexibility, fault-tolerance and maintainability. As a prelude to the second part, this chapter discusses the fundamental requirements that are imposed independently of a particular application.

First of all, Section 6.1 contrasts the characteristics of various kinds of networks in order to emphasise the shape of a storage network. Section 6.2 introduces various possibilities in the storage network for device sharing and data sharing among several servers. The final part of the chapter deals with the two fundamental requirements of IT systems: availability of data (fault-tolerance, Section 6.3) and adaptability (Section 6.4).

In the following chapters we will be discussing the major topics of network backup (Chapter 7), archiving (Chapter 8), business continuity (Chapter 9) and the management of storage networks (chapters 10 and 11).

6.1 DEFINITION OF THE TERM 'STORAGE NETWORK'

In our experience, ambiguities regarding the definition of the various transmission techniques for storage networks crop up again and again. This section therefore illustrates

Storage Networks Explained: Basics and Application of Fibre Channel SAN, NAS, iSCSI, InfiniBand and FCoE, Second Edition
U. Troppens R. Erkens W. Müller-Friedt N. Haustein R. Wolafka © 2009 John Wiley & Sons, Ltd

networks and storage networks once again from various points of view. It considers the layering of the various protocols and transmission techniques (Section 6.1.1), discusses once again at which points in the I/O path networks can be implemented (Section 6.1.2) and it again defines the terms LAN (Local Area Network), MAN (Metropolitan Area Network), WAN (Wide Area Network) and SAN (Storage Area Network) (Section 6.1.3).

6.1.1 Layering of the transmission techniques and protocols

If we greatly simplify the Open System Interconnection (OSI) reference model, then we can broadly divide the protocols for storage networks into three layers that build upon one another: network technologies, transport protocols and application protocols (Figure 6.1). The transmission techniques provide the necessary physical connection between several end devices. Building upon these, transport protocols facilitate the data exchange between end devices via the underlying networks. Finally, the application protocols determine which type of data the end participants exchange over the transport protocol.

Transmission techniques represent the necessary prerequisite for data exchange between several participants. In addition to the already established Ethernet, the first part of the book introduces Fibre Channel and InfiniBand. They all define a medium (cable, radio

File-oriented	Block-oriented	Memory oriented
• NFS	• SCSI	• Remote Direct Memory
• CIFS	• Fibre Channel (FCP)	Access (RDMA)
• HTTP	• Internet SCSI (iSCSI)	
• DAFS	• Internet FCP (iFCP)	

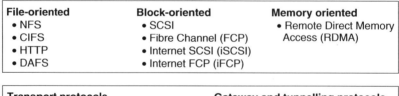

Transport protocols	Gateway and tunnelling protocols
• Fibre Channel (FC-2, FC-3)	• IP over Fibre Channel (IPFC)
• TCP/IP	• Fibre Channel over IP (FCIP)
• UDP/IP	
• Virtual Interface Architecture (VIA)	
Remote Direct Memory Access (RDMA)	

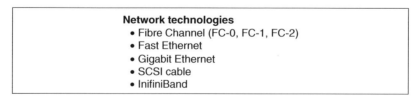

Network technologies
- Fibre Channel (FC-0, FC-1, FC-2)
- Fast Ethernet
- Gigabit Ethernet
- SCSI cable
- InifiniBand

Figure 6.1 The communication techniques introduced in the first part of the book can be divided into transmission techniques, transport protocols and application protocols.

frequency) and the encoding of data in the form of physical signals, which are transmitted over the medium.

Transport protocols facilitate the exchange of data over a network. In addition to the use of the tried and tested and omnipresent TCP protocol the first part of the book introduces Fibre Channel and the Virtual Interface Architecture (VIA). Transport protocols can either be based directly upon a transmission technique such as, for example, Virtual Interfaces over Fibre Channel, InfiniBand or Ethernet or they can use an alternative transport protocol as a medium. Examples are Fibre Channel over IP (FCIP) and IP over Fibre Channel (IPFC). Additional confusion is caused by the fact that Fibre Channel defines both a transmission technique (FC-0, FC-1, FC-2) and a transport protocol (FC-2, FC-3) plus various application protocols (FC-4).

Application protocols define the type of data that is transmitted over a transport protocol. With regard to storage networks we differentiate between block-oriented and file-oriented application protocols. SCSI is the mother of all block-oriented application protocols for block-oriented data transfer. All further block-oriented application protocols such as FCP, iFCP and iSCSI were derived from the SCSI protocol. File-oriented application protocols transmit files or file fragments. Examples of file-oriented application protocols discussed in this book are NFS, CIFS, FTP, HTTP and DAFS.

6.1.2 Networks in the I/O path

The logical I/O path offers a second point of view for the definition of transmission techniques for storage networks. Figure 6.2 illustrates the logical I/O path from the disk to the application and shows at which points in the I/O path networks can be used. Different application protocols are used depending upon location. The same transport protocols and transmission techniques can be used regardless of this.

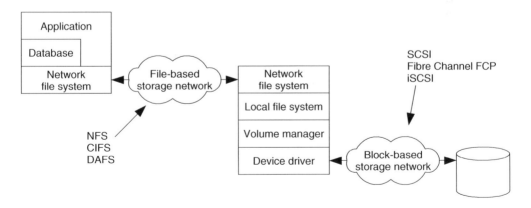

Figure 6.2 Storage networks in the I/O path.

Below the volume manager, block-oriented application protocols are used. Depending upon technique these are SCSI and SCSI offshoots such as FCP, iFCP, iSCSI and FCoE. Today, block-oriented storage networks are found primarily between computers and storage systems. However, within large disk subsystems too the SCSI cable is increasingly being replaced by a network transmission technique (Figure 6.3).

Above the volume manager and file system, file-oriented application protocols are used. Here we find application protocols such as NFS, CIFS, HTTP, FTP and DAFS. In Chapter 4 three different fields of application for file-oriented application protocols were discussed: traditional file sharing, high-speed LAN file sharing and the World Wide Web. Shared disk file systems, which realise the network within the file system, should also be mentioned as a special case.

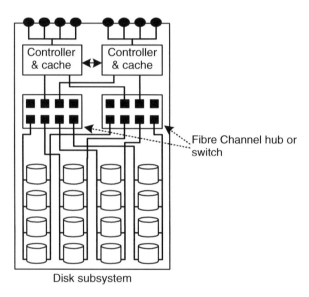

Figure 6.3   A storage network can be hidden within a disk subsystem.

6.1.3   Data networks, voice networks and storage networks

Today we can differentiate between three different types of communication networks: data networks, voice networks and storage networks. The term 'voice network' refers to the omnipresent telephone network. Data networks describe the networks developed in the 1990s for the exchange of application data. Data networks are subdivided into LAN, MAN and WAN, depending upon range. Storage networks were defined in the first chapter as networks that are installed in addition to the existing LAN and are primarily used for data exchange between computers and storage devices.

In introductions to storage networks, storage networks are often called SANs and compared with conventional LANs (a term for data networks with low geographic extension). Fibre Channel technology is often drawn upon as a representative for the entire category of storage networks. This is clearly because Fibre Channel is currently the dominant technology for storage networks. Two reasons lead us to compare LAN and SAN: first, LANs and Fibre Channel SANs currently have approximately the same geographic range. Second, quite apart from capacity bottlenecks, separate networks currently have to be installed for LANs and SANs because the underlying transmission technologies (Ethernet or Fibre Channel) are incompatible.

We believe it is very likely that the three network categories – storage networks, data networks and voice networks – will converge in the future, with TCP/IP, or at least Ethernet, being the transport protocol jointly used by all three network types. We discussed the economic advantages of storage networks over Ethernet in Sections 3.5.2 and 3.7.1. We see it as an indication of the economic advantages of voice transmission over IP (Voice over IP, VoIP) that more and more reputable network manufacturers are offering VoIP devices.

6.2 STORAGE SHARING

In Part I of the book you heard several times that one advantage of storage networks is that several servers can share storage resources via the storage network. In this context, storage resources mean both storage devices such as disk subsystems and tape libraries and also the data stored upon them. This section discusses various variants of storage device sharing and data sharing based upon the examples of disk storage pooling (Section 6.2.1), dynamic tape library sharing (Section 6.2.2) and data sharing (Section 6.2.3).

6.2.1 Disk storage pooling

Disk storage pooling describes the possibility that several servers share the capacity of a disk subsystem. In a server-centric IT architecture each server *possesses* its own storage: Figure 6.4 shows three servers with their own storage. Server 2 needs more storage space, but the free space in the servers 1 and 3 cannot be assigned to server 2. Therefore, further storage must be purchased for server 2, even though free storage capacity is available on the other servers.

In a server-centric IT architecture the storage capacity available in the storage network can be assigned much more flexibly. Figure 6.5 shows the same three servers as Figure 6.4. The same storage capacity is installed in the two figures. However, in Figure 6.5 only one storage system is present, which is shared by several servers (disk storage pooling). In this arrangement, server 2 can be assigned additional storage capacity by the reconfiguration

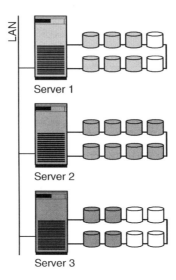

Figure 6.4 Inflexible: storage assignment in server-centric systems. Server 2 cannot use the free storage space of servers 1 and 3.

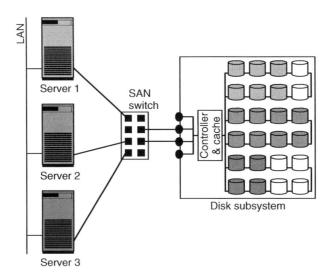

Figure 6.5 Better: storage pooling in storage-centric IT systems. Free storage capacity can be assigned to all servers.

of the disk subsystem without the need for changes to the hardware or even the purchase of a new disk subsystem.

In Section 5.2.2 ('Implementation-related limitations of storage networks') we discussed that storage pooling across several storage devices from various manufacturers in many cases limited. The main reason for this is the incompatibility of the device drivers for various disk subsystems. In the further course of Chapter 5 we showed how virtualisation in the storage network can help to overcome these incompatibilities and, in addition, further increase the efficiency of the storage pooling.

6.2.2   Dynamic tape library sharing

Tape libraries, like disk subsystems, can be shared among several servers. In tape library sharing we distinguish between static partitioning of the tape library and dynamic tape library sharing. In static partitioning the tape library is broken down into several virtual tape libraries; each server is assigned its own virtual tape library (Figure 6.6). Each tape drive and each tape in the tape library are unambiguously assigned a virtual tape library; all virtual tape libraries share the media changer that move the tapes cartridges back and for between slots and tape drives. The reconfiguration of the assignment of tapes and tape drives to a certain virtual tape library is expensive in comparison to tape library sharing.

Tape library sharing offers greater flexibility (Figure 6.7). In this approach, the various servers dynamically negotiate which servers use which tape drives and tapes. To this

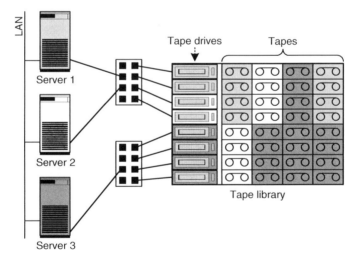

Figure 6.6   The partitioning of a tape library breaks down the tape library statically into several virtual tape libraries. Each server is assigned its own virtual tape library. The servers only share the media changer for the moving of the tapes.

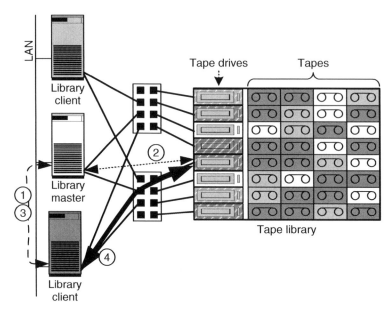

Figure 6.7 Tape library sharing: The library client informs the library master that it wants to write to a free tape (1). The library master places a free tape in a free drive (2) and sends the corresponding information back to the library client (3). The library client then writes directly via the storage network (4).

end one server acts as library master, all others as library clients. The library master co-ordinates access to the tapes and tape drives of the tape library. If a library client wishes to write data to a new tape then it first of all requests a free tape from the library master. The library master selects a tape from the store of free tapes (scratch pool) and places it in a free drive. Then it makes a note in its database that this tape is now being used by the library client and it informs the library client which drive the tape is in. Finally, the library client can write the data directly to the tape via the storage network.

In contrast to the conventional partitioning of tape libraries in which each tape and each tape drive are statically assigned to a certain server, the advantage of the tape library sharing is that tapes and drives are dynamically assigned. Depending upon requirements, sometimes one and sometimes another server can use a tape or a tape drive.

This requires that all servers access the tape library according to the same rules. Tape library sharing is currently an important component of network backup over storage networks (Section 7.8.6). But today, different applications – either on the same server or on different servers – cannot participate in the tape library sharing of a network backup system because there is no common synchronisation protocol. However, some tape libraries with built-in library managers support dynamic tape library sharing, but this requires that the applications integrate the proprietary API of the respective library manager. In the long term, these proprietary APIs need to be standardised to reduce the costs for library sharing across heterogeneous application and heterogeneous tape library boundaries. The IEEE 1244 Standard for Removable Media Management (Section 11.5) specifies exactly such an interface for tape library sharing.

6.2.3   Data sharing

In contrast to device sharing (disk storage pooling, tape library partitioning and tape library sharing) discussed earlier, in which several servers share a storage device at block level, data sharing is the use of a common data set by several applications. In data sharing we differentiate between data copying and real-time data sharing.

In data copying, as the name suggests, data is copied. This means that several versions of the data are kept, which is fundamentally a bad thing: each copy of the data requires storage space and care must be taken to ensure that the different versions are copied according to the requirements of the applications at the right times. Errors occur in particular in the maintenance of the various versions of the data set, so that subsequent applications repeatedly work with the wrong data.

Despite these disadvantages, data copying is used in production environments. The reasons for this can be:

- *Generation of test data*
  Copies of the production data are helpful for the testing of new versions of applications and operating systems and for the testing of new hardware. In Section 1.3 we used the example of a server upgrade to show how test data for the testing of new hardware could be generated using instant copies in the disk subsystem. The important point here was that the applications are briefly suspended so that the consistency of the copied data is guaranteed. As an alternative to instant copies the test data could also be generated using snapshots in the file system.

- *Data protection (backup)*
  The aim of data protection is to keep up-to-date copies of data at various locations as a precaution to protect against the loss of data by hardware or operating errors. Data protection is an important application in the field of storage networks. It is therefore dealt with separately in Chapter 7.

- *Archiving*
  The goal of archiving is to freeze the current version of data so that it can be used further at a later time. Many years or decades can pass between the archiving and the further usage of the data. Digital archiving is becoming more and important due to the increase in the digital storage of information. We therefore address this separately in Chapter 8.

- *Data replication*
  Data replication is the name for the copying of data for access to data on computers that are far apart geographically. The objective of data replication is to accelerate data access

and save network capacity. There are many applications that automate the replication of data. Within the World Wide Web the data is replicated at two points: first, every web browser caches local data in order to accelerate access to pages called up frequently by an individual user. Second, many Internet Service Providers (ISPs) install a so-called proxy server. This caches the contents of web pages that are accessed by many users. Other examples of data replication are the mirroring of FTP servers (FTP mirror), replicated file sets in the Andrew File System (AFS) or the Distributed File System (DFS) of the Distributed Computing Environment (DCE), and the replication of mail databases.

- *Conversion into more efficient data formats*
  It is often necessary to convert data into a different data format because certain calculations are cheaper in the new format. In the days before the pocket calculator logarithms were often used for calculations because, for example, the addition of logarithms yielded the same result as the multiplication in the origin space only more simply. For the same reasons, in modern IT systems data is converted to different data formats. In data mining, for example, data from various sources is brought together in a database and converted into a data format in which the search for regularities in the data set is simpler.

- *Conversion of incompatible data formats*
  A further reason for the copying of data is the conversion of incompatible data formats. A classic example is when applications originally developed independently of one another are being brought together over time.

Real-time data sharing represents an alternative to data copying. In real-time data sharing all applications work on a single data set. Real-time data sharing saves storage space, avoids the cost and errors associated with the management of several data versions and all applications work on the up-to-date data set. For the reasons mentioned above for data copying it is particularly important to replace the conversion of incompatible data sets by real-time data sharing.

The logical separation of applications and data is continued in the implementation. In general, applications and data in the form of file systems and databases are installed on different computers. This physical separation aids the adaptability, and thus the maintainability, of overall systems. Figure 6.8 shows several applications that work on a single data set, with applications and data being managed independently of one another. This has the advantage that new applications can be introduced without existing applications having to be changed.

However, in the configuration shown in Figure 6.8 the applications may generate so much load that a single data server becomes a bottleneck and the load has to be divided amongst several data servers. There are two options for resolving this bottleneck without data copying: first, the data set can be partitioned (Figure 6.9) by splitting it over several data servers. If this is not sufficient, then several parallel access paths can be established to the same data set (Figure 6.10). Parallel databases and shared disk file systems such as

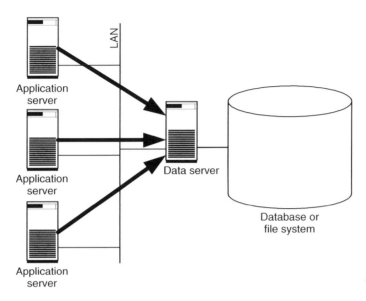

Figure 6.8   Real-time data sharing leads to the separation of applications and data: several applications work on the same data set.

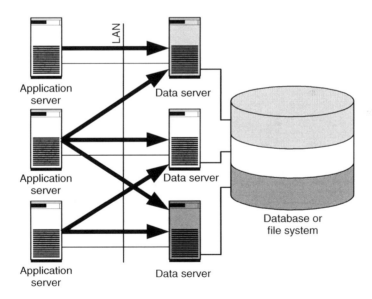

Figure 6.9   Static load balancing of the data servers by the partitioning of the data set.

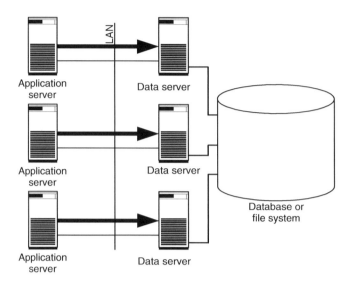

Figure 6.10   Dynamic load balancing of the data server by parallel databases and shared disk file systems.

the General Parallel File System (GPFS) introduced in Section 4.3.1 provide the functions necessary for this.

6.3   AVAILABILITY OF DATA

Nowadays, the availability of data is an important requirement made of IT systems. This section discusses how the availability of data and applications can be maintained in various error conditions. Individually, the following will be discussed: the failure of an I/O bus (Section 6.3.1), the failure of a server (Section 6.3.2), the failure of a disk subsystem (Section 6.3.3), and the failure of a storage virtualisation instance which is placed in the storage network (Section 6.3.4). The case study 'protection of an important database' discusses a scenario in which the protective measures that have previously been discussed are combined in order to protect an application against the failure of an entire data centre (Section 6.3.5).

6.3.1   Failure of an I/O bus

Protection against the failure of an I/O bus is relatively simple and involves the installation of several I/O buses between server and storage device. Figure 6.11 shows a scenario for

SCSI and Figure 6.12 shows one for Fibre Channel. In Figure 6.12 protection against the failure of an I/O bus is achieved by two storage networks that are independent of one another. Such separate storage networks are also known as a 'dual storage network' or 'dual SAN'.

The problem here: operating systems manage storage devices via the triple host bus adapter, SCSI target ID and SCSI logical unit number (LUN). If, for example, there are two connections from a server to a disk subsystem, the operating system recognises the same disk twice (Figure 6.13).

So-called multipathing software recognises that a storage device can be reached over several paths. Figure 6.14 shows how multipathing software reintegrates the disk found twice in Figure 6.13 to form a single disk again. Multipathing software can act at various points depending upon the product:

- in the volume manager (Figure 6.14, right);
- as an additional virtual device driver between the volume manager and the device driver of the disk subsystem (Figure 6.14, left);
- in the device driver of the disk subsystem;
- in the device driver of the host bus adapter card.

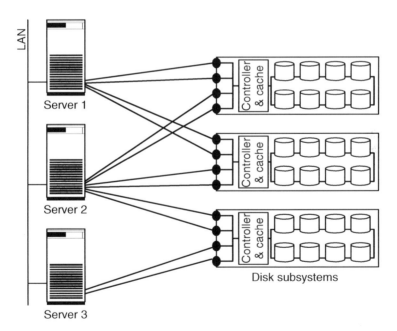

Figure 6.11 Redundant SCSI cable between server and disk subsystem protects against the failure of a SCSI cable, a SCSI host bus adapter or a connection port in the disk subsystem.

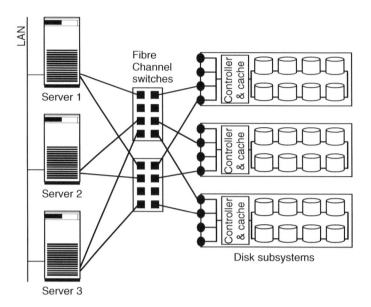

Figure 6.12 So-called dual storage networks (dual SAN) expand the idea of redundant SCSI cable to storage networks.

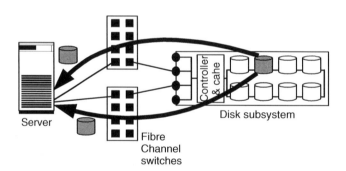

Figure 6.13 Problem: As a result of the redundant I/O path the operating system recognises a single physical or virtual disk several times.

Fibre Channel plans to realise this function in the FC-3 layer. However, this part of the Fibre Channel standard has not yet been realised in real products. We believe it is rather unlikely that these functions will ever actually be realised within the Fibre Channel Protocol stack. In the past the principle of keeping the network protocol as simple as possible and realising the necessary intelligence in the end devices has prevailed in networks.

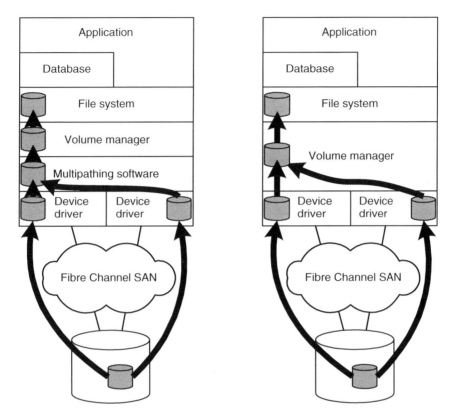

Figure 6.14 Solution: Multipathing software brings the multiply recognised disks back together. Multipathing software can be realised at various points in the operating system.

The multipathing software currently available on the market differs in the mode in which it uses redundant I/O buses:

- *Active/passive mode*
  In active/passive mode the multipathing software manages all I/O paths between server and storage device. Only one of the I/O paths is used for actual data traffic. If the active I/O path fails, the multipathing software activates one of the other I/O paths in order to send the data via this one instead.

- *Active/active mode*
  In active/active mode the multipathing software uses all available I/O paths between server and storage device. It distributes the load evenly over all available I/O channels. In addition, the multipathing software continuously monitors the availability of the individual I/O paths; it activates or deactivates the individual I/O paths depending upon their availability.

It is obvious that the active/active mode utilises the underlying hardware better than the active/passive mode, since it combines fault-tolerance with load distribution.

6.3.2 Failure of a server

Protection against the failure of an entire server is somewhat trickier. The only thing that can help here is to provide a second server that takes over the tasks of the actual application server in the event of its failure. So-called cluster software monitors the state of the two computers and starts the application on the second computer if the first computer fails.

Figure 6.15 shows a cluster for a file server, the disks of which are connected over Fibre Channel SAN. Both computers have access to the disks, but only one computer actively accesses them. The file system stored on the disks is exported over a network file system such as NFS or CIFS. To this end a virtual IP address (vip) is configured for the cluster. Clients access the file system via this virtual IP address.

If the first computer fails, the cluster software automatically initiates the following steps:

1. Activation of the disks on the stand-by computer.
2. File system check of the local file system stored on the disk subsystem.
3. Mounting of the local file system on the stand-by computer.
4. Transfer of the virtual cluster IP address.
5. Export of the local file system via the virtual cluster IP address.

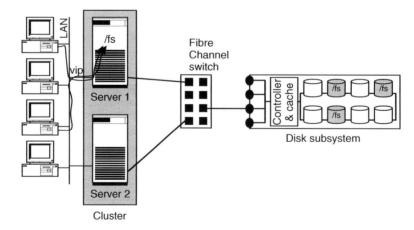

Figure 6.15 Server cluster to protect against the failure of a server: Server 1 exports the file system '/fs' over the virtual IP address 'vip'.

This process is invisible to clients of the file server apart from the fact that they cannot access the network file system for a brief period so file accesses may possibly have to be repeated (Figure 6.16).

Server clustering and redundant I/O buses are two measures that are completely independent of each other. In practice, as shown in Figure 6.17, the two measures are nevertheless combined. The multipathing software reacts to errors in the I/O buses significantly more quickly than the cluster software so the extra cost of the redundant I/O buses is usually justified.

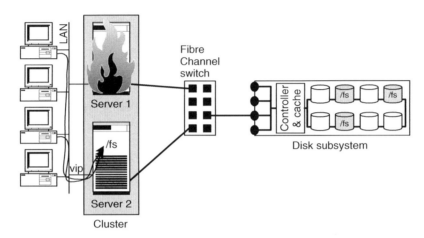

Figure 6.16   Failure of the primary server. The stand-by server activates the disks, checks the consistency of the local file system and exports this over the virtual IP address 'vip'.

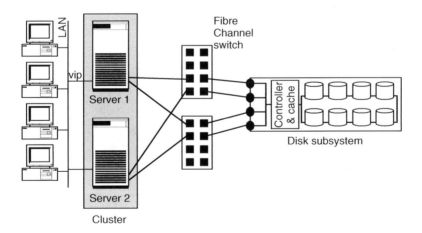

Figure 6.17   Server cluster and redundant I/O buses are independent concepts.

6.3.3 Failure of a disk subsystem

In Chapter 2 we discussed how disk subsystems implement a whole range of measures to increase their own fault-tolerance. Nevertheless, disk subsystems can sometimes fail, for example in the event of physical impairments such as fire or water damage or due to faults that should not happen at all according to the manufacturer. The only thing that helps in the event of faults in the disk subsystem is to mirror the data on two disk subsystems.

Mirroring (RAID 1) is a form of virtualisation, for which various realisation locations were discussed in Section 5.1. In contrast to classical RAID 1 within the disk subsystem for protection against its failure, the data is mirrored on two different disk subsystems, which are wherever possible separated by a fire protection wall and connected to two independent electric circuits. From the point of view of reducing the load on the server, the realisation of the mirroring by the disk subsystem in the form of remote mirroring is optimal (Figure 6.18, cf. also Section 2.7.2 and Section 5.1.)

From the point of view of fault-tolerance, however, remote mirroring through the disk subsystem represents a single point of failure: if the data in the disk subsystem is falsified on the way to the disk subsystem (controller faults, connection port faults), the copy of the data is also erroneous. Therefore, from the point of view of fault-tolerance, mirroring in the volume manager or in the application itself is optimal (Figure 6.19). In this approach the data is written to two different disk subsystems via two different physical I/O paths.

A further advantage of volume manager mirroring compared to remote mirroring is due to the way the two variants are integrated into the operating system. Volume manager mirroring is a solid component of every good volume manager: the volume manager reacts automatically to the failure and the restarting of a disk subsystem. On the other hand, today's operating systems in the Open System world are not yet good at handling copies of disks created by a disk subsystem. Switching to such a copy generally requires manual support. Although, technically, an automated reaction to the failure or the restarting of

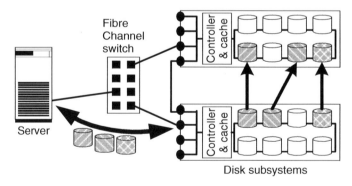

Figure 6.18 Remote mirroring of the disk subsystems protects against the failure of a disk subsystem.

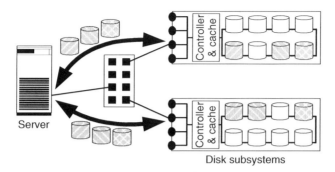

Disk subsystems

Figure 6.19 Volume manager mirroring is often a favourable alternative to disk subsystem remote mirroring for protection against the failure of a disk subsystem.

a disk subsystem is possible, however due to a missing integration into the operating system additional scripts are required.

On the other hand, there are some arguments in favour of remote mirroring. In addition to the performance benefits discussed above, we should also mention the fact that remote mirroring is supported over greater distances than volume manager mirroring. As a rule of thumb, volume manager can be used up to a maximum distance of six to ten kilometres between server and disk subsystem; for greater distances remote mirroring currently has to be used. Furthermore, remote mirroring offers advantages in respect of avoiding data loss when major failures occur. We will be taking a closer look at this aspect within the context of business continuity (Section 9.5.5).

For business continuity we will work out the very different characteristics of volume manager mirroring and remote mirroring with regard to fail-safe and loss-free operations (Section 9.4.2). The requirements for high availability and loss-free operation are often the deciding factors when determining whether volume manager mirroring or remote mirroring should be used. Later on we will show that certain requirements can only be fulfilled if both techniques are combined together (Section 9.5.6).

Figure 6.20 shows how volume manager mirroring, server clustering and redundant I/O buses can be combined. In this configuration the management of the disks is somewhat more complicated: each server sees each disk made available by the disk subsystem four times because each host bus adapter finds each disk over two connection ports of the disk subsystem. In addition, the volume manager mirrors the data on two disk subsystems. Figure 6.21 shows how the software in the server brings the disks recognised by the operating system back together again: the file system writes the data to a logical disk provided by the volume manager. The volume manager mirrors the data on two different virtual disks, which are managed by the multipathing software. The multipathing software also manages the four different paths of the two disks. It is not visible here whether the disks exported from the disk subsystem are also virtualised within the disk subsystem.

The configuration shown in Figure 6.20 offers good protection against the failure of various components, whilst at the same time providing a high level of availability of

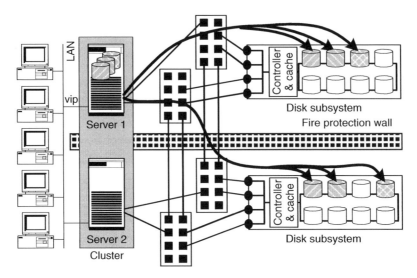

Figure 6.20   Combined use of server clustering, volume manager mirroring and redundant I/O buses.

data and applications. However, this solution comes at a price. Therefore, in practice, sometimes one and sometimes another protective measure is dispensed with for cost reasons. Often, for example, the following argument is used: 'The data is mirrored within the disk subsystem by RAID and additionally protected by means of network backup. That should be enough.'

6.3.4   Failure of virtualisation in the storage network

Virtualisation in the storage network is an important technology for the consolidation of storage resources in large storage networks, with which the storage resources of several disk subsystems can be centrally managed (Chapter 5). However, it is necessary to be clear about the fact that precisely such a central virtualisation instance represents a single point of failure. Even if the virtualisation instance is protected against the failure of a single component by measures such as clustering, the data of an entire data centre can be lost as a result of configuration errors or software errors in the virtualisation instance, since the storage virtualisation aims to span all the storage resources of a data centre.

Therefore, the same considerations apply for the protection of a virtualisation instance positioned in the storage network (Section 5.7) against the failure as the measures to protect against the failure of a disk subsystem discussed in the previous section. Therefore, the mirroring of important data from the server via two virtualisation instances should also be considered in the case of virtualisation in the storage network.

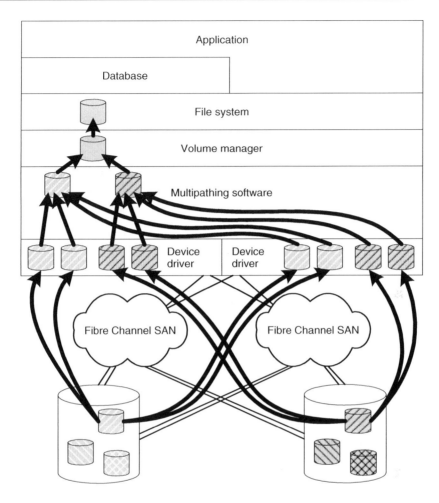

Figure 6.21   The logical I/O path for volume manager mirroring with redundant I/O buses.

6.3.5   Failure of a data centre based upon the case study 'protection of an important database'

The measures of server clustering, redundant I/O buses and disk subsystem mirroring (volume manager mirroring or remote mirroring) discussed above protect against the failure of a component within a data centre. However, these measures are useless in the event of the failure of a complete data centre (fire, water damage). To protect against the failure of a data centre it is necessary to duplicate the necessary infrastructure in a backup data centre to continue the operation of the most important applications.

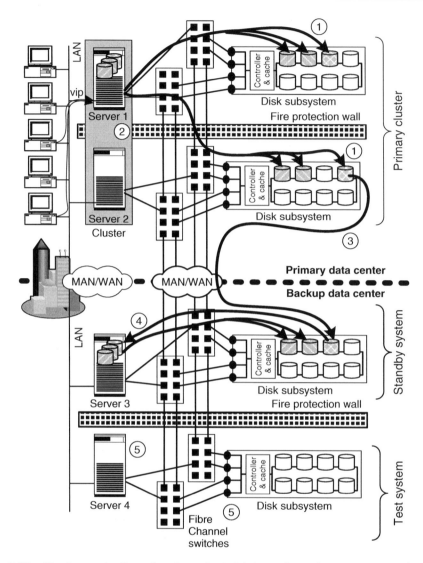

Figure 6.22 For the protection of an important database the volume manager mirrors data files and log files on two different disk subsystems (1). The database server and the I/O buses between database servers and disk subsystems are designed with built-in redundancy (2). In the backup data centre only the log files are mirrored (3). There, the log files are integrated into the data set with a delay (4). In normal operation the second computer and the second disk subsystem in the backup data centre are used for other tasks (5).

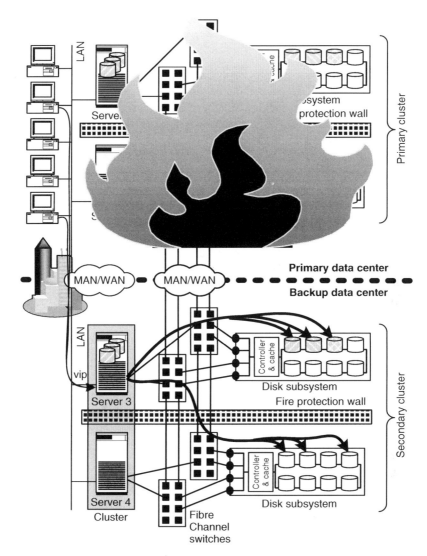

Figure 6.23   In the event of failure of the primary data centre the two computers in the backup data centre are configured as a cluster. Likewise, the data of the database is mirrored to the second disk subsystem via the volume manager. Thus the same configuration is available in the backup data centre as in the primary data centre.

Figure 6.22 shows the interaction between the primary data centre and its backup data centre based upon the case study 'protection of an important database'. In the case study, all the measures discussed in this section for protection against the failure of a component are used.

In the primary data centre all components are designed with built-in redundancy. The primary server is connected via two independent Fibre Channel SANs (Dual SAN) to two disk subsystems, on which the data of the database is stored. Dual SANs have the advantage that even in the event of a serious fault in a SAN (defective switch, which floods the SAN with corrupt frames), the connection via the other SAN remains intact. The redundant paths between servers and storage devices are managed by appropriate multipathing software.

Each disk subsystem is configured using a RAID procedure so that the failure of individual physical disks within the disk subsystem in question can be rectified. In addition, the data is mirrored in the volume manager so that the system can withstand the failure of a disk subsystem. The two disk subsystems are located at a distance from one another in the primary data centre. They are separated from one another by a fire protection wall.

Like the disk subsystems, the two servers are spatially separated by a fire protection wall. In normal operation the database runs on one server; in the meantime the second server is used for other, less important tasks. If the primary server fails, the cluster software automatically starts the database on the second computer. It also terminates all other activities on the second computer, thus making all its resources fully available to the main application.

Remote mirroring takes place via an IP connection. Mirroring utilises knowledge of the data structure of the database: in a similar manner to journaling in file systems (Section 4.1.2), databases write each change into a log file before then integrating it into the actual data set. In the example, only the log files are mirrored in the backup data centre. The complete data set was only transferred to the backup data centre once at the start of mirroring. Thereafter this data set is only ever adjusted with the aid of the log files.

This has two advantages: a powerful network connection between the primary data centre and the remote backup data centre is very expensive. The necessary data rate for this connection can be halved by only transferring the changes to the log file. This cuts costs.

In the backup data centre the log files are integrated into the data set after a delay of two hours. As a result, a copy of the data set that is two hours old is always available in the backup data centre. This additionally protects against application errors: if a table space is accidentally deleted in the database then the user has two hours to notice the error and interrupt the copying of the changes in the backup data centre.

A second server and a second disk subsystem are also operated in the backup data centre, which in normal operation can be used as a test system or for other, less time-critical tasks such as data mining. If the operation of the database is moved to the backup data centre, these activities are suspended (Figure 6.23). The second server is configured as a stand-by server for the first server in the cluster; the data of the first disk subsystem is mirrored to the second disk subsystem via the volume manager. Thus a completely redundant system is available in the backup data centre.

The realisation of the case study discussed here is possible with current technology. However, it comes at a price; for most applications this cost will certainly not be justified. The main point of the case study is to highlight the possibilities of storage networks. In practice you have to decide how much failure protection is necessary and how much this may cost. At the end of the day, protection against the loss of data or the tempo-rary non-availability of applications must cost less than the data loss or the temporary non-availability of applications itself.

The cost-effective protection of applications from failures is an important aspect of planning IT systems. Chapter 9 takes an in-depth look at the requirements for availability of applications and delineates between solutions that are technically the best and solutions that are considered necessary from the business perspective and therefore are the most sensible economically.

6.4 ADAPTABILITY AND SCALABILITY OF IT SYSTEMS

A further requirement of IT systems is that of adaptability and scalability: successful companies have to adapt their business processes to new market conditions in ever shorter cycles. Along the same lines, IT systems must be adapted to new business processes so that they can provide optimal support for these processes. Storage networks are also required to be scalable: on average the storage capacity required by a company doubles in the course of each year. This means that anyone who has 1 terabyte of data to manage today will have 32 terabytes in five years time. A company with only 250 gigabytes today will reach 32 terabytes in seven years time.

This section discusses the adaptability and scalability of IT systems on the basis of clusters for load distribution (Section 6.4.1), the five-tier architecture for web application servers (Section 6.4.2) and the case study 'structure of a travel portal' (Section 6.4.3).

6.4.1 Clustering for load distribution

The term 'cluster' is very frequently used in information technology, but which is not clearly defined. The meaning of the term 'cluster' varies greatly depending upon context. As the greatest common denominator we can only state that a cluster is a combination of components or servers that perform a common function in one form or another.

This section expands the cluster concept for protection against the failure of a server introduced in Section 6.3.2 to include clustering for load distribution. We discuss three different forms of clusters based upon the example of a file server. The three different forms of cluster are comparable to the modes of multipathing software.

The starting point is the so-called shared-null configuration (Figure 6.24). The compo-nents are not designed with built-in redundancy. If a server fails, the file system itself is no longer available, even if the data is mirrored on two different disk subsystems and redundant I/O buses are installed between server and disk subsystems (Figure 6.25).

In contrast to the shared-null configuration, shared-nothing clusters protect against the failure of a server. The basic form of the shared-nothing cluster was discussed in Section 6.3.2 in relation to the protection of a file server against the failure of a server. Figure 6.26 once again shows two shared-nothing clusters each with two servers.

Shared-nothing clusters can be differentiated into active/active and active/passive configurations. In the active/active configuration, applications run on both computers; for example, the computers 'server 1' and 'server 2' in Figure 6.26 each export a file system. If one of the two computers fails, the other computer takes over the tasks of the failed computer in addition to its own (Figure 6.27, top).

This taking over of the applications of the failed server can lead to performance bottlenecks in active/active configurations. The active/passive configuration can help in this situation. In this approach the application runs only on the primary server, the second computer in the cluster (stand-by server) does nothing in normal operation. It is exclusively

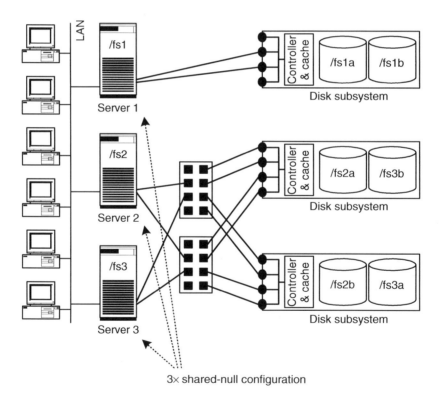

3× shared-null configuration

Figure 6.24 Shared-null configuration: The server is not designed with built-in redundancy.

there to take over the applications of the primary server if this fails. If the primary server fails, the stand-by server takes over its tasks (Figure 6.27, bottom).

The examples in Figures 6.26 and 6.27 show that shared-nothing clusters with only two servers are relatively inflexible. More flexibility is offered by shared-nothing clusters with more than two servers, so-called enhanced shared-nothing clusters. Current shared-nothing cluster software supports shared-nothing clusters with several dozens of computers.

Figures 6.28 and 6.29 show the use of an enhanced shared-nothing cluster for static load balancing: during the daytime when the system is busy, three different servers each export two file systems (Figure 6.28). At night, access to the data is still needed; however, a single server can manage the load for the six file systems (Figure 6.28). The two other servers are freed up for other tasks in this period (data mining, batch processes, backup, maintenance).

One disadvantage of the enhanced shared-nothing cluster is that it can only react to load peaks very slowly. Appropriate load balancing software can, for example, move the file system '/fs2' to one of the other two servers even during the day if the load on the

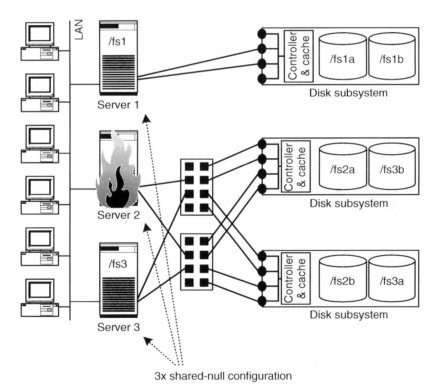

Figure 6.25   Shared-null configuration: In the event of the failure of a server, the file system '/fs2' can no longer be accessed despite redundant disk subsystems and redundant I/O paths.

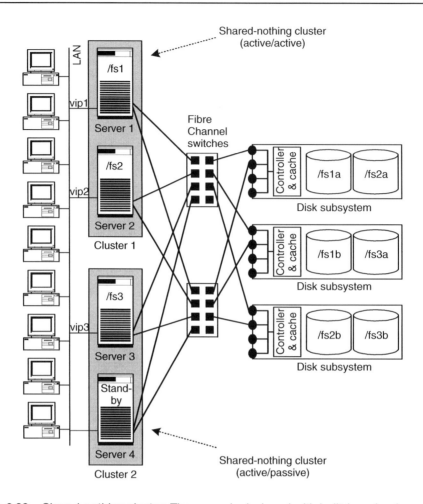

Figure 6.26 Shared-nothing cluster: The server is designed with built-in redundancy.

file system '/fs1' is higher. However, this takes some time, which means that this process is only worthwhile for extended load peaks.

A so-called shared-everything cluster offers more flexibility in comparison to enhanced shared-nothing clusters. For file servers, shared disk file systems are used as local file systems here, so that all servers can access the data efficiently over the storage network. Figure 6.30 shows a file server that is configured as a shared-everything cluster with three servers. The shared disk file system is distributed over several disk subsystems. All three servers export this file system to the clients in the LAN over the same virtual IP address by means of a conventional network file system such as NFS or CIFS. Suitable load balancing software distributes new incoming accesses on the network file system equally

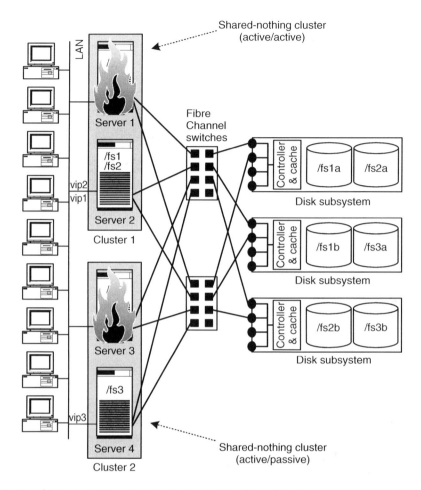

Figure 6.27 Shared-nothing cluster: In the event of the failure of a server, the other server takes over the applications of the failed server.

amongst all three servers. If the three servers are not powerful enough, a fourth server can simply be linked to the cluster.

The shared-everything cluster also offers advantages in the event of the failure of a single server. For example, the file server in Figure 6.30 is realised in the form of a distributed application. If one server fails, as in Figure 6.31, recovery measures are only necessary for those clients that have just been served by the failed computer. Likewise, recovery measures are necessary for the parts of the shared disk file system and the network file system have just been managed by the failed computer. None of the other clients of the file server notice the failure of a computer apart from a possible reduction in performance.

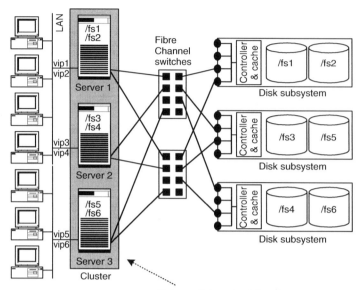

Figure 6.28 Enhanced shared-nothing cluster: The servers are designed with built-in redundancy and the server cluster consists of more than two computers.

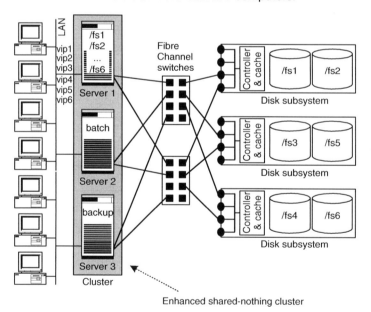

Figure 6.29 Enhanced shared-nothing cluster: For the purpose of load balancing, applications from one server can be moved to another server.

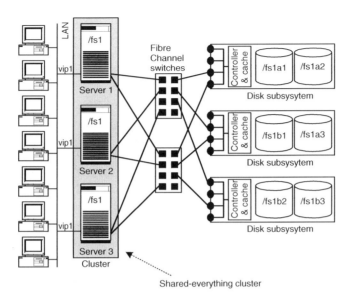

Figure 6.30   Shared-everything cluster: Parallel applications run on several computers. Incoming requests from the client are dynamically and uniformly distributed to all computers in the cluster.

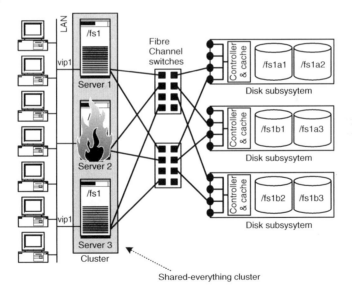

Figure 6.31   Shared-everything cluster: in the event of the failure of a server only the parts of the local shared disk file system and the network file system that have just been managed by the failed computer have to be re-synchronised. None of the other clients notice the failure of the computer, apart from possible reductions in performance.

Despite their advantages, shared-everything clusters are very seldom used. The reason for this is quite simply that this form of cluster is the most difficult to realise, so most cluster products and applications only support the more simply realised variants of shared-nothing or enhanced shared-nothing.

6.4.2 Web architecture

In the 1990s the so-called three-tier architecture established itself as a flexible architecture for IT systems (Figure 6.32). The three-tier architecture isolates the tasks of data management, applications and representation into three separate layers. Figure 6.33 shows a possible implementation of the three-tier architecture.

Individually the three layers have the following tasks:

- *Data*
  Information in the form of data forms the basis for the three-tier architecture. Databases and file systems store the data of the applications on block-oriented disks or disk subsystems. In addition, the data layer can provide interfaces to external systems and legacy applications.

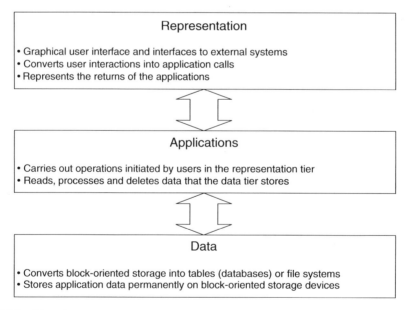

Figure 6.32   The three-tier architecture divides IT systems into data, applications and representation. Each layer can be modified or expanded without this being visible to another layer.

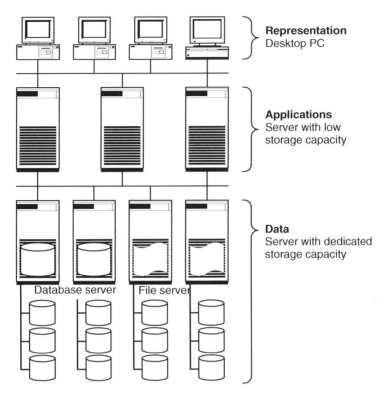

Figure 6.33   A separate LAN can be installed between application servers and data servers for the implementation of a three-tier architecture, since the clients (representation layer) communicate only with the application servers.

- *Applications*
  Applications generate and process data. Several applications can work on the same databases or file systems. Depending upon changes to the business processes, existing applications are modified and new applications added. The separation of applications and databases makes it possible for no changes, or only minimal changes, to have to be made to the underlying databases or file systems in the event of changes to applications.

- *Representation*
  The representation layer provides the user interface for the end user. In the 1990s the user interface was normally realised in the form of the graphical interface on a PC. The corresponding function calls of the application are integrated into the graphical interface so that the application can be controlled from there.

Currently, the two outer layers can be broken down into sub-layers so that the three-tier architecture is further developed into a five-tier architecture Figure 6.34 and Figure 6.35:

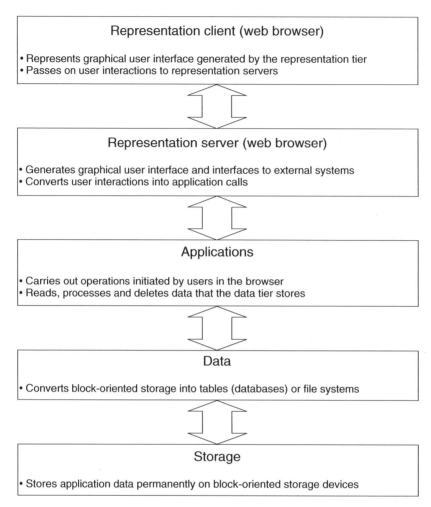

Figure 6.34 In the five-tier architecture the representation layer is split up into representation server and representation client and the data layer is split into data management and storage devices.

- *Splitting of the representation layer*
  In recent years the representation layer has been split up by the World Wide Web into web servers and web browsers. The web servers provide statically or dynamically generated websites that are represented in the browsers. Websites with a functional scope comparable to that of conventional user interfaces can currently be generated using Java and various script languages.

  The arrival of mobile end devices such as mobile phones and Personal Digital Assistants (PDAs) has meant that web servers had to make huge modifications to websites

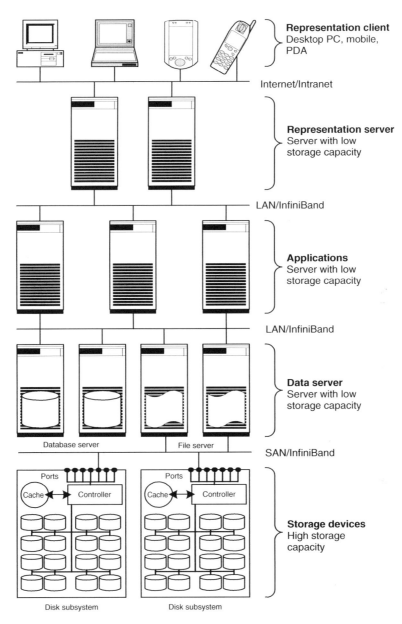

Figure 6.35   The web architecture (five-tier architecture) supports numerous different end devices. Furthermore, a storage network is inserted between the data management and the storage devices so that the IT architecture is transformed from a server-centric to a storage-centric architecture.

to bring them into line with the properties of the new end devices. In future there will be user interfaces that are exclusively controlled by means of the spoken word – for example navigation systems for use in the car, that are connected to the Internet for requesting up-to-date traffic data.

- *Splitting of the data layer*
  In the 1990s, storage devices for data were closely coupled to the data servers (storage-centric IT architecture). In the previous chapters storage networks were discussed in detail, so at this point of the book it should be no surprise to learn that the data layer is split into the organisation of the data (databases, file servers) and the storage space for data (disk subsystems).

6.4.3   Web applications based upon the case study 'travel portal'

This section uses the 'travel portal' case study to demonstrate the implementation of a so-called web application. The case study is transferable to web applications for the support of business processes. It thus shows the possibilities opened up by the Internet and highlights the potential and the change that stand before us with the transformation to e-business. Furthermore, the example demonstrates once again how storage networks, server clusters and the five-tier architecture can fulfil requirements such as the fault-tolerance, adaptability and scalability of IT systems.

Figure 6.36 shows the realisation of the travel portal in the form of a web application. Web application means that users can use the information and services of the travel portal from various end devices such as PC, PDA and mobile phone if these are connected to the Internet. The travel portal initially supports only editorially prepared content (including film reports, travel catalogues, transport timetable information) and content added by the users themselves (travel tips and discussion forums), which can be called up via conventional web browsers. Figure 6.37 shows the expansion of the travel portal by further end devices such as mobile phones and PDAs and by further services.

To use the travel portal, users first of all build up a connection to the representation server by entering the URL. Depending upon its type, the end device connects to a web server (HTTP server) or, for example, to a Wireless Application Protocol (WAP) server. The end user only perceives the web server as being a single web server. In fact, a cluster of representation servers is working in the background. The load balancer of the representation server accepts the request to build up a connection and passes it on to the computer with the lowest load.

Once a connection has been built up the web browser transfers the user identifier, for example, in the form of a cookie or the mobile number, and the properties of the end device (for example, screen resolution). The web server calls up the user profile from the user management. Using this information the web server dynamically generates websites (HTML, WML, iMode) that are optimally oriented towards the requirements of the user.

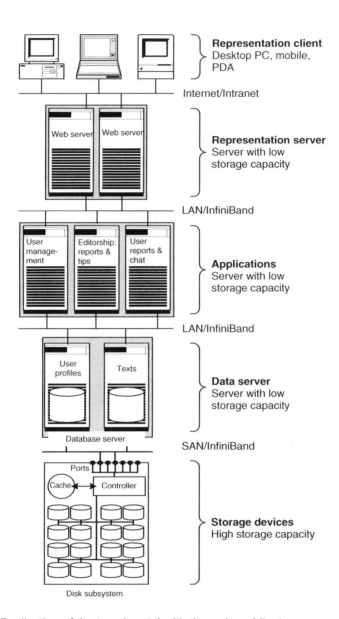

Figure 6.36   Realisation of the travel portal with the web architecture.

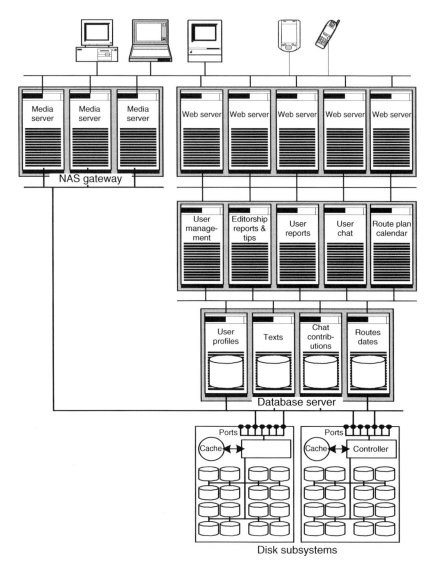

Figure 6.37 The five-tier architecture facilitates the flexible expansion of the travel portal: each layer can be expanded independently without this having major consequences for the other layers.

Thus the representation of content can be adjusted to suit the end device in use at the time. Likewise, content, adverts and information can be matched to the preferences of the user; one person may be interested in the category of city tips for good restaurants, whilst another is interested in museums.

The expansion of the travel portal to include the new 'hotel tips' application takes place by the linking of the existing 'city maps' and 'hotel directory' databases (Figure 6.37). The application could limit the selection of hotels by a preferred price category stored in the user profile or the current co-ordinates of the user transmitted by a mobile end device equipped with GPS.

Likewise, the five-tier architecture facilitates the support of new end devices, without the underlying applications having to be modified. For example, in addition to the conventional web browsers and WAP phones shown in Figure 6.37 you could also implement mobile PDAs (low resolution end devices) and a pure voice interface for car drivers.

All server machines are connected together via a fast network such as Gigabit Ethernet and InfiniBand. With the aid of appropriate cluster software, applications can be moved from one computer to another. Further computers can be added to the cluster if the overall performance of the cluster is not sufficient.

Storage networks bring with them the flexibility needed to provide the travel portal with the necessary storage capacity. The individual servers impose different requirements on the storage network:

- *Databases*
  The databases require storage space that meets the highest performance requirements. To simplify the administration of databases the data should not be stored directly upon raw devices, instead it should be stored within a file system in files that have been specially formatted for the database. Nowadays (2009), only disk subsystems connected via Fibre Channel are considered storage devices. NAS servers cannot yet be used in this situation due to the lack of availability of standardised high-speed network file systems such as RDMA enabled NFS. In future, it will be possible to use storage virtualisation on file level here.

- *Representation server and media servers*
  The representation servers augment the user interfaces with photos and small films. These are stored on separate media servers that the end user's web browser can access directly over the Internet. As a result, the media do not need to travel through the internal buses of the representation servers, thus freeing these up. Since the end users access the media over the Internet via comparatively slow connections, NAS servers are very suitable. Depending upon the load upon the media servers, shared-nothing or shared-everything NAS servers can be used. Storage virtualisation on file level again offers itself as an alternative here.

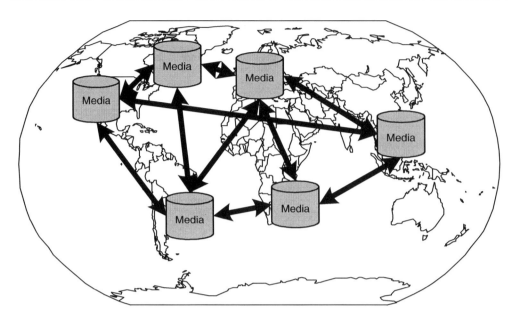

Figure 6.38   The content of the media servers is replicated at various locations in order to save network capacity.

- *Replication of the media servers*
  The users of the travel portal access it from various locations around the globe. Therefore, it is a good idea to store pictures and films at various sites around the world so that the large data quantities are supplied to users from a server located near the user (Figure 6.38). This saves network capacity and generally accelerates the transmission of the data. The data on the various cache servers is synchronised by appropriate replication software. Incidentally, the use of the replication software is independent of whether the media servers at the various sites are configured as shared-nothing NAS servers, shared-everything NAS servers, or as a storage virtualisation on the file level.

6.5 SUMMARY

In the first part of the book the building blocks of storage networks were introduced. Building upon these, this chapter has explained the fundamental principles of the use of storage networks and shown how storage networks help to increase the availability and the adaptability of IT systems.

As an introduction to the use of storage networks, we elaborated upon the characteristics of storage networks by illustrating the layering of the techniques for storage networks, investigated various forms of storage networks in the I/O path and defined storage networks in relation to data networks and voice networks. Storage resource sharing was introduced as a first application of storage networks. Individually, disk storage pooling, tape library partitioning, tape library sharing and data sharing were considered. We described the redundant I/O buses and multipathing software, redundant server and cluster software, redundant disk subsystems and volume manager mirroring or disk subsystem remote mirroring to increase the availability of applications and data, and finally redundant storage virtualisation. Based upon the case study 'protection of an important database' we showed how these measures can be combined to protect against the failure of a data centre. With regard to adaptability and scalability, the term 'cluster' was expanded to include the property of load distribution. Individually, shared-null configurations, shared-nothing clusters, enhanced shared-nothing clusters and shared-everything clusters were introduced. We then introduced the five-tier architecture – a flexible and scalable architecture for IT systems. Finally, based upon the case study 'travel portal', we showed how clusters and the five-tier architecture can be used to implement flexible and scalable web applications.

As other important applications of storage networks, the following chapters discuss network backup (Chapter 7) and archiving (Chapter 8). A flexible and adaptable architecture for data protection is introduced and we show how network backup systems can benefit from the use of disk subsystems and storage networks.

7

Network Backup

Network backup systems can back up heterogeneous IT environments incorporating several thousands of computers largely automatically. In the classical form, network backup systems move the data to be backed up via the local area network (LAN); this is where the name 'network backup' comes from. This chapter explains the basic principles of network backup and shows typical performance bottlenecks for conventional server-centric IT architectures. Finally, it shows how storage networks and intelligent storage systems help to overcome these performance bottlenecks.

Before getting involved in technical details, we will first discuss a few general conditions that should be taken into account in backup (Section 7.1). Then the backup, archiving and hierarchical storage management (HSM) services will be discussed (Section 7.2) and we will show which components are necessary for their implementation (Sections 7.3 and 7.4). This is followed by a summary of the measures discussed up to this point that are available to network backup systems to increase performance (Section 7.5). Then, on the basis of network backup, further technical boundaries of server-centric IT architectures will be described (Section 7.6) that are beyond the scope of Section 1.1, and we will explain why these performance bottlenecks can only be overcome to a limited degree within the server-centric IT architecture (Section 7.7). Then we will show how data can be backed up significantly more efficiently with a storage-centric IT architecture (Section 7.8). Building upon this, the protection of file servers (Section 7.9) and databases (Section 7.10) using storage networks and network backup systems will be discussed. Finally, organisational aspects of data protection will be considered (Section 7.11).

Storage Networks Explained: Basics and Application of Fibre Channel SAN, NAS, iSCSI, InfiniBand and FCoE, Second Edition
U. Troppens R. Erkens W. Müller-Friedt N. Haustein R. Wolafka © 2009 John Wiley & Sons, Ltd

7.1 GENERAL CONDITIONS FOR BACKUP

Backup is always a headache for system administrators. Increasing amounts of data have to be backed up in ever shorter periods of time. Although modern operating systems come with their own backup tools, these tools only represent isolated solutions, which are completely inadequate in the face of the increasing number and heterogeneity of systems to be backed up. For example, there may be no option for monitoring centrally whether all backups have been successfully completed overnight or there may be a lack of overall management of the backup media.

Changing preconditions represent an additional hindrance to data protection. There are three main reasons for this:

1. As discussed in Chapter 1, installed storage capacity doubles every 4–12 months depending upon the company in question. The data set is thus often growing more quickly than the infrastructure in general (personnel, network capacity). Nevertheless, the ever-increasing quantities of data still have to be backed up.
2. Nowadays, business processes have to be adapted to changing requirements all the time. As business processes change, so the IT systems that support them also have to be adapted. As a result, the daily backup routine must be continuously adapted to the ever-changing IT infrastructure.
3. As a result of globalisation, the Internet and e-business, more and more data has to be available around the clock: it is no longer feasible to block user access to applications and data for hours whilst data is backed up. The time window for backups is becoming ever smaller.

Network backup can help us to get to grips with these problems.

7.2 NETWORK BACKUP SERVICES

Network backup systems such as Arcserve (Computer Associates), NetBackup (Symantec/Veritas), Networker (EMC/Legato) and Tivoli Storage Manager (IBM) provide the following services:

- Backup
- Archive
- Hierarchical Storage Management (HSM)

The main task of network backup systems is to back data up regularly. To this end, at least one up-to-date copy must be kept of all data, so that it can be restored after a hardware or application error ('file accidentally deleted or destroyed by editing', 'error in the database programming').

The goal of archiving is to freeze a certain version of data so that precisely this version can be retrieved at a later date. For example, at the end of a project the data that was used can be archived on a backup server and then deleted from the local hard disk. This releases local disk space and accelerates the backup and restore processes, because only the data currently being worked on needs to be backed up or restored. Data archiving has become so important during the last few years that we treat it as a separate topic in the next chapter (Chapter 8).

HSM finally leads the end user to believe that any desired size of hard disk is present. HSM moves files that have not been accessed for a long time from the local disk to the backup server; only a directory entry remains in the local file server. The entry in the directory contains meta-information such as file name, owner, access rights, date of last modification and so on. The metadata takes up hardly any space in the file system compared to the actual file contents, so space is actually gained by moving the file content from the local disk to the backup server.

If a process accesses the content of a file that has been moved in this way, HSM blocks the accessing process, copies the file content back from the backup server to the local file system and only then gives clearance to the accessing process. Apart from the longer access time, this process remains completely hidden to the accessing processes and thus also to end users. Older files can thus be automatically moved to cheaper media (tapes) and, if necessary, fetched back again without the end user having to alter his behaviour.

Strictly speaking, HSM and backup and archive are separate concepts. However, HSM is a component of many network backup products, so the same components (media, software) can be used both for backup, archive and also for HSM. When HSM is used, the backup software used must at least be HSM-capable: it must back up the metadata of the moved files and the moved files themselves, without moving the file contents back to the client. HSM-capable backup software can speed up backup and restore processes because only the meta-information of the moved files has to be backed up and restored, not their file contents.

A network backup system realises the above-mentioned functions of backup, archive and HSM by the coordination of backup server and a range of backup clients (Figure 7.1). The server provides central components such as the management of backup media that are required by all backup clients. However, different backup clients are used for different operating systems and applications. These are specialised in the individual operating systems or applications in order to increase the efficiency of data protection or the efficiency of the movement of data.

The use of terminology regarding network backup systems is somewhat sloppy: the main task of network backup systems is the backup of data. Server and client instances of network backup systems are therefore often known as the backup server and backup client, regardless of what tasks they perform or what they are used for. A particular server

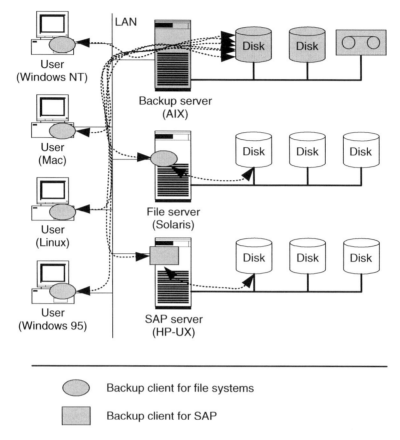

Figure 7.1 Network backup systems can automatically back up heterogeneous IT environments via the LAN. A platform-specific backup client must be installed on all clients to be backed up.

instance of a network backup system could, for example, be used exclusively for HSM, so that this instance should actually be called a HSM server – nevertheless this instance would generally be called a backup server. A client that provides the backup function usually also supports archive and the restore of backups and archives – nevertheless this client is generally just known as a backup client. In this book we follow the general, untidy conventions, because the phrase 'backup client' reads better than 'backup-archive-HSM and restore client'.

The two following sections discuss details of the backup server (Section 7.3) and the backup client (Section 7.4). We then turn our attention to the performance and the use of network backup systems.

7.3 COMPONENTS OF BACKUP SERVERS

Backup servers consist of a whole range of component parts. In the following we will discuss the main components: job scheduler (Section 7.3.1), error handler (Section 7.3.2), metadata database (Section 7.3.3) and media manager (Section 7.3.4).

7.3.1 Job scheduler

The job scheduler determines what data will be backed up when. It must be carefully configured; the actual backup then takes place automatically.

With the aid of job schedulers and tape libraries many computers can be backed up overnight without the need for a system administrator to change tapes on site. Small tape libraries have a tape drive, a magazine with space for around ten tapes and a media changer that can automatically move the various tapes back and forth between magazine and tape drive. Large tape libraries have several dozen tape drives, space for several thousands of tapes and a media changer or two to insert the tapes in the drives.

7.3.2 Error handler

If a regular automatic backup of several systems has to be performed, it becomes difficult to monitor whether all automated backups have run without errors. The error handler helps to prioritise and filter error messages and generate reports. This avoids the situation in which problems in the backup are not noticed until a backup needs to be restored.

7.3.3 Metadata database

The metadata database and the media manager represent two components that tend to be hidden to end users. The metadata database is the brain of a network backup system. It contains the following entries for every backup up object: name, computer of origin, date of last change, date of last backup, name of the backup medium, etc. For example, an entry is made in the metadata database for every file to be backed up.

The cost of the metadata database is worthwhile: in contrast to backup tools provided by operating systems, network backup systems permit the implementation of the incremental-forever strategy in which a file system is only fully backed up in the first backup. In subsequent backups, only those files that have changed since the previous backup are backed up. The current state of the file system can then be calculated on the backup server from database operations from the original full backup and from all subsequent incremental backups, so that no further full backups are necessary. The calculations in the metadata database are generally performed faster than a new full backup.

Even more is possible: if several versions of the files are kept on the backup server, a whole file system or a subdirectory dated three days ago, for example, can be restored (point-in-time restore) – the metadata database makes it possible.

7.3.4   Media manager

Use of the incremental-forever strategy can considerably reduce the time taken by the backup in comparison to the full backup. The disadvantage of this is that over time the backed up files can become distributed over numerous tapes. This is critical for the restoring of large file systems because tape mounts cost time. This is where the media manager comes into play. It can ensure that only files from a single computer are located on one tape. This reduces the number of tape mounts involved in a restore process, which means that the data can be restored more quickly.

A further important function of the media manager is so-called tape reclamation. As a result of the incremental-forever strategy, more and more data that is no longer needed is located on the backup tapes. If, for example, a file is deleted or changed very frequently over time, earlier versions of the file can be deleted from the backup medium. The gaps on the tapes that thus become free cannot be directly overwritten using current techniques. In tape reclamation, the media manager copies the remaining data that is still required from several tapes, of which only a certain percentage is used, onto a common new tape. The tapes that have thus become free are then added to the pool of unused tapes.

There is one further technical limitation in the handling of tapes: current tape drives can only write data to the tapes at a certain speed. If the data is transferred to the tape drive too slowly this interrupts the write process, the tape rewinds a little and restarts the write process. The repeated rewinding of the tapes costs performance and causes unnecessary wear to the tapes so they have to be discarded more quickly. It is therefore better to send the data to the tape drive quickly enough so that it can write the data onto the tape in one go (streaming).

The problem with this is that in network backup the backup clients send the data to be backed up via the LAN to the backup server, which forwards the data to the tape drive. On the way from backup client via the LAN to the backup server there are repeated fluctuations in the transmission rate, which means that the streaming of tape drives is repeatedly interrupted. Although it is possible for individual clients to achieve streaming by additional measures (such as the installation of a separate LAN between backup client and backup server) (Section 7.7), these measures are expensive and technically not scalable at will, so they cannot be realised economically for all clients.

The solution: the media manager manages a storage hierarchy within the backup server. To achieve this, the backup server must be equipped with hard disks and tape libraries. If a client cannot send the data fast enough for streaming, the media manager first of all stores the data to be backed up to hard disk. When writing to a hard disk it makes no difference what speed the data is supplied at. When enough of the data to be backed up has been temporarily saved to the hard disk of the backup server, the media manager automatically moves large quantities of data from the hard disk of the backup server to its tapes. This process only involves recopying the data within the backup server, so that streaming is guaranteed when writing the tapes.

This storage hierarchy is used, for example, for the backup of user PCs (Figure 7.2). Many user PCs are switched off overnight, which means that backup cannot be guaranteed overnight. Therefore, network backup systems often use the midday period to back up user PCs. Use of the incremental-forever strategy means that the amount of data to be backed up every day is so low that such a backup strategy is generally feasible. All user PCs are first of all backed up to the hard disk of the backup server in the time window from 11:15 to 13:45. The media manager in the backup server then has a good

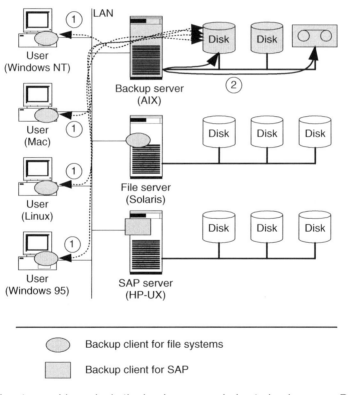

Figure 7.2 The storage hierarchy in the backup server helps to back up user PCs efficiently. First of all, all PCs are backed up to the hard disks of the backup server (1) during the midday period. Before the next midday break the media manager copies the data from the hard disks to tapes (2).

20 hours to move the data from the hard disks to tapes. Then the hard disks are once again free so that the user PCs can once again be backed up to hard disk in the next midday break.

In all operations described here the media manager checks whether the correct tape has been placed in the drive. To this end, the media manager writes an unambiguous signature to every tape, which it records in the metadata database. Every time a tape is inserted the media manager compares the signature on the tape with the signature in the metadata database. This ensures that no tapes are accidentally overwritten and that the correct data is written back during a restore operation.

Furthermore, the media manager monitors how often a tape has been used and how old it is, so that old tapes are discarded in good time. If necessary, it first copies data that is still required to a new tape. Older tape media formats also have to be wound back and forwards now and then so that they last longer; the media manager can also automate the winding of tapes that have not been used for a long time.

A further important function of the media manager is the management of data in a so-called off-site store. To this end, the media manager keeps two copies of all data to be backed up. The first copy is always stored on the backup server, so that data can be quickly restored if it is required. However, in the event of a large-scale disaster (fire in the data centre) the copies on the backup server could be destroyed. For such cases the media manager keeps a second copy in an off-site store that can be several kilometres away. The media manager supports the system administrator in moving the correct tapes back and forwards between backup server and off-site store. It even supports tape reclamation for tapes that are currently in the off-site store.

7.4 BACKUP CLIENTS

A platform-specific client (backup agent) is necessary for each platform to be backed up. The base client can back up and archive files and restores them if required. The term platform is used here to mean the various operating systems and the file systems that they support. Furthermore, some base clients offer HSM for selected file systems.

The backup of file systems takes place at file level as standard. This means that each changed file is completely re-transferred to the server and entered there in the metadata database. By using backup at volume level and at block level it is possible to change the granularity of the objects to be backed up.

When backup is performed at volume level, a whole volume is backed up as an individual object on the backup server. We can visualise this as the output of the Unix command 'dd' being sent to the backup server. Although this has the disadvantage that free areas, on which no data at all has been saved, are also backed up, only very few metadata database operations are necessary on the backup server and on the client side it is not necessary to spend a long time comparing which files have changed since the last backup. As a result, backup and restore operations can sometimes be performed more quickly at volume level than they can at file level. This is particularly true when restoring large file systems with a large number of small files.

Backup on block level optimises backup for members of the external sales force, who only connect up to the company network now and then by means of a laptop via a dial-up line or the Internet. In this situation the performance bottleneck is the low transmission capacity between the backup server and the backup client. If only one bit of a large file is changed, the whole file must once again be forced down the network. When backing up on block level the backup client additionally keeps a local copy of every file backed up. If a file has changed, it can establish which parts of the file have changed. The backup client sends only the changed data fragments (blocks) to the backup server. This can then reconstruct the complete file. As is the case for backup on file level, each file backed up is entered in the metadata database. Thus, when backing up on block level the quantity of data to be transmitted is reduced at the cost of storage space on the local hard disk.

In addition to the standard client for file systems, most network backup systems provide special clients for various applications. For example, there are special clients for Microsoft Exchange or IBM Lotus Domino that make it possible to back up and restore individual documents. We will discuss the backup of file systems and NAS servers (Section 7.9) and databases (Section 7.10) in more detail later on.

7.5   PERFORMANCE GAINS AS A RESULT OF NETWORK BACKUP

The underlying hardware components determine the maximum throughput of network backup systems. The software components determine how efficiently the available hardware is actually used. At various points of this chapter we have already discussed how network backup systems can help to better utilise the existing infrastructure:

- *Performance increase by the archiving of data*
  Deleting data that has already been archived from hard disks can accelerate the daily backup because there is less data to back up. For the same reason, file systems can be restored more quickly.

- *Performance increase by HSM*
  By moving file contents to the HSM server, file systems can be restored more quickly. The directory entries of files that have been moved can be restored comparatively quickly; the majority of the data, namely the file contents, do not need to be fetched back from the HSM server.

- *Performance increase by the incremental-forever strategy*
  After the first backup, only the data that has changed since the last backup is backed up. On the backup server the metadata database is used to calculate the latest state of the

data from the first backup and all subsequent incremental backups, so that no further full backups are necessary. The backup window can thus be significantly reduced.

- *Performance increase by reducing tape mounts*
  The media manager can ensure that data that belongs together is only distributed amongst a few tapes. The number of time-consuming tape mounts for the restoring of data can thus be reduced.

- *Performance increase by streaming*
  The efficient writing of tapes requires that the data is transferred quickly enough to the tape drive. If this is not guaranteed the backup server can first temporarily store the data on a hard drive and then send the data to the tape drive in one go.

- *Performance increase by backup on volume level or on block level*
  As default, file systems are backed up on file level. Large file systems with several hundreds of thousands of files can sometimes be backed up more quickly if they are backed up at volume level. Laptops can be backed up more quickly if only the blocks that have changed are transmitted over a slow connection to the backup server.

7.6   PERFORMANCE BOTTLENECKS OF NETWORK BACKUP

At some point, however, the technical boundaries for increasing the performance of backup are reached. When talking about technical boundaries, we should differentiate between application-specific boundaries (Section 7.6.1) and those that are determined by server-centric IT architecture (Section 7.6.2).

7.6.1   Application-specific performance bottlenecks

Application-specific performance bottlenecks are all those bottlenecks that can be traced back to the 'network backup' application. These performance bottlenecks play no role for other applications.

The main candidate for application-specific performance bottlenecks is the metadata database. A great deal is demanded of this. Almost every action in the network backup system is associated with one or more operations in the metadata database. If, for example, several versions of a file are backed up, an entry is made in the metadata database for each version. The backup of a file system with several hundreds of thousands of files can thus be associated with a whole range of database operations.

A further candidate for application-specific performance bottlenecks is the storage hierarchy: when copying the data from hard disk to tape the media manager has to load the data from the hard disk into the main memory via the I/O bus and the internal buses, only to forward it from there to the tape drive via the internal buses and I/O bus. This means that the buses can get clogged up during the copying of the data from hard disk to tape. The same applies to tape reclamation.

7.6.2 Performance bottlenecks due to server-centric IT architecture

In addition to these two application-specific performance bottlenecks, some problems crop up in network backup that are typical of a server-centric IT architecture. Let us mention once again as a reminder the fact that in a server-centric IT architecture storage devices only exist in relation to servers; access to storage devices always takes place via the computer to which the storage devices are connected. The performance bottlenecks described in the following apply for all applications that are operated in a server-centric IT architecture.

Let us assume that a backup client wants to back data up to the backup server (Figure 7.3). The backup client loads the data to be backed up from the hard disk into the main memory of the application server via the SCSI bus, the PCI bus and the system bus, only to forward it from there to the network card via the system bus and the PCI bus. On the backup server the data must once again be passed through the buses twice. In backup, large quantities of data are generally backed up in one go. During backup, therefore, the buses of the participating computers can become a bottleneck, particularly if the application server also has to bear the I/O load of the application or the backup server is supposed to support several simultaneous backup operations.

The network card transfers the data to the backup server via TCP/IP and Ethernet. Previously the data exchange via TCP/IP was associated with a high CPU load. However, the CPU load caused by TCP/IP data traffic can be reduced using TCP/IP offload engines (TOE) (Section 3.5.2).

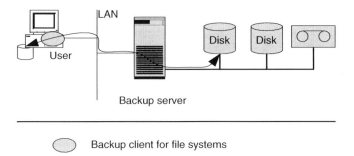

Figure 7.3 In network backup, all data to be backed up must be passed through both computers. Possible performance bottlenecks are internal buses, CPU and the LAN.

7.7 LIMITED OPPORTUNITIES FOR INCREASING PERFORMANCE

Backup is a resource-intensive application that places great demands upon storage devices, CPU, main memory, network capacity, internal buses and I/O buses. The enormous amount of resources required for backup is not always sufficiently taken into account during the planning of IT systems. A frequent comment is 'the backup is responsible for the slow network' or 'the slow network is responsible for the restore operation taking so long'. The truth is that the network is inadequately dimensioned for end user data traffic and backup data traffic. Often, data protection is the application that requires the most network capacity. Therefore, it is often sensible to view backup as the primary application for which the IT infrastructure in general and the network in particular must be dimensioned.

In every IT environment, most computers can be adequately protected by a network backup system. In almost every IT environment, however, there are computers – usually only a few – for which additional measures are necessary in order to back them up quickly enough or, if necessary, to restore them. In the server-centric IT architecture there are three approaches to taming such data monsters: the installation of a separate LAN for the network backup between backup client and backup server (Section 7.7.1), the installation of several backup servers (Section 7.7.2) and the installation of backup client and backup server on the same physical computer (Section 7.7.3).

7.7.1 Separate LAN for network backup

The simplest measure to increase backup performance more of heavyweight backup clients is to install a further LAN between backup client and backup server in addition to the existing LAN and to use this exclusively for backup (Figure 7.4). An expensive, but powerful, transmission technology such as leading edge Gigabit Ethernet generations can also help here.

The concept of installing a further network for backup in addition to the existing LAN is comparable to the basic idea of storage networks. In contrast to storage networks, however, in this case only computers are connected together; direct access to all storage devices is not possible. All data thus continues to be passed via TCP/IP and through application server and backup server which leads to a blockage of the internal bus and the I/O buses.

Individual backup clients can thus benefit from the installation of a separate LAN for network backup. This approach is, however, not scalable at will: due to the heavy load on the backup server this cannot back up any further computers in addition to the backup of one individual heavyweight client.

Despite its limitations, the installation of a separate backup LAN is sufficient in many environments. With Fast-Ethernet you can still achieve a throughput of over 10 MByte/s.

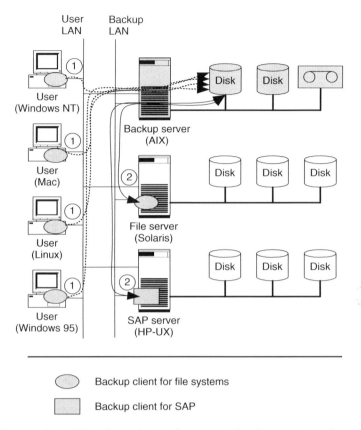

Figure 7.4   Approach 1: The throughput of network backup can be increased by the installation of a second LAN. Normal clients are still backed up via the User LAN (1). Only heavyweight clients are backed up via the second LAN (2).

The LAN technique is made even more attractive by Gigabit Ethernet, 10Gigabit Ethernet and the above-mentioned TCP/IP offload engines that free up the server CPU significantly with regard to the TCP/IP data traffic.

7.7.2   Multiple backup servers

Installing multiple backup servers distributes the load of the backup server over more hardware. For example, it would be possible to assign every heavyweight backup client a special backup server installed exclusively for the backup of this client (Figure 7.5). Furthermore, a further backup server is required for the backup of all other backup clients. This approach is worthwhile in the event of performance bottlenecks in the metadata

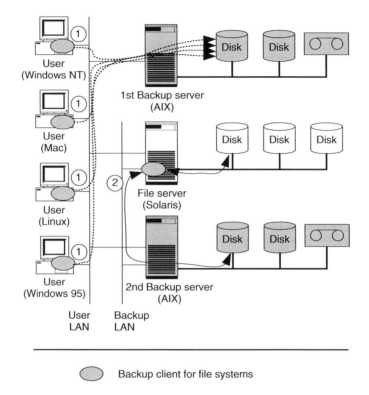

Backup client for file systems

Figure 7.5 Approach 2: Dedicated backup servers can be installed for heavyweight backup clients. Normal clients continue to be backed up on the first backup server (1). Only the heavyweight client is backed up on its own backup server over the separate LAN (2).

database or in combination with the first measure, the installation of a separate LAN between the heavyweight backup client and backup server.

The performance of the backup server can be significantly increased by the installation of multiple backup servers and a separate LAN for backup. However, from the point of view of the heavyweight backup client the problem remains that all data to be backed up must be passed from the hard disk into the main memory via the buses and from there must again be passed through the buses to the network card. This means that backup still heavily loads the application server. The resource requirement for backup could be in conflict with the resource requirement for the actual application.

A further problem is the realisation of the storage hierarchy within the individual backup server since every backup server now requires its own tape library. Many small tape libraries are more expensive and less flexible than one large tape library. Therefore, it would actually be better to buy a large tape library that is used by all servers. In a server-centric IT architecture it is, however, only possible to connect multiple computers to the same tape library to a very limited degree.

7.7.3   Backup server and application server on the same physical computer

The third possible way of increasing performance is to install the backup server and application server on the same physical computer (Figure 7.6). This results in the backup client also having to run on this computer. Backup server and backup client communicate over Shared Memory (Unix), Named Pipe or TCP/IP Loopback (Windows) instead of via LAN. Shared Memory has an infinite bandwidth in comparison to the buses, which means that the communication between backup server and backup client is no longer the limiting factor.

However, the internal buses continue to get clogged up: the backup client now loads the data to be backed up from the hard disk into the main memory via the buses. The backup server takes the data from the main memory and writes it, again via the buses, to the backup medium. The data is thus once again driven through the internal bus twice.

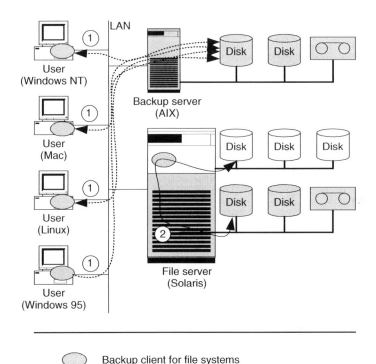

⬭   Backup client for file systems

Figure 7.6   Approach 3: Application server, backup server and backup client are installed on one computer. Normal clients continue to be backed up on the first backup server (1). Only the heavyweight client is backed up within the same computer (2).

Tape reclamation and any copying operations within the storage hierarchy of the backup server could place an additional load on the buses.

Without further information we cannot more precisely determine the change to the CPU load. Shared Memory communication (or Named Pipe or TCP/IP Loopback) dispenses with the CPU-intensive operation of the network card. On the other hand, a single computer must now bear the load of the application, the backup server and the backup client. This computer must incidentally possess sufficient main memory for all three applications.

One problem with this approach is the proximity of production data and copies on the backup server. SCSI permits a maximum cable length of 25 m. Since application and backup server run on the same physical computer, the copies are a maximum of 50 m away from the production data. In the event of a fire or comparable damage, this is disastrous. Therefore, either a SCSI extender should be used or the tapes taken from the tape library every day and placed in an off-site store. The latter goes against the requirement of largely automating data protection.

7.8   NEXT GENERATION BACKUP

Storage networks open up new possibilities for getting around the performance bottlenecks of network backup described above. They connect servers and storage devices, so that during backup production data can be copied directly from the source hard disk to the backup media, without passing it through a server (server-free backup, Section 7.8.1). LAN-free backup (Section 7.8.2) and LAN-free backup with shared disk file systems (Section 7.8.3) are two further alternative methods of accelerating backup using storage networks. The use of instant copies (Section 7.8.4) and remote mirroring (Section 7.8.5) provide further possibilities for accelerating backup and restore operations. The introduction of storage networks also has the side effect that several backup servers can share a tape library (Section 7.8.6).

7.8.1   Server-free backup

The ultimate goal of backup over a storage network is so-called server-free backup (Figure 7.7). In backup, the backup client initially determines which data has to be backed up and then sends only the appropriate metadata (file name, access rights, etc.) over the LAN to the backup server. The file contents, which make up the majority of the data quantity to be transferred, are then written directly from the source hard disk to the backup medium (disk, tape, optical) over the storage network, without a server being connected in between. The network backup system coordinates the communication between source hard disk and backup medium. A shorter transport route for the backup of data is not yet in sight with current storage techniques.

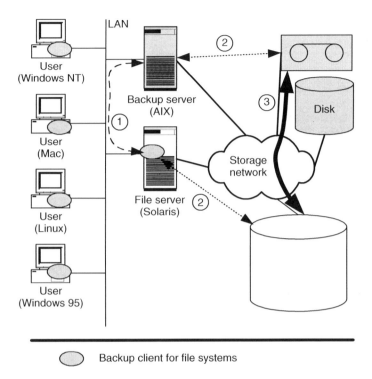

Backup client for file systems

Figure 7.7   In server-free backup, backup server and backup client exchange lightweight metadata via the LAN (1). After it has been determined which data blocks have to be backed up, the network backup system can configure the storage devices for the data transfer via the storage network (2). The heavyweight file contents are then copied directly from the source hard disk to the backup medium via the storage network (3).

   The performance of server-free backup is predominantly determined by the performance of the underlying storage systems and the connection in the storage network. Shifting the transport route for the majority of the data from the LAN to the storage network without a server being involved in the transfer itself means that the internal buses and the I/O buses are freed up on both the backup client and the backup server. The cost of coordinating the data traffic between source hard disk and backup medium is comparatively low.

   A major problem in the implementation of server-free backup is that the SCSI commands have to be converted en route from the source hard disk to the backup medium. For example, different blocks are generally addressed on source medium and backup medium. Or, during the restore of a deleted file in a file system, this file has to be restored to a different area if the space that was freed up is now occupied by other files. In the backup from hard disk to tape, even the SCSI command sets are slightly different. Therefore, software called 3rd-Party SCSI Copy Command is necessary for the protocol conversion. It can be realised at various points: in a switch of the storage network in a box specially connected to the storage network that is exclusively responsible for the protocol conversion, or in one of the two participating storage systems themselves.

According to our knowledge, server-free backup at best made it to the demonstration centres of the suppliers' laboratories. Some suppliers were involved in aggressive marketing around 2002 and 2003, claiming that their network backup products support server-free backup. In our experience, server-free backup is basically not being used in any production environments today although it has been available for some time. In our opinion this confirms that server-free backup is still very difficult to implement, configure and operate with the current state of technology.

7.8.2 LAN-free backup

LAN-free backup dispenses with the necessity for the 3rd-Party SCSI Copy Command by realising comparable functions within the backup client (Figure 7.8). As in server-free backup, metadata is sent via the LAN. File contents, however, no longer go through the backup server: for backup the backup client loads the data from the hard disk into the main memory via the appropriate buses and from there writes it directly to the backup medium via the buses and the storage network. To this end, the backup client must be able to access the backup server's backup medium over the storage network. Furthermore, backup server and backup client must synchronise their access to common devices. This is easier to realise than server-free backup and thus well proven in production environments.

In LAN-free backup the load on the buses of the backup server is reduced but not the load on those of the backup client. This can impact upon other applications (databases, file and web servers) that run on the backup client at the same time as the backup.

LAN-free backup is already being used in production environments. However, the manufacturers of network backup systems only support LAN-free backup for certain applications (databases, file systems, e-mail systems), with not every application being supported on every operating system. Anyone wanting to use LAN-free backup at the moment must take note of the manufacturer's support matrix (see Section 3.4.6). It can be assumed that in the course of the next years the number of the applications and operating systems supported will increase further.

7.8.3 LAN-free backup with shared disk file systems

Anyone wishing to back up a file system now for which LAN-free backup is not supported can sometimes use shared disk file systems to rectify this situation (Figure 7.9). Shared disk file systems are installed upon several computers. Access to data is synchronised over the LAN; the individual file accesses, on the other hand, take place directly over the storage network (Section 4.3). For backup the shared disk file system is installed on the

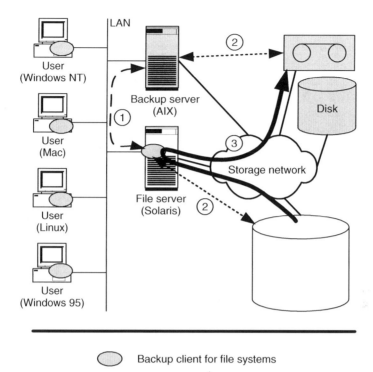

Backup client for file systems

Figure 7.8 In LAN-free backup too, backup servers and backup clients exchange lightweight metadata over the LAN (1). The backup server prepares its storage devices for the data transfer over the storage network and then hands control of the storage devices over to the backup client (2). This then copies heavyweight file contents directly to the backup medium via the storage network (3).

file server and the backup server. The prerequisite for this is that a shared disk file system is available that supports the operating systems of backup client and backup server. The backup client is then started on the same computer on which the backup server runs, so that backup client and backup server can exchange the data via Shared Memory (Unix) or Named Pipe or TCP/IP Loopback (Windows).

In LAN-free backup using a shared disk file system, the performance of the backup server must be critically examined. All data still has to be passed through the buses of the backup server; in addition, the backup client and the shared disk file system run on this machine. LAN data traffic is no longer necessary within the network backup system; however, the shared disk file system now requires LAN data traffic for the synchronisation of simultaneous data accesses. The data traffic for the synchronisation of the shared disk file system is, however, comparatively light. At the end of the day, you have to measure whether backup with a shared disk file system increases performance for each individual case.

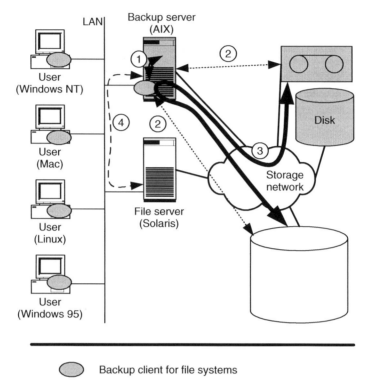

⬭   Backup client for file systems

Figure 7.9 When backing up using shared disk file systems, backup server and backup client run on the same computer (1). Production data and backup media are accessed on the backup server (2), which means that the backup can take place over the storage network (3). The shared disk file system requires a LAN connection for the lightweight synchronisation of parallel data accesses (4).

The performance of LAN-free backup using a shared disk subsystem is not as good as the performance of straight LAN-free backup. However, it can be considerably better than backup over a LAN. Therefore, this approach has established itself in production environments, offering an attractive workaround until LAN-free (or even server-free) backup is available. It can also be assumed that this form of data backup will become important due to the increasing use of shared disk file systems.

7.8.4   Backup using instant copies

Instant copies can virtually copy even terabyte-sized data sets in a few seconds, and thus freeze the current state of the production data and make it available via a second access path. The production data can still be read and modified over the first access path, so that the operation of the actual application can be continued, whilst at the same time the frozen state of the data can be backed up via the second access path.

Instant copies can be realised on three different levels:

1. *Instant copy in the block layer (disk subsystem or block-based virtualisation)*
Instant copy in the disk subsystem was discussed in detail in Section 2.7.1: intelligent disk subsystems can virtually copy all data of a virtual disk onto a second virtual disk within a few seconds. The frozen data state can be accessed and backed up via the second virtual disk.

2. *Instant copy in the file layer (local file system, NAS server or file-based virtualisation)*
Many file systems also offer the possibility of creating instant copies. Instant copies on file system level are generally called snapshots (Section 4.1.3). In contrast to instant copies in the disk subsystem the snapshot can be accessed via a special directory path.

3. *Instant copy in the application*
Finally, databases in particular offer the possibility of freezing the data set internally for backup, whilst the user continues to access it (hot backup, online backup).

Instant copies in the local file system and in the application have the advantage that they can be realised with any hardware. Instant copies in the application can utilise the internal data structure of the application and thus work more efficiently than file systems. On the other hand, applications do not require these functions if the underlying file system already provides them. Both approaches consume system resources on the application server that one would sometimes prefer to make available to the actual application. This is the advantage of instant copies in external devices (e.g., disk subsystem, NAS server, network-based virtualisation instance): although it requires special hardware, application server tasks are moved to the external device thus freeing up the application server.

Backup using instant copy must be synchronised with the applications to be backed up. Databases and file systems buffer write accesses in the main memory in order to increase their performance. As a result, the data on the disk is not always in a consistent state. Data consistency is the prerequisite for restarting the application with this data set and being able to continue operation. For backup it should therefore be ensured that an instant copy with consistent data is first generated. The procedure looks something like this:

1. Shut down the application.

2. Perform the instant copy.

3. Start up the application again.

4. Back up the data of the instant copy.

Despite the shutting down and restarting of the application the production system is back in operation very quickly.

Data protection with instant copies is even more attractive if the instant copy is controlled by the application itself: in this case the application must ensure that the data on disk is consistent and then initiate the copying operation. The application can then continue operation after a few seconds. It is no longer necessary to stop and restart the application.

Instant copies thus make it possible to backup business-critical applications every hour with only very slight interruptions. This also accelerates the restoring of data after application errors ('accidental deletion of a table space'). Instead of the time-consuming restore of data from tapes, the frozen copy that is present in the storage system can simply be put back.

With the aid of instant copies in the disk subsystem it is possible to realise so-called application server-free backup. In this, the application server is put at the side of a second server that serves exclusively for backup (Figure 7.10). Both servers are directly

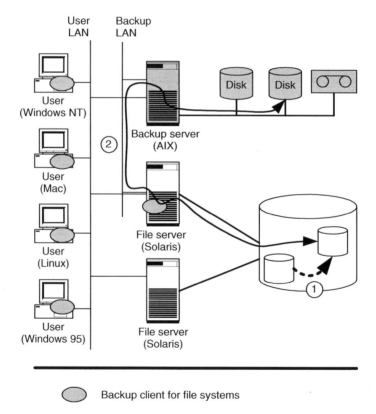

Figure 7.10 Application server-free backup utilises the functions of an intelligent disk subsystem. To perform a backup the application is operated for a short period in such a manner as to create a consistent data state on the disks, so that data can be copied by means of instant copy (1). The application can immediately switch back to normal operation; in parallel to this the data is backed up using the instant copy (2).

connected to the disk subsystem via SCSI; a storage network is not absolutely necessary. For backup the instant copy is first of all generated as described above: (1) shut down application; (2) generate instant copy; and (3) restart application. The instant copy can then be accessed from the second computer and the data is backed up from there without placing a load on the application server. If the instant copy is not deleted in the disk subsystem, the data can be restored using this copy in a few seconds in the event of an error.

7.8.5 Data protection using remote mirroring

Instant copies help to quickly restore data in the event of application or operating errors; however, they are ineffective in the event of a catastrophe: after a fire the fact that there are several copies of the data on a storage device does not help. Even a power failure can become a problem for a 24×7 operation.

The only thing that helps here is to mirror the data by means of remote mirroring on two disk subsystems, which are at least separated by a fire protection barrier. The protection of applications by means of remote mirroring has already been discussed in detail in Sections 6.3.3 and 6.3.5.

Nevertheless, the data still has to be backed up: in synchronous remote mirroring the source disk and copy are always identical. This means that if data is destroyed by an application or operating error then it is also immediately destroyed on the copy. The data can be backed up to disk by means of instant copy or by means of classical network backup to tapes. Since storage capacity on disk is more expensive than storage capacity on tape, only the most important data is backed up using instant copy and remote mirroring. For most data, backup to tape is still the most cost effective. In Section 9.5.5 we will take another but more detailed look at the combination of remote mirroring and backup.

7.8.6 Tape library sharing

It is sometimes necessary for more than one backup server to be used. In some data centres there is so much data to be backed up that, despite the new techniques for backup using storage networks presented here, several ten of backup servers are needed. In other cases, important applications are technically partitioned off from the network so that they have better protection from outside attacks. Ideally, everyone would like to have a separate backup server in every partitioned network segment with each backup server equipped with its own small tape library. However, having many small libraries is expensive in terms of procurement cost and is also more difficult to manage than large tape libraries. Therefore, a large tape library is frequently acquired and shared by all the backup servers over a storage network using tape library sharing (Section 6.2.2).

Figure 7.11 shows the use of tape library sharing for network backup: one backup server acts as library master, all others as library clients. If a backup client backs up data to a backup server that is configured as a library client, then this first of all requests a free tape from the library master. The library master selects the tape from its pool of free tapes and places it in a free drive. Then it notes in its metadata database that this tape is now being used by the library client and it informs the library client of the drive that the tape is in. Finally, the backup client can send the data to be backed up via the LAN to the backup server, which is configured as the library client. This then writes the data directly to tape via the storage network.

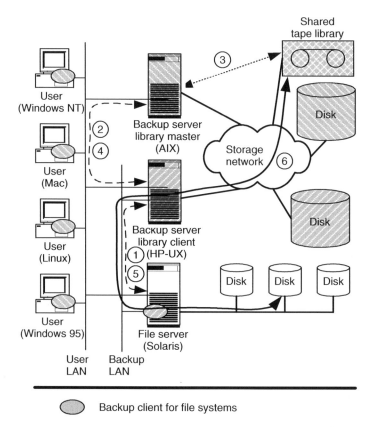

    Backup client for file systems

Figure 7.11   In tape library sharing two backup servers share a large tape library. If a client wants to back data up directly to tape with the second backup server (library client) (1) then this initially requests a tape and a drive from the library master (2). The library master places a free tape in a free drive (3) and returns the information in question to the library client (4). The library client now informs the backup client (5) that it can back the data up (6).

7.9 BACKUP OF FILE SYSTEMS

Almost all applications store their data in file systems or in databases. Therefore, in this section we will examine the backup of file servers (Section 7.9) and in the next section we will look more closely at that of databases (Section 7.10). The chapter concludes with organisational aspects of network backup (Section 7.11).

This section first of all discusses fundamental requirements and problems in the backup of file servers (Section 7.9.1). Then a few functions of modern file systems will be introduced that accelerate the incremental backup of file systems (Section 7.9.2). Limitations in the backup of NAS servers will then be discussed (Section 7.9.3). We will then introduce the Network Data Management Protocol (NDMP), a standard that helps to integrate the backup of NAS servers into an established network backup system (Section 7.9.4).

7.9.1 Backup of file servers

We use the term file server to include computers with a conventional operating system such as Windows or Unix that exports part of its local file systems via a network file system (NFS, CIFS) or makes it accessible as service (Novell, FTP, HTTP). The descriptions in this section can be transferred to all types of computers, from user PCs through classical file servers to the web server.

File servers store three types of information:

- Data in the form of files;
- Metadata on these files such as file name, creation date and access rights; and
- Metadata on the file servers such as any authorised users and their groups, size of the individual file systems, network configuration of the file server and names, components and rights of files or directories exported over the network.

Depending upon the error situation, different data and metadata must be restored. The restore of individual files or entire file systems is relatively simple: in this case only the file contents and the metadata of the files must be restored from the backup server to the file server. This function is performed by the backup clients introduced in Section 7.4.

Restoring an entire file server is more difficult. If, for example, the hardware of the file server is irreparable and has to be fully replaced, the following steps are necessary:

1. Purchasing and setting up of appropriate replacement hardware.
2. Basic installation of the operating system including any necessary patches.
3. Restoration of the basic configuration of the file server including LAN and storage network configuration of the file server.

4. If necessary, restoration of users and groups and their rights.

5. Creation and formatting of the local file systems taking into account the necessary file system sizes.

6. Installation and configuration of the backup client.

7. Restoration of the file systems with the aid of the network backup system.

This procedure is very labour-intensive and time-consuming. The methods of so-called Image Restore (also known as Bare Metal Restore) accelerate the restoration of a complete computer: tools such as 'mksysb' (AIX), 'Web Flash Archive' (Solaris) or various disk image tools for Windows systems create a complete copy of a computer (image). Only a boot diskette or boot CD and an appropriate image is needed to completely restore a computer without having to work through steps 2–7 described above. Particularly advantageous is the integration of image restore in a network backup system: to achieve this the network backup system must generate the appropriate boot disk. Furthermore, the boot diskette or boot CD must create a connection to the network backup system.

7.9.2 Backup of file systems

For the classical network backup of file systems, backup on different levels (block level, file level, file system image) has been discussed in addition to the incremental-forever strategy. The introduction of storage networks makes new methods available for the backup of file systems such as server-free backup, application server-free backup, LAN-free backup, shared disk file systems and instant copies.

The importance of the backup of file systems is demonstrated by the fact that manufacturers of file systems are providing new functions specifically targeted at the acceleration of backups. In the following we introduce two of these new functions – the so-called archive bit and block level incremental backup.

The archive bit supports incremental backups at file level such as, for example, the incremental-forever strategy. One difficulty associated with incremental backups is finding out quickly which files have changed since the previous backup. To accelerate this decision, the file system adds an archive bit to the metadata of each file: the network backup system sets this archive bit immediately after it has backed a file up on the backup server. Thus the archive bits of all files are set after a full backup. If a file is altered, the file system automatically clears its archive bit. Newly generated files are thus not given an archive bit. In the next incremental backup the network backup system knows that it only has to back up those files for which the archive bits have been cleared.

The principle of the archive bit can also be applied to the individual blocks of a file system in order to reduce the cost of backup on block level. In Section 7.4 a comparatively expensive procedure for backup on block level was introduced: the cost of the copying and comparing of files by the backup client is greatly reduced if the file system manages the quantity of altered blocks itself with the aid of the archive bit for blocks and the network backup system can call this up via an API.

Unfortunately, the principle of archive bits cannot simply be combined with the principle of instant copies: if the file system copies uses instant copy to copy within the disk subsystem for backup (Figure 7.10), the network backup system sets the archive bit only on the copy of the file system. In the original data the archive bit thus remains cleared even though the data has been backed up. Consequently, the network backup system backs this data up at the next incremental backup because the setting of the archive bit has not penetrated through to the original data.

7.9.3 Backup of NAS servers

NAS servers are preconfigured file servers; they consist of one or more internal servers, preconfigured disk capacity and usually a stripped-down or specific operating system (Section 4.2.2). NAS servers generally come with their own backup tools. However, just like the backup tools that come with operating systems, these tools represent an isolated solution (Section 7.1). Therefore, in the following we specifically consider the integration of the backup of NAS servers into an existing network backup system.

The optimal situation would be if there were a backup client for a NAS server that was adapted to suit both the peculiarities of the NAS server and also the peculiarities of the network backup system used. Unfortunately, it is difficult to develop such a backup client in practice.

If the NAS server is based upon a specific operating system the manufacturers of the network backup system sometimes lack the necessary interfaces and compilers to develop such a client. Even if the preconditions for the development of a specific backup client were in place, it is doubtful whether the manufacturer of the network backup system would develop a specific backup client for all NAS servers: the necessary development cost for a new backup client is still negligible in comparison to the testing cost that would have to be incurred for every new version of the network backup system and for every new version of the NAS server.

Likewise, it is difficult for the manufacturers of NAS servers to develop such a client. The manufacturers of network backup systems publish neither the source code nor the interfaces between backup client and backup server, which means that a client cannot be developed. Even if such a backup client already exists because the NAS server is based upon on a standard operating system such as Linux, Windows or Solaris, this does not mean that customers may use this client: in order to improve the Plug&Play-capability of NAS servers, customers may only use the software that has been tested and certified by the NAS manufacturer. If the customer installs non-certified software, then he can lose support for the NAS server. Due to the testing cost, manufacturers of NAS servers may be able to support some, but certainly not all network backup systems.

Without further measures being put in place, the only possibility that remains is to back the NAS server up from a client of the NAS server (Figure 7.12). However, this approach, too, is doubtful for two reasons.

First, this approach is only practicable for smaller quantities of data: for backup the files of the NAS server are transferred over the LAN to the NFS or CIFS client on which the backup client runs. Only the backup client can write the files to the backup medium using advanced methods such as LAN-free backup.

Second, the backup of metadata is difficult. If an NAS server supports the export of the local file system both via CIFS and also via NFS then the backup client only accesses one of the two protocols on the files – the metadata of the other protocol is lost. NAS servers would thus have to store their metadata in special files so that the network backup system

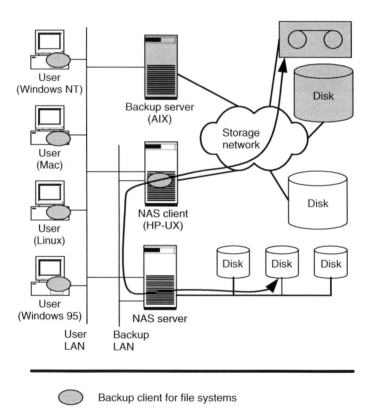

Backup client for file systems

Figure 7.12 When backing up a NAS server over a network file system, the connection between the NAS server and backup client represents a potential performance bottleneck. Backup over a network file system makes it more difficult to back up and restore the metadata of the NAS server.

can back these up. There then remains the question of the cost for the restoring of a NAS server or a file system. The metadata of NAS servers and files has to be re-extracted from these files. It is dubious whether network backup systems can automatically initiate this process.

As a last resort for the integration of NAS servers and network backup systems, there remains only the standardisation of the interfaces between the NAS server and the network backup system. This would mean that manufacturers of NAS servers would only have to develop and test one backup client that supports precisely this interface. The backup systems of various manufacturers could then back up the NAS server via this interface. In such an approach the extensivity of this interface determines how well the backup of NAS servers can be linked into a network backup system. The next section introduces a standard for such an interface – the Network Data Management Protocol (NDMP).

7.9.4   The Network Data Management Protocol (NDMP)

The Network Data Management Protocol (NDMP) defines an interface between NAS servers and network backup systems that makes it possible to back up NAS servers without providing a specific backup client for them. More and more manufacturers – both of NAS servers and network backup systems – are supporting NDMP. The current version of NDMP is Version 4; Version 5 is in preparation.

More and more manufacturers of NAS servers and network backup systems support NDMP, making it a de facto standard. An Internet draft of NDMP Version 4 has been available for some time. Furthermore, a requirements catalogue exists for NDMP Version 5. NDMP Version 4 has some gaps, such as the backup of snapshots that some manufacturers deal with through proprietary extensions. These vendor-specific extensions relate to NDMP Version 4. However, NDMP has received widespread acceptance as a technology for the integration of NAS servers and network backup systems.

NDMP uses the term 'data management operations' to describe the backup and restoration of data. A so-called data management application (DMA) – generally a backup system – initiates and controls the data management operations, with the execution of a data management operation generally being called an NDMP session. The DMA cannot directly access the data; it requires the support of so-called NDMP services (Figure 7.13). NDMP services manage the current data storage, such as file systems, backup media and tape libraries. The DMA creates an NDMP control connection for the control of every participating NDMP service; for the actual data flow between source medium and backup medium a so-called NDMP data connection is established between the NDMP services in question. Ultimately, the NDMP describes a client-server architecture, with the DMA taking on the role of the NDMP client. An NDMP server is made up of one or more NDMP services. Finally, the NDMP host is the name for a computer that accommodates one or more NDMP servers.

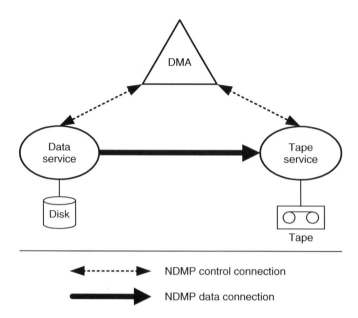

Figure 7.13   NDMP standardises the communication between the data management appli-
cation (DMA) – generally a backup system – and the NDMP services (NDMP data service,
NDMP tape service), which represent the storage devices. The communication between the
NDMP services and the storage devices is not standardised.

NDMP defines different forms of NDMP services. All have in common that they only
manage their local state. The state of other NDMP services remains hidden to an NDMP
service. Individually, NDMP Version 4 defines the following NDMP services:

- *NDMP data service*
  The NDMP data service forms the interface to primary data such as a file system
  on a NAS server. It is the source of backup operations and the destination of restore
  operations. To backup a file system, the NDMP data service converts the content of
  the file system into a data stream and writes this in an NDMP data connection, which
  is generally created by means of a TCP/IP connection. To restore a file system it reads
  the data stream from an NDMP data connection and from this reconstructs the content
  of a file system. The data service only permits the backup of complete file systems; it
  is not possible to back up individual files. By contrast, individual files or directories
  can be restored in addition to complete file systems.
    The restore of individual files or directories is also called 'direct access recovery'. To
  achieve this, the data service provides a so-called file history interface, which it uses to
  forward the necessary metadata to the DMA during the backup. The file history stores
  the positions of the individual files within the entire data stream. The DMA cannot

read this so-called file locator data, but it can forward it to the NDMP tape service in the event of a restore operation. The NDMP tape service then uses this information to wind the tape to the appropriate position and read the files in question.

- *NDMP tape service*
  The NDMP tape service forms the interface to the secondary storage. Secondary storage, in the sense of NDMP, means computers with connected tape drive, connected tape library or a CD burner. The tape service manages the destination of a backup or the source of a restore operation. For a backup, the tape service writes an incoming data stream to tape via the NDMP data connection; for a restore it reads the content of a tape and writes this as a data stream in a NDMP data connection. The tape service has only the information that it requires to read and write, such as tape size or block size. It has no knowledge of the format of the data stream. It requires the assistance of the DMA to mount tapes in a tape drive.

- *NDMP SCSI pass through service*
  The SCSI pass through service makes it possible for a DMA to send SCSI commands to a SCSI device that is connected to a NDMP server. The DMA requires this service, for example, for the mounting of tapes in a tape library.

The DMA holds the threads of an NDMP session together: it manages all state information of the participating NDMP services, takes on the management of the backup media and initiates appropriate recovery measures in the event of an error. To this end the DMA maintains an NDMP control connection to each of the participating NDMP services, which – like the NDMP data connections – are generally based upon TCP/IP. Both sides – DMA and NDMP services – can be active within an NDMP session. For example, the DMA sends commands for the control of the NDMP services, whilst the NDMP services for their part send messages if a control intervention by the DMA is required. If, for example, an NDMP tape service has filled a tape, it informs the DMA. This can then initiate a tape unmount by means of an NDMP SCSI pass through service.

The fact that both NDMP control connections and NDMP data connections are based upon TCP/IP means that flexible configuration options are available for the backup of data using NDMP. The NDMP architecture supports backup to a locally connected tape drive (Figure 7.14) and likewise to a tape drive connected to another computer, for example a second NAS server or a backup server (Figure 7.15). This so-called remote backup has the advantage that smaller NAS servers do not need to be equipped with a tape library. Further fields of application of remote backup are the replication of file systems (disk-to-disk remote backup) and of backup tapes (tape-to-tape remote backup).

In remote backup the administrator comes up against the same performance bottlenecks as in conventional network backup over the LAN (Section 7.6). Fortunately, NDMP local backup and LAN-free backup of network backup systems complement each other excellently: a NAS server can back up to a tape drive available in the storage network, with the network backup system coordinating access to the tape drive outside of NDMP by means of tape library sharing (Figure 7.16).

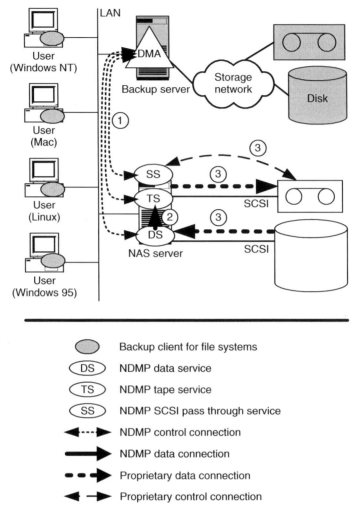

Figure 7.14 NDMP data service, NDMP tape service and NDMP SCSI pass through service all run on the same computer in a local backup using NDMP. NDMP describes the protocols for the NDMP control connection (1) and the NDMP data connection (2). The communication between the NDMP services and the storage devices is not standardised (3).

NDMP Version 4 offers the possibility of extending the functionality of the protocol through extensions. This option is being used by some manufacturers to provide some important functions that NDMP is lacking – for example, the backup and management of snapshots. Unfortunately, these extensions are vendor specific. This means, for example, that it is important in each individual case to check carefully that a specific network backup system can back up the snapshots of a certain NAS server.

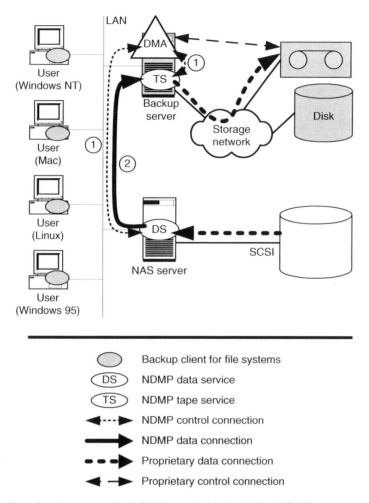

	Backup client for file systems
DS	NDMP data service
TS	NDMP tape service
◀---▶	NDMP control connection
➡	NDMP data connection
●-●-▶	Proprietary data connection
◀ ―▶	Proprietary control connection

Figure 7.15 In a backup over the LAN (remote backup) the NDMP tape service runs on the computer to which the backup medium is connected. The communication between the remote services is guaranteed by the fact that NDMP control connections (1) and NDMP data connections (2) are based upon TCP/IP. The backup server addresses the tape library locally, which means that the NDMP SCSI pass through service is not required here.

In Version 5, NDMP will have further functions such as multiplexing, compressing and encryption. To achieve this, NDMP Version 5 expands the architecture to include the so-called translator service (Figure 7.17). Translator services process the data stream (data stream processor): they can read and change one or more data streams. The implementation of translator services is in accordance with that of previous NDMP services. This means that the control of the translator service lies with the DMA; other participating NDMP

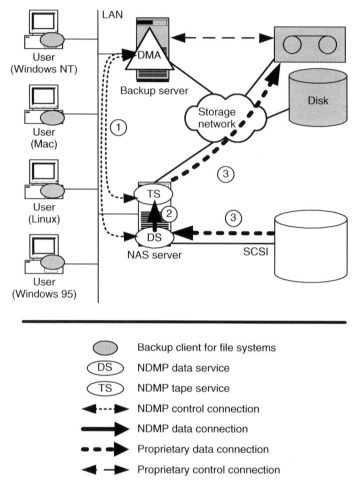

Figure 7.16 NDMP local backup can be excellently combined with the LAN-free backup of network backup systems.

services cannot tell whether an incoming data stream was generated by a translator service or a different NDMP service. NDMP Version 5 defines the following translator services:

- *Data stream multiplexing*
  The aim of data stream multiplexing is to bundle several data streams into one data stream (N:1-multiplexing) or to generate several data streams from one (1:M-multiplexing). Examples of this are the backup of several small, slower file systems onto a faster tape drive (N:1-multiplexing) or the parallel backup of a large file system onto several tape drives (1:M-multiplexing).

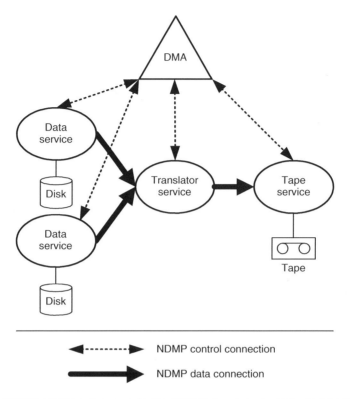

Figure 7.17   NDMP Version 5 expands the NDMP services to include translator services, which provide functions such as multiplexing, encryption and compression.

- *Data stream compression*
  In data stream compression the translator service reads a data stream, compresses it and sends it back out. Thus the data can be compressed straight from the hard disk, thus freeing up the network between it and the backup medium.

- *Data stream encryption*
  Data stream encryption works on the same principle as data stream compression, except that it encrypts data instead of compressing it. Encryption is a good idea, for example, for the backup of small NAS servers at branch offices to a backup server in a data centre via a public network.

NDMP offers many opportunities to connect NAS servers to a network backup system. The prerequisite for this is NDMP support on both sides. NDMP data services cover approximately the functions that backup clients of network backup systems provide. One weakness of NDMP is the backup of the NAS server metadata, which makes the restoration of a NAS server after the full replacement of hardware significantly more difficult (Section 7.9.1). Furthermore, there is a lack of support for the backup of file systems with the aid of snapshots or instant copies. Despite these missing functions NDMP has established itself as a standard and so we believe that it is merely a matter of time before NDMP is expanded to include these functions.

7.10 BACKUP OF DATABASES

Databases are the second most important organisational form of data after the file systems discussed in the previous section. Despite the measures introduced in Section 6.3.5, it is sometimes necessary to restore a database from a backup medium. The same questions are raised regarding the backup of the metadata of a database server as for the backup of file servers (Section 7.9.1). On the other hand, there are clear differences between the backup of file systems and databases. The backup of databases requires a fundamental understanding of the operating method of databases (Section 7.10.1). Knowledge of the operating method of databases helps us to perform both the conventional backup of databases without storage networks (Section 7.10.2) and also the backup of databases with storage networks and intelligent storage subsystems (Section 7.10.3) more efficiently.

7.10.1 Functioning of database systems

One requirement of database systems is the atomicity of transactions, with transactions bringing together several write and read accesses to the database to form logically coherent units. Atomicity of transactions means that a transaction involving write access should be performed fully or not at all.

Transactions can change the content of one or more blocks that can be distributed over several hard disks or several disk subsystems. Transactions that change several blocks are problematic for the atomicity. If the database system has already written a few of the blocks to be changed to disk and has not yet written others and then the database server goes down due to a power failure or a hardware fault, the transaction has only partially been performed. Without additional measures the transaction can neither be completed nor undone after a reboot of the database server because the information necessary for this is no longer available. The database would therefore be inconsistent.

The database system must therefore store additional information regarding transactions that have not yet been concluded on the hard disk in addition to the actual database. The database system manages this information in so-called log files. It first of all notes every pending change to the database in a log file before going on to perform the changes to the blocks in the database itself. If the database server fails during a transaction, the database system can either complete or undo incomplete transactions with the aid of the log file after the reboot of the server.

Figure 7.18 shows a greatly simplified version of the architecture of database systems. The database system fulfils the following two main tasks:

- *Database: storing the logical data structure to block-oriented storage*
  First, the database system organises the data into a structure suitable for the applications and stores this on the block-oriented disk storage. In modern database systems the relational data model, which stores information in interlinked tables, is the main model used for this. To be precise, the database system stores the logical data directly onto the disk, circumventing a file system, or it stores it to large files. The advantages and disadvantages of these two alternatives have already been discussed in Section 4.1.1.

- *Transaction machine: changing the database*
  Second, the database system realises methods for changing the stored information. To this end, it provides a database language and a transaction engine. In a relational database the users and applications initiate transactions via the database language SQL and thus call up or change the stored information. Transactions on the logical, application-near data structure thus bring about changes to the physical blocks on the disks. The transaction system ensures, amongst other things, that the changes to the data

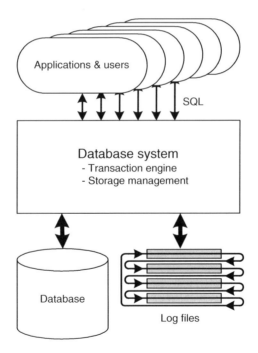

Figure 7.18 Users start transactions via the database language (SQL) in order to read or write data. The database system stores the application data in block-oriented data (database) and it uses log files to guarantee the atomicity of the transactions.

set caused by a transaction are either completed or not performed at all. As described above, this condition can be guaranteed with the aid of log files even in the event of computer or database system crashes.

The database system changes blocks in the data area, in no specific order, depending on how the transactions occur. The log files, on the other hand, are always written sequentially, with each log file being able to store a certain number of changes. Database systems are generally configured with several log files written one after the other. When all log files have been fully written, the database system first overwrites the log file that was written first, then the next, and so on.

A further important function for the backup of databases is the backup of the log files. To this end, the database system copies full log files into a file system as files and numbers these sequentially: logfile 1, logfile 2, logfile 3, etc. These copies of the log files are also called archive log files. The database system must be configured with enough log files that there is sufficient time to copy the content of a log file that has just been fully written into an archive log file before it is once again overwritten.

7.10.2 Classical backup of databases

As in all applications, the consistency of backed up data also has to be ensured in databases. In databases, consistency means that the property of atomicity of the transactions is maintained. After the restore of a database it must therefore be ensured that only the results of completed transactions are present in the data set. In this section we discuss various backup methods that guarantee precisely this. In the next section we explain how storage networks and intelligent storage systems help to accelerate the backup of databases (Section 7.10.3).

The simplest method for the backup of databases is the so-called cold backup. For cold backup, the database is shut down so that all transactions are concluded, and then the files or volumes in question are backed up. In this method, databases are backed up in exactly the same way as file systems. In this case it is a simple matter to guarantee the consistency of the backed up data because no transactions are taking place during the backup.

Cold backup is a simple to realise method for the backup of databases. However, it has two disadvantages. First, in a 24×7 environment you cannot afford to shut down databases for backup, particularly as the backup of large databases using conventional methods can take several hours. Second, without further measures all changes since the last backup would be lost in the event of the failure of a disk subsystem. For example, if a database is backed up overnight and the disk subsystem fails on the following evening all changes from the last working day are lost.

With the aid of the archive log file the second problem, at least, can be solved. The latest state of the database can be recreated from the last backup of the database, all archive log files backed up since and the active log files. To achieve this, the last backup of the database must first of all be restored from the backup medium – in the example above the backup from the previous night. Then all archive log files that have been created since the last backup are applied to the data set, as are all active log files. This procedure, which is also called forward recovery of databases, makes it possible to restore the latest state even a long time after the last backup of the database. However, depending upon the size of the archive log files this can take some time.

The availability of the archive log files is thus an important prerequisite for the successful forward recovery of a database. The file system for the archive log files should, therefore, be stored on a different hard disk to the database itself (Figure 7.19) and additionally protected by RAID. Furthermore, the archive log files should be backed up regularly.

Log files and archive log files form the basis of two further backup methods for databases: hot backup and fuzzy backup. In hot backup, the database system writes pending changes to the database to the log files only. The actual database remains unchanged

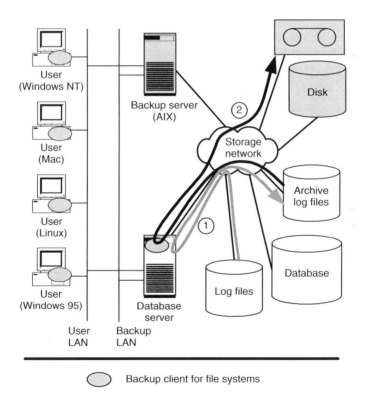

Figure 7.19 The database system copies the archive log files into a file system (1) located on a different storage system to the database and its log files. From there, the archive log files can be backed up using advanced techniques such as LAN-free backup.

at this time, so that the consistency of the backup is guaranteed. After the end of the backup, the database system is switched back into the normal state. The database system can then incorporate the changes listed in the log files into the database.

Hot backup is suitable for situations in which access to the data is required around the clock. However, hot backup should only be used in phases in which a relatively low number of write accesses are taking place. If, for example, it takes two hours to back up the database and the database is operating at full load, the log files must be dimensioned so that they are large enough to be able to save all changes made during the backup. Furthermore, the system must be able to complete the postponed transactions after the backup in addition to the currently pending transactions. Both together can lead to performance bottlenecks.

Finally, fuzzy backup allows changes to be made to the database during its backup so that an inconsistent state of the database is backed up. The database system is nevertheless capable of cleaning the inconsistent state with the aid of archive log files that have been written during the backup.

With cold backup, hot backup and fuzzy backup, three different methods are available for the backup of databases. Network backup systems provide backup clients for databases, which means that all three backup methods can be automated with a network backup system. According to the principle of keeping systems as simple as possible, cold backup or hot backup should be used whenever possible.

7.10.3 Next generation backup of databases

The methods introduced in the previous section for the backup of databases (cold backup, hot backup and fuzzy backup) are excellently suited for use in combination with storage networks and intelligent storage subsystems. In the following we show how the backup of databases can be performed more efficiently with the aid of storage networks and intelligent storage subsystems.

The integration of hot backup with instant copies is an almost perfect tool for the backup of databases. Individually, the following steps should be performed:

1. Switch the database over into hot backup mode so that there is a consistent data set in the storage system.
2. Create the instant copy.
3. Switch the database back to normal mode.
4. Back up the database from the instant copy.

This procedure has two advantages: first, access to the database is possible throughout the process. Second, steps 1–3 only take a few seconds, so that the database system only has to catch up comparatively few transactions after switching back to normal mode.

Application server-free backup expands the backup by instant copies in order to additionally free up the database server from the load of the backup (Section 7.8.5). The concept shown in Figure 7.10 is also very suitable for databases. Due to the large quantity of data involved in the backup of databases, LAN-free backup is often used – unlike in the figure – in order to back up the data generated using instant copy.

In the previous section (Section 7.10.2) we explained that the time of the last backup is decisive for the time that will be needed to restore a database to the last data state. If the last backup was a long time ago, a lot of archive log files have to be reapplied. In order to reduce the restore time for a database it is therefore necessary to increase the frequency of database backups.

The problem with this approach is that large volumes of data are moved during a complete backup of databases. This is very time-consuming and uses a lot of resources, which means that the frequency of backups can only be increased to a limited degree. Likewise, the delayed copying of the log files to a second system (Section 6.3.5) and the holding of several copies of the data set on the disk subsystem by means of instant copy can only seldom be economically justified due to the high hardware requirement and the associated costs.

In order to nevertheless increase the backup frequency of a database, the data volume to be transferred must therefore be reduced. This is possible by means of an incremental backup of the database on block level. The most important database systems offer backup tools for this by means of which such database increments can be generated. Many network backup systems provide special adapters (backup agents) that are tailored to the backup tools of the database system in question. However, the format of the increments is unknown to the backup software, so that the incremental-forever strategy cannot be realised in this manner. This would require manufacturers of database systems to publish the format of the increments.

The backup of databases using the incremental-forever strategy therefore requires that the backup software knows the format of the incremental backups, so that it can calculate the full backups from them. To this end, the storage space of the database must be provided via a file system that can be incrementally backed up on block level using the appropriate backup client. The backup software knows the format of the increments so the incremental-forever strategy can be realised for databases via the circuitous route of file systems.

7.11 ORGANISATIONAL ASPECTS OF BACKUP

In addition to the necessary technical resources, the personnel cost of backing data up is also often underestimated. We have already discussed (1) how the backup of data has to be continuously adapted to the ever-changing IT landscape; and (2) that it is necessary to continuously monitor whether the backup of data is actually performed according to plan. Both together quite simply take time, with the time cost for these activities often being underestimated.

As is the case for any activity, human errors cannot be avoided in backup, particularly if time is always short due to staff shortages. However, in the field of data protection these human errors always represent a potential data loss. The costs of data loss can be enormous: for example, Marc Farley (Building Storage Networks, 2000) cites a figure of US$ 1000 per employee as the cost for lost e-mail databases. Therefore, the personnel requirement for the backup of data should be evaluated at least once a year. As part of this process, personnel costs must always be compared to the cost of lost data.

The restore of data sometimes fails due to the fact that data has not been fully backed up, tapes have accidentally been overwritten with current data or tapes that were already worn and too old have been used for backups. The media manager can prevent most of these problems.

However, this is ineffective if the backup software is not correctly configured. One of the three authors can well remember a situation in the early 1990ies in which he was not able to restore the data after a planned repartitioning of a disk drive. The script for the backup of the data contained a single typing error. This error resulted in an empty partition being backed up instead of the partition containing the data.

The restore of data should be practised regularly so that errors in the backup are detected before an emergency occurs, in order to practise the performance of such tasks and in order to measure the time taken. The time taken to restore data is an important cost variable: for example, a multi-hour failure of a central application such as SAP can involve significant costs.

Therefore, staff should be trained in the following scenarios, for example:

- Restoring an important server including all applications and data to equivalent hardware;
- Restoring an important server including all applications and data to new hardware;
- Restoring a subdirectory into a different area of the file system;
- Restoring an important file system or an important database;
- Restoring several computers using the tapes from the off-site store;
- Restoring old archives (are tape drives still available for the old media?).

The cost in terms of time for such exercises should be taken into account when calculating the personnel requirement for the backup of data.

7.12 SUMMARY

Storage networks and intelligent storage subsystems open up new possibilities for solving the performance problems of network backup. However, these new techniques are significantly more expensive than classical network backup over the LAN. Therefore, it is first necessary to consider at what speed data really needs to be backed up or restored. Only then is it possible to consider which alternative is the most economical: the new techniques will be used primarily for heavyweight clients and for 24×7 applications. Simple clients will continue to be backed up using classical methods of network backup and for medium-sized clients there remains the option of installing a separate LAN for the backup of data. All three techniques are therefore often found in real IT systems nowadays.

Data protection is a difficult and resource-intensive business. Network backup systems allow the backup of data to be largely automated even in heterogeneous environments. This automation takes the pressure off the system administrator and helps to prevent errors such as the accidental overwriting of tapes. The use of network backup systems is indispensable in large environments. However, it is also worthwhile in smaller environments. Nevertheless, the personnel cost of backup must not be underestimated.

This chapter started out by describing the general conditions for backup: strong growth in the quantity of data to be backed up, continuous adaptation of backup to ever-changing IT systems and the reduction of the backup window due to globalisation. The transition to network backup was made by the description of the backup, archiving and hierarchical storage management (HSM). We then discussed the server components necessary for the implementation of these services (job scheduler, error handler, media manager and meta-data database) plus the backup client. At the centre was the incremental-forever strategy and the storage hierarchy within the backup server. Network backup was also considered from the point of view of performance: we first showed how network backup systems can contribute to using the existing infrastructure more efficiently. CPU load, the clogging of the internal buses and the inefficiency of the TCP/IP/Ethernet medium were highlighted as performance bottlenecks. Then, proposed solutions for increasing performance that are possible within a server-centric IT architecture were discussed, including their limitations. This was followed by proposed solutions to overcome the performance bottlenecks in a storage-centric IT architecture. Finally, the backup of large file systems and databases was described and organisational questions regarding network backup were outlined.

The next chapter (Chapter 8) on archiving considers another important field of application for storage networks. In contrast to backup, archiving is about 'freezing' data and keeping it for years or even decades. In the subsequent chapter (Chapter 9) we will discuss backup as an important building block in business continuity.

8

Archiving

The goal of archiving is to preserve data for long periods of time. With the increasing digitalisation of information and data, digital archiving is quickly gaining in importance. People want to preserve digital photos, music and films for following generations. Companies are obligated to archive business documents for years, decades or even longer in order to fulfil regulatory requirements. This chapter presents an overview of the basic requirements and concepts of digital archiving.

To begin our discussion on archiving we will first differentiate between the terms that are used in the archiving environment (Section 8.1). We then introduce reasons, basic conditions and requirements for archiving (Section 8.2). This leads us to implementation considerations (Section 8.3) and the necessity for interfaces between different components of archive systems (Section 8.4). Lastly, we present archive solutions for different areas of application (Section 8.5) and look at the organisational and operational aspects of archiving (Section 8.6).

8.1 TERMINOLOGY

We will explain the basic terms and relationships of archiving for the purposes of developing a uniform vocabulary for this chapter. We first differentiate between 'information' and 'data' (Section 8.1.1) and define the terms 'archiving' (Section 8.1.2) and 'digital archiving' (Section 8.1.3). This is followed by a reference architecture for digital archive systems (Section 8.4). Lastly, we also make a distinction between archiving (Section 8.1.5)

Storage Networks Explained: Basics and Application of Fibre Channel SAN, NAS, iSCSI, InfiniBand and FCoE, Second Edition
U. Troppens R. Erkens W. Müller-Friedt N. Haustein R. Wolafka © 2009 John Wiley & Sons, Ltd

and backup (Section 8.1.5) and explain archiving in the context of Information Lifecycle Management (ILM) (Section 8.1.6).

8.1.1 Differentiating between information and data

In the following sections we differentiate between 'information' and 'data.' Information is always something that is viewed in a business context and is generated, analysed, related, altered and deleted there. For storage and processing in computers, information must first be transferred into a format that the computer understands. For example, information is stored in files as text that is represented as zeros and ones on storage media. The zeros and the ones themselves actually have no special significance. However, when the content of the file, thus the zeros and ones, is transferred through programs into a readable form for people, then the data becomes information. Information is therefore an interpretation of data.

8.1.2 Archiving

Archiving is a process in which information, and therefore data, which will usually not change again and is to be retained for a long period of time, is transferred to and stored in an archive. An assurance is required that access will be granted for a long period of time and that the data will be protected. For example, the archiving process determines when which data is to be archived and where the best location is to store it. What is essential with archiving is that the information and the data can be retrieved and made available for processing – otherwise the whole archiving process is a waste of time.

Archiving information is already a very old concept if one thinks about the cave paintings of Grotte Chauvet in Vallon Pont d'Arc in France. These are estimated to be around 31,000 years old – and may constitute the first recorded information provided by mankind. The techniques for passing down information for posterity have continued to improve since then. First information was chiselled into stone and burned into clay; then in around 4,000 BC the Egyptians discovered papyrus, the predecessor to the paper we use today. The invention of book printing by, among others, Johannes Gutenberg from Mainz in the 15th century provided an easy method for information to be duplicated. This resulted in many interesting written records being passed on through the ages and, of course, also to an enormous growth in archived information.

8.1.3 Digital archiving

Digital archiving means that a digital computer-supported archive system stores and manages information and data on digital media. Digital media is storage media, such as hard disk, solid state disk (flash memory), optical storage devices and magnetic tapes, which allow data to be recorded.

In the past, most data and information was stored on paper or microfilm (microfiche) and kept in an archive room. It was common for these archives to be provided with digital computer-supported indices. However, due to the long and complicated management and search mechanisms involved, this traditional approach is no longer up to date. Furthermore, because of the growth in data quantities and the space required for it, paper archives would sooner or later become too expensive. Our discussion therefore does not consider archives that store information on paper or microfiche and use a digital index as being digital archives.

Today a large amount of information is available digitally. Consequently, many companies have already decided to archive information digitally. The advantages of digital archiving are summarised below:

- Offers fast access to information
- Permits simple linking of data and information
- Allows efficient search for content of information
- Requires less space
- Allows automatic duplication and filing of data
- Regulations support digital archiving

The digital archiving of information, photos, films and letters has also been popular for some time in the private sphere. Many people depend on removable hard disks, solid state disks (such as flash memory), compact disks (CD) and digital versatile disks (DVD) for storing information long term. Magnetic tapes are less common in the private sphere. For the management of their information many users rely on the structure of a file system or use additional management programs.

8.1.4 Reference architecture for digital archive systems

In our experience a three-layer architecture (Figure 8.1), consisting of applications (layer 1), archive management (layer 2) and archive storage (layer 3), has proven to be effective to describe digital archive systems. In practice, a number of different systems and architectures that deviate to some degree from this model are also used (Section 8.5). For example, some archiving solutions lack the middle layer of archive management. In this case, the other layers (application and archive storage) then usually contain the required functions of the middle layer. For a structural representation of the discussion we will orientate ourselves throughout this chapter to the three-layer architecture illustrated in Figure 8.1.

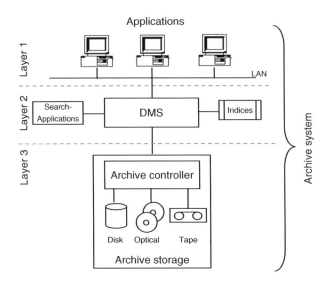

Figure 8.1 The general architecture of a digital archive system comprises three layers, consisting of the applications where the data is created, a document management system (DMS) and an archive storage system. We also refer to this architecture as the 'Reference Architecture for Archive Systems'.

At the top layer (layer 1) a digital archive system consists of applications that run on computer systems and generate, process and analyse information and then store this information in the form of data. These can be standard applications such as email systems. However, they can also be the workplaces of staff that generate and archive information and data, such as invoices, documents, drawings, damage reports and patent specifications, on a daily basis. Special devices, such as those that automatically digitalise (scan) incoming invoices and feed them into a digital archive system, can also be found on layer 1.

The applications in layer 1 communicate over an interface (discussed in detail in Section 8.4.1) with archive management (layer 2) and use it to transmit the data being archived. Archive management is usually a software system that is run on a separate computer system. Archive management is commonly referred to as a 'Document Management System (DMS)'.

Sometimes the term Enterprise Content Management (ECM) is used instead of DMS although ECM is not used consistently in the literature or by vendors. Sometimes ECM is used synonymously with DMS. However, the Association for Information and Image Management (AIIM) interprets ECM as techniques for collecting, managing, storing, retaining and providing information and documents for the support of organisational processes in enterprises. In this definition a DMS is part of an ECM solution that manages documents. In this book we concern ourselves with the 'archiving side' of DMS and ECM so that both concepts overlap and we therefore make no further distinction between them.

A DMS provides a number of services to applications. It processes and manages information and data in a variety of different ways. For example, it ensures that information is categorised and then handled and stored in accordance with requirements. A DMS can also link data and information from different applications and thus map complete business processes. For example, it is possible for data from an Enterprise Resource Planning system (ERP system) to be combined with files from a file system over a relational database. A DMS can also index data and store the resulting metadata in its own database. A clear example of indexing is a full-text index that greatly simplifies later searches for information based on content. A DMS usually has suitable applications for the search of archived data. This can be a tool within the DMS or an interface that is also used by the applications in layer 1.

The DMS transfers the data being archived to archive storage (layer 3). Another interface that we discuss in detail in Section 8.4.3 is located between the DMS and archive storage. Archive storage contains the storage media and organises the storage of the data. It also ensures that data cannot be altered or deleted before expiration of the retention period. More information on the functions and implementations of archive storage systems appears in Section 8.3.

8.1.5 Differentiating between archiving and backup

Backup (Chapter 7) and archiving are two different fields of application in data processing. In practice, there is often no clear-cut distinction made between the two. As a result, sometimes backups are made and copies are saved in the belief that the requirements for archiving are being met. However, this can be a misconception. Table 8.1 compares

Table 8.1 The table compares the characteristics of backup (Chapter 7) and archiving. Backup generates multiple copies of data that are retained for a short lifetime whereas archiving retains the data for long periods of time.

Feature	Backup	Archiving
Purpose	Data protection	Data retention and preservation
Retention times	Usually short: 1 day to 1 year	Usually long: 5, 10, 30 years and longer
Regulatory compliance	Usually not required	Usually mandatory
Original data	Usually exist in addition to the backup data	Is the archive data
Number of versions	Multiple	One
Media	Rapid data transfer, reusable, cost-effective	Robust, WORM media, in some cases cost-effective
Write access	High throughput	Medium throughput
Read access	High throughput	Fast response time, medium throughput
Search access	Based on file names	Based on content

the main differences between archiving and backup. We will explain these differences in detail.

The aim of backup is to generate one or more copies of data that can be used in the event of a failure to reconstruct the original data state. Backup copies are therefore produced and stored in addition to the original data, usually at a different location. Backup is therefore a method that ensures loss-free operation in production systems.

Backup copies usually have a short lifetime because the data being backed up frequently changes. When the next backup is made, the old version is overwritten or a new version of the data is produced as a copy. The original data remains in the original location, such as in the production system. The management of the different versions is an important function of backup systems. The storage of backup copies is normally not subject to legal requirements.

In contrast, with archiving the aim is to retain data that will not be altered again or only altered very infrequently for a long period of time and to protect it from being changed. At the same time efficient search and read access must be provided. There are different reasons for archiving data (Section 8.2.1). Some of them are based on legal requirements and standards (Section 8.2.2). Another reason is to relieve the burden on production storage systems and applications by transferring data to archive systems. Keeping data and information for future generations is another important reason for archiving.

Unlike data backup, the life cycle of archived data usually spans long periods of time – 5, 10, 30 years and longer – depending on regulatory requirements and goals. There are some who would like to keep their data forever. As archived data is never changed again, only one version is available. The ability to manage different versions is therefore not a primary requirement of archive systems – even though it might be appropriate for certain kinds of data.

As a result, backup requires an infrastructure that is different from the one for archiving. With backup, the original data is still available in primary storage and can be accessed by the application (Figure 8.2). Usually a separate backup system is required. In some cases backup is also integrated into the application. The backup system initiates the copying process and manages the metadata as well as the relationship between the original and the copy and between versions.

Unlike backup, the primary aim of archiving is not to create copies of data for disaster recovery but instead to transfer original data for long retention in an archive system (Figure 8.3). This takes place either directly through an application or through a special archiving process. According to the reference architecture for archive systems (Section 8.1.4), the archiving process is a service of a DMS (layer 2). The archiving process transfers data from the application computer's primary storage to archive storage. To improve the retrieval of the data, the archiving process can also index the data and provide search functions. The archiving process provides access to the archived data; this access comes either directly from the original application or from a special archive application.

Due to the long retention times involved, archiving has requirements for storage media that differ from those for backup. With archiving the priority is stability, regulatory compliance and cost – all of which have to be considered over long periods of time. Backup,

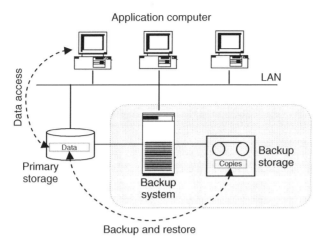

Figure 8.2 The general infrastructure for backup comprises application computers that store their data on a primary storage system and a backup system that copies the data to the appropriate backup storage media.

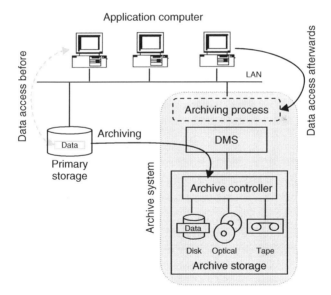

Figure 8.3 The general infrastructure for archiving comprises application computers that store their data on a primary storage system and an archive system where data is transferred during archiving. The archiving process transfers the data from primary storage through the DMS for indexing to the archive storage for long-term storage.

on the other hand, needs media that permits fast access and high data throughput and is also reusable.

Backup and archiving can easily be combined in an archive system (Figure 8.4). In this case backup copies are generated from the archived data within archive storage. This is a normal and sensible procedure because it contributes towards safeguarding continuous and loss-free operation in an archive system (Sections 8.3.6 and 8.3.7). The archived data as well as the metadata of the DMS is backed up conceivably using the same backup system.

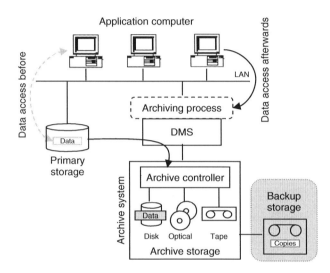

Figure 8.4   The general infrastructure for archiving can be expanded to include the backup of the archive data. This additionally requires an infrastructure for the backup, including backup media, to copy the archived data and its metadata.

8.1.6   Differentiating between archiving and ILM

Information Lifecycle Management (ILM) relates to a large number of applications in data processing, whereas backup, hierarchical storage management (HSM) and archiving are typical ILM techniques. In our experience, the concepts 'archiving' and 'ILM' are often confused, so we would like to distinguish clearly between them.

The goal of Information Lifecycle Management is the cost-optimal storage of data in accordance with the current requirements of a user or an enterprise. ILM is not a single product or service; instead, it encompasses processes, techniques and infrastructure. One standard definition of ILM states that the cost of storing data is in relationship to the 'value' of the data. The 'value' is best described by the requirements for data storage. For example, let us say a company wants to store very important data in a highly available storage system and also have copies of this data. These are concrete data storage requirements that can ultimately be translated into cost.

As time goes by, the requirements for the data may change. Figure 8.5 shows a typical example of the changing requirements for a certain type of data. Intensive work is done on a few files or data sets during the creation and collaboration phases. For example, a presentation has to be prepared and a contract drawn up on short notice for a proposal for a client. During this time the data undergoes many changes and so it has to be stored in a highly available way – depending on what is needed, maybe even on fast storage devices. At a certain point the documents are finalised and the proposal is sent to the client. This might be the starting point for the archiving phase. The data is then retained for legal and reference purposes and the requirements for high availability and fast data access decrease. The requirements for data storage therefore also decrease so that the data can be transferred to more cost-effective but slower media. When the data reaches its expiration date, it can then be deleted from the archive and the storage capacity reused.

The requirements for backup data are similar to those shown in the typical example in Figure 8.5. Current production data – such as the proposal to the client mentioned above – is the most important data. If this data is lost, the most current backup copy should be restored as quickly as possible. Backup that was carried out one week ago has limited usefulness. A current backup copy is thus considerably more valuable than one that is already a few days old.

The requirements for archive data are somewhat different. For example, if data is archived for legal reasons (Section 8.2.1), the 'value' of that data remains constant for the entire duration of the compulsory retention period because of the threat of fines in

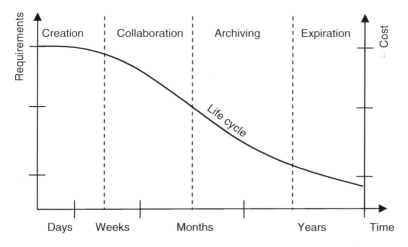

Figure 8.5 In view of the classical life cycle of information, the requirements for data are the highest during the creation and collaboration phases, and, consequently, so are the costs of data storage. As time goes by the requirements decrease and a more cost-effective storage is possible.

the event of any data loss. However, the access speed requirements for the data change. After a long time these can suddenly increase again because a second collaboration phase is inserted at the end of the archiving period (Figure 8.6). For example, during the negotiation of a loan agreement a large amount of data is collected until the contract is signed and sealed. During this time access to the contract data is frequent. In the years that follow the contract remains filed and there is less access to the contract data. The requirements for the data storage therefore decrease. When the terms are renegotiated shortly before the expiration of the loan agreement, the requirements for the data increase again, which means there is a demand for fast data access. In this example, the contract data can be transferred back to a fast and highly available storage system during the first year. After that the data can be transferred to a slower and more cost-efficient storage medium. When the contract is close to expiration, the data might be transferred again to a faster storage system because of the necessity for frequent and speedy access. This enables the costs for the storage to be adapted to requirements over the entire life cycle.

Archiving and ILM are two different concepts that can be combined together. With archiving the priority is the long-term retention of data, whereas with ILM it is the cost-effective storage of data over its entire life cycle. With the large quantities of data involved and the long periods of inactivity of archived data, ILM methods can be used to provide considerable cost reductions to archive solutions. Therefore, requirements for the ILM area should be incorporated into the planning of archive systems. At the same time it is important to separate business process-related requirements for archiving from purely cost-related requirements (Section 8.2).

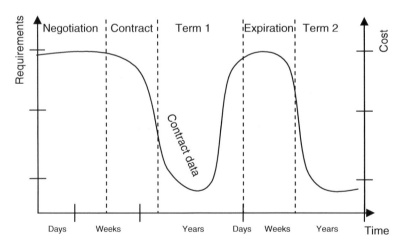

Figure 8.6 The life cycle of archive data is characterised by long retention times and changing requirements. These changing requirements must be taken into account during the storage process so that costs can be optimised.

8.2 MOTIVATION, CONDITIONS AND REQUIREMENTS

There are multiple reasons for archiving (Section 8.2.1) with regulatory requirements being a frequent one for digital archiving (Section 8.2.2). In comparison to the retention periods for archive data required for legal reasons, technical progress is rapid (Section 8.2.3) and, consequently, a great deal of attention has to be paid to the robustness of archive systems (Section 8.2.4). Environmental factors pose further risks (Section 8.2.5), leading to the need for these systems to be adaptable and scalable (Section 8.2.6). This section ends with a consideration of other operational requirements (Section 8.2.7), the requirement for cost-effective archiving (Section 8.2.8) and closing remarks (Section 8.2.9).

8.2.1 Reasons for archiving

Regulatory requirements are a key reason for archiving data. For example, the US Securities and Exchange Commission (SEC) Regulations 17 CFR 240.17a-3 and 17 CFR 240.17a-4 stipulate the records retention requirements for the securities broker-dealer industry. According to these regulations, every member, broker and dealer is required to preserve records subject to rules 17a-3 for a period of up to 6 years.

Reducing costs is another important reason for archiving. For example, if all historical data on business transactions were directly kept in a production system to provide a large data basis for data mining in the future, there would be a tremendous accumulation of data in the production databases. The existence of large quantities of data in a production database system can in turn have a negative impact on performance. Furthermore, the performance and availability requirements for production data are frequently higher than for archived data, thus usually making storage in production systems more expensive than in archive systems. Archiving allows data that will not change again to be transferred from a production system to an archive system, thereby freeing up the load on the production system.

The preservation of information is a further important reason for archiving data. Companies want to preserve the knowledge they have accumulated to secure intellectual property and competitive advantage. Saving information for future generations is an equally important social reason for archiving. This also pertains to the general public. Many people archive their digital photos and films to keep them for future generations or of course just for themselves. Social services and governmental offices also retain information such as birth certificates and family trees.

8.2.2 Legal requirements

Regulatory-compliant archiving refers to archiving that is subject to certain legal conditions and requirements. In practice, the term 'regulatory compliance' is used with different meanings. We take regulatory-compliant to mean that the legal requirements for the protection of data from destruction (deletion) and manipulation will be adhered to for a specified retention period.

In this context compliance is seen as being connected to auditing. Audits are generally carried out by authorised staff or institutes, such as authorised auditors and the tax office. An audit of regulatory compliance is based on legal regulations and requirements.

In many areas the law recognises digital archiving as regulatory-compliant but nevertheless makes special stipulations on how the data is to be archived. Germany also has something called 'Grundsätze zum Datenzugriff und zur Prüfbarkeit digitaler Unterlagen (GDPdU),' which provides basic information on data access and the auditing of digital documents. German companies must ensure that all tax-related data is identified, unaltered and in its complete form for a period of 10 years. GDPdU itself is a regulation that is based on German tax and trade laws.

A study of the individual laws and regulations would extend beyond the scope of this chapter. However, most legal requirements can be generalised. The following is a list of the most important ones:

- The type of data requiring archiving is specified – for example, invoices, contracts and receipts. This data must be identified and retained in accordance with legal requirements.

- The retention periods for different types of data is specified. In Germany the trade law states that accounting and trade information must be archived for a period between 6 and 10 years. The American licensing authorities for medicine (Food and Drug Administration, FDA) require that clinical data is saved until after the death of the children of trial patients. This can amount to more than 100 years.

- Data must be protected from manipulation or deletion during the retention period. It must at all times correspond to its original state. No alteration to the data is permissible.

- Access to archived data must be granted to an auditor at all times. In most cases the time frame for the availability of the data vacillates between one hour and 14 days.

- Data access shall only be allowed for authorised personnel and must be implemented in such a way that non-authorised personnel have no way of gaining access. Systems that store the data of more than one client must provide an assurance that clients only have access to their own data.

- The systems and processes used for archiving data must be documented. In some cases audit trails must be produced of all instances of data access. Guidelines and standards that specify the form of the documentation are usually available.

- A request is sometimes made regarding the existence of copies of data in distinct fire sections.

- A demand for the deletion of data or data sets is made upon expiration of the retention period.

Only very few regulatory requirements refer in detail to the type of storage technology that should be used for the actual archiving. The French standard NFZ42-013 tends to be an exception with its demand that only optical media is permissible for data storage. The freedom to choose a storage media gives vendors and users alike a high degree of flexibility as well as responsibility.

Regulatory requirements can also extend across other countries and continents. For example, a German company that trades on the New York Stock Exchange (NYSE) has to comply with the rules of the SEC. The same applies to pharmaceutical companies. Companies that want to operate in the American market must adhere to the rules of the FDA and the Health Insurance Portability and Accountability Act (HIPAA). This means that legal requirements vary from area and country and a company must check to make sure its way of archiving data is in compliance with these requirements.

8.2.3 Technical progress

The legally required periods for data retention are an eternity compared to how quickly technical, corporate and social changes take place. "If technology were to develop with the speed of a jet airplane, the corresponding business models would move ahead at Porsche speed, people would follow with a bicycle and politics would adapt to the legal framework at a snail's pace. Now try to teach a snail to move as fast as a Porsche." This quotation by David Bosshart, a futurologist at the Gottlieb Duttweiler Institute for Economics in Zurich, knocks the nail on its head: Technology is developing at a galloping speed today and archived data usually outlives the technology.

With archiving, technical progress is a two-edged sword. On one hand, this kind of progress produces improvements in components and systems. On the other hand, these advances do not come without associated risks. Due to the long retention periods for archive data there is a danger that the components (software, hardware, storage media) will become obsolete. For example, access to archived data can be lost if a component is obsolete and no longer compatible with other new components. Despite data integrity, data can sometimes no longer be read, such as when a particular data format cannot be interpreted anymore. For example, many operating systems, data processing programs and table calculations that were popular in the 1990s are no longer being sold and are rarely available.

Technical progress affects all layers of an archive system (see Figure 8.1): applications, DMS and archive storage. Infrastructure components, such as the computers and the operating systems on which applications are run, are also affected by advances in technology. This applies as well to storage networks, including their protocols and interfaces. Considerable risks are associated with the aging and changing of the components in archive systems and this is something that definitely requires attention. One appropriate way for coping with the technical progress is logical and physical migration of systems, software and data (Sections 8.3.8 and 8.3.9).

8.2.4 Requirement for stability

Technical progress plays an important role when it comes to the long retention periods for archived data with special attention given to the long-term stability of an archive system. The durability of an archive solution is an elementary requirement that must be conceptualised and given a high priority during the planning and procurement of these systems.

Data formats for archived data should be selected in such a way that they can still be read and represented several years or even decades down the line. Data formats must also satisfy the requirements for regulatory compliance. Standardised data formats supported by the applications of many different vendors are available for this purpose. This ensures that the data can still be read even if certain vendors have left the market.

The stability of a storage technology can be viewed from two angles: storage medium and storage device. The stability or durability of storage media signifies how long data is physically readable on a particular medium. Due to their physical properties, some storage technologies are suitable for long-term storage; these include optical media (Section 8.3.1). With this technology data bits are 'burned' into a medium and are very durable. In contrast, with magnetic recording techniques, such as disks and magnetic tape, the data bits are produced through magnetic means. However, magnetic fields deteriorate over time and this can reduce the readability of the data.

Environmental influences also play a big role with technology. In the case of magnetic tapes, moisture and strong magnetic fields can affect the durability of the stored data. The lifespan of optical media is reduced due to UV light and pollution. Disks are sensitive to mechanical stress, heat and strong magnetic fields.

With storage devices, stability refers to the length of time a device is available on the market. Some technologies suddenly simply disappear, such as the 5 1/4″ diskettes and corresponding disk drives. Another danger is that the interface for a storage device is no longer supported, such as with the token-ring technology in data networks.

Another important aspect of stability relates to the vendors and their archive systems. This is linked to such questions as: Will the vendor and the archive systems and components still be around several years hence or, in other words, will the vendor still be supporting the system over many – 10, 20, 30 or more – years? When seeking answers to these questions and choosing archive system components, it is also important to examine the background of the respective vendor (Section 8.3.10).

A component-based architecture with standardised and open interfaces (Section 8.4) can also offer a high degree of stability. Open and standardised interfaces are a prerequisite for the adaptability of IT systems in general and archive systems in particular. They are the basis for the replacement of archive system components independently of the original vendors, for the seamless migration of archived data to new media and they usually have a long lifetime (Section 8.3.8).

8.2.5   Risks from the environment and from society

Like all IT systems, archive systems are exposed to a variety of risks from the environment. Possible causes of failures and system outages range all the way from natural catastrophes, such as earthquakes, flooding and whirlwinds, to widespread power failures as well as to socio-political risks (Section 9.1). A key example of the latter was the attack on September 11, 2001, in the United States. Archive systems must be equipped to cope with disruptions of this type and at the same time guarantee loss-free and continuous operation.

Other risk factors result from economic changes, such as company takeovers and mergers that affect archive systems and data. Usually a company merger will entail an integration of archived data into an existing archive system. Of course, this must be carried out in compliance with the regulatory requirements. Likewise relevant archived data must be separated and transferred to new archive systems when part of a company is split off.

Politics may move at a snail's pace . . . , but archiving periods are so long that it is likely the relevant legal requirements will have changed in the meantime. These changes may have to be incorporated in respect of the archived data. For example, the law can demand that data be archived for a shorter or longer period than originally planned. The processes and parameters of the archive systems must therefore be flexible to meet these requirements.

8.2.6   Requirement for adaptability and scalability

Due to the long retention periods required for archived data, an equal amount of attention must be given to environmental changes and to technical progress. In addition to stability, the adaptability and scalability of an archive system must be considered as a basic requirement during procurement and planning. Scalability mainly relates to growth in data quantities and data throughput. Adaptability refers to the integration of new components and requirements into an existing archive system as well as the splitting off of some data and components.

Numerous studies have shown that the volume of archive data balloons to the same degree as general data. An added factor is that the retention periods for archived data are usually very long, which over the years leads to an accumulative effect in required capacity. Therefore, the capacity of archive systems must be highly scalable.

Equally, the amount of data processed by an archive system during a particular unit of time also increases. The result is an increase in the number of parallel write and read access operations to an archive system. To deal with these requirements, an archive system must be scalable for data throughput.

The requirement for adaptability results from company and legal changes that occur. It must be possible for data from multiple archives to be integrated into a central archive;

alternatively, it should be possible to split the data from one archive into two separate and independent archives. At the same time attention must be given to the legal requirements involved, such as the logical separation of data for different clients or companies.

The components of an archive system must also be adaptable. It should be possible to add other applications to an archive system, to replace existing components or to use new technologies. Multiple DMS systems should also be able to work together as a group (Section 8.5.5) to enable an easier integration of existing archive systems and an expansion of the archive system from the standpoint of scalability.

Not only stability but also scalability and adaptability benefit from a component-based architecture with standardised and open interfaces supported by multiple vendors (Section 8.4). This is a prerequisite for ensuring that a variety of different applications, DMS and archive systems can communicate with one another. Standardised interfaces also help to simplify the process of transferring data and replacing the components of archive systems – for example, with equivalent components from another vendor.

8.2.7 Operational requirements

Operational requirements represent another type of requirement for archive systems. These deal with the different operational characteristics, such as access speed, transparency of data access, data throughput and continuous and loss-free operation.

Access speed defines the maximum time it should take for accessing data in an archive system. Data access can constitute reading, writing or querying data. Querying refers to queries sent to the DMS or to the archive storage. For a guarantee of scalability there should be no restriction to the amount of parallel access in an archive system.

Transparency in data access is sometimes an important requirement from a user's view and contributes to efficient processes within a company. Users may want to access archived data just as they access data on a production system. As the archive system may use slower storage media such as tape, a requirement should exist to notify a user if data is being accessed on such a slow medium.

Requirements for data throughput define how much data can be processed during one unit of time by an archive system. This operational requirement therefore refers to the quantity of data of a certain size that can be written and read in an archive system during a specific period of time. There are two archive system components and processes to consider: one is the indexing capabilities of the DMS and the other is the throughput that can be handled by the archive storage system.

Continuous operation (Chapter 9) relates to requirements on how an archive system is designed to cope with failures. The operating time of archive systems is impacted by maintenance work and repairs as well as by fault occurrences. Loss-free operation should ensure that neither data nor metadata is lost during such occurrences.

The operational requirements mentioned can change for the user as time goes by (Section 8.1.6). Therefore, it is important that possible changes in requirements for archived data are automatically anticipated in the respective component of the archive system.

There is an increasing trend towards data being archived immediately without first being kept in production storage. This may further increase the operational requirements of archive data. Of course, this is only appropriate for data that will not change anymore, such as scanned invoices. The advantage is that production storage is not overloaded with data that is to be archived. This reduces the administrative effort required for operations such as backup and restore as well as the resource management of production storage.

8.2.8 Cost-related requirements

Cost is another important criterion for archive systems. This applies not only to procurement costs but also to the operating costs over long periods of archiving. The general rule of thumb is that the procurement cost of an archive system should amount to 30% of the total cost for an operational period of 5–7 years. This means that more than 70% of the total cost must be applied towards the operation of the archive system. The operational costs include system administration and maintenance as well as power consumption and space rental costs. It also includes the cost of data and system migration.

Power consumption is a major operational cost factor. The production of energy is becoming more and more expensive, which is leading to a constant increase in energy prices. Removable media is becoming a real advantage now as it only consumes power and produces heat when data is being accessed. A study titled 'Is Tape Really Cheaper than Disk' published by the Data Mobility Group in 2005 confirms that over a 7-year period tape storage media is considerably cheaper than disks.

8.2.9 Conclusion: Archive systems as a strategic investment

The legal requirements for archiving with the aspects of stability, scalability and adaptability can be used to determine which basic elements must be considered in the planning and implementation of archive systems.

In practice, however, these long-term aspects are often neglected during the planning and procurement stage. Instead short-term aspects, such as operational criteria and procurement costs, become the deciding factors in purchase decisions. In today's age where decisions are often driven by quarterly results, the procurement costs are the only thing taken into account, whereas operating costs and other aspects are sometimes ignored.

When an archive system is viewed as a strategic investment, it is important to incorporate archiving into the general processes of risk management (Section 9.1.9). Due to the long life of the archived data relevant measures should be taken to guard against the risks resulting from technical progress (Section 8.2.3), environmental risks and company changes (Section 8.2.5). The goal should be to avoid a proliferation of different archiving solutions within a company. One solution could be to establish a central archive (Section 8.5.5) and to use components with standardised interfaces (Section 8.4).

Companies should view archive systems as strategic investments that help to support their long-term success. This is the only way in which long-term requirements such as stability, scalability and adaptability can be properly incorporated. The following sections on implementation considerations (Section 8.3) and interfaces for archive systems (Section 8.4) offer further ideas for realising long-term requirements.

8.3 IMPLEMENTATION CONSIDERATIONS

In this section we introduce storage technologies that are very important for the implementation of archive systems, in particular for archive storage. We therefore start by demonstrating how the requirements described earlier (Section 8.2) can be implemented. These considerations should provide the reader with the criteria necessary for the selection of an archive storage system.

The section begins with a comparison of the different write once read many (WORM) storage technologies (Section 8.3.2). This is followed by implementation considerations for data security (Section 8.3.2), data integrity (Section 8.3.3), proof of regulatory compliance (Section 8.3.4), data deletion (Section 8.3.5) and continuous (Section 8.3.6) and loss-free (Section 8.3.7) operation. The section concludes with suggestions for the implementation of storage hierarchies (Section 8.3.8) and technology-independent data storage (Section 8.3.9), along with considerations for making a strategic selection of vendors and components (Section 8.3.10).

8.3.1 WORM storage technologies

The concept 'WORM storage technology' (write once, read many) refers to storage media and systems in which data that has once been stored cannot be modified or deleted prior to expiration. WORM technologies are usually implemented on the storage medium in archive storage. It is important to note that the use of WORM storage technology is usually not sufficient on its own to make an archive system regulatory-compliant. For example, it is not helpful if data is protected in WORM-protected archive storage but the references to the data and the metadata can simply be deleted in a DMS. For an assurance of regulatory compliance the entire archive system must be protected from data deletion and modification. Appropriate measures for this kind of protection are discussed in Sections 8.3.2 and 8.3.3.

There are three basic WORM storage technologies: hard disks and disk subsystems (also referred to as disk), optical storage and magnetic tapes. These technologies have difference characteristics in terms of scalability (Section 8.2.6), cost (Section 8.2.8) and operational requirements (Section 8.2.7).

Disk-based WORM storage

WORM technologies have been implemented on disks for some time now. In the discussion that follows we refer to this as disk-based WORM storage. Here data is stored on conventional disks that do not have immanent WORM functionality, meaning that the disk medium itself is rewritable. Instead the control software used for storing the data on the disks has functions added to prevent the data being deleted or changed once it is stored. This control software can be embedded in the disk drive firmware itself or in a disk subsystem controlling a plurality of disk drives. This produces WORM characteristics similar to those of traditional (optical) WORM media.

The advantage of this approach is obvious: Disk subsystems allow fast data access, can be designed for high availability (Section 2.8) and thus satisfy high operational requirements. These systems also allow storage space to be reused when data is deleted after expiration of the retention period. This is not possible with other WORM technologies.

A disadvantage of disk-based WORM storage is that users and auditors generally do not accept the regulatory compliance of such systems unless an audit has been undertaken. Therefore vendors of disk-based WORM storage systems often engage an auditor or the like for an assessment of the regulatory compliance of their disk-based WORM storage. Another disadvantage of these systems is that disks constantly consume power and produce heat. Over long periods of time the operating costs can be quite substantial (Section 8.2.8). The overall robustness of disks is also limited because some of their components are constantly in motion and therefore subject to wear and tear.

Some vendors reduce power consumption by enabling disks within the disk subsystem to be switched off automatically when data is not being accessed. As a consequence, when data is accessed, the disks first have to start rotating again, thus data access takes longer. When for reasons of redundancy data is striped over multiple disks (RAID), multiple disks then have to be switched on each time data is accessed. This results in high power consumption and increases the overall time for data access because the system has to wait for the slowest disk to become ready. Additionally, this constant switching on and off of the disks shortens their lifespan.

The type of disks selected for archiving have a direct impact on the reliability of disk-based WORM storage. Today inexpensive Serial ATA (SATA) disks are often used for archiving to reduce cost. However, it is this type of disk that has a high failure rate (Section 2.5.5). The loss of archived data through disk failure is almost inevitable unless additional measures, such as RAID protection, are undertaken. Additional protection such as backup (Section 8.3.6) or data mirroring (8.3.7) should therefore be provided to disk-based systems to prevent data error and failures.

Optical WORM media

Traditional WORM media, such as CDs, DVDs, ultra-density optical (UDO) and BluRay DVD, is based on optical technology. When data is written, the associated bits are 'burned' onto this media. This makes it physically impossible to overwrite or alter the data.

Optical media is traditionally used for regulatory-compliant digital archiving and is generally accepted by users and auditors alike. Access to data on an optical medium is random, which ensures that access is fast. Another advantage of optical WORM media is that it is removable and therefore does not consume power continuously; nor does it constantly have to be cooled.

The disadvantage of the optical technologies available today is low storage capacity, which lags behind other technologies by a factor of 5 to 10. Optical storage media currently (2009) provides a capacity of 25–60 GBytes (UDO and BluRay) per medium. Vendors are planning capacities of 100–200 GBytes for the future. Another disadvantage compared to disks is longer access times. A physical movement of the removable media is usually required to access the data on an optical medium in an automated library, which usually takes 10–15 seconds. This is not optimal when a user requires frequent online access. Write throughput for optical media is relatively low compared to other techniques. The write speed of UDO disk drives is currently at 5–10 Mbytes/s. Once optical WORM media has been written it cannot be used again, and therefore has to be disposed of when all data on it has expired.

WORM magnetic tapes

Magnetic tape systems have been offering WORM functionality for some time. These are also referred to as WORM tapes. A magnetic WORM tape cassette usually contains a special internal marker that identifies it as a WORM medium. The firmware of the tape drives ensures that each spot on a WORM tape can only be written once although the tape material is physically writable more than once.

Magnetic tapes offer a number of advantages for archiving: They have a high capacity and allow high data throughput. The capacity of the WORM tapes available today is 1 TByte with a throughput of 160 MBytes/s. Tape libraries are also easily expandable. Over the years WORM tapes can be added in stages, thereby enabling the storage capacity of archive storage to cope well with increasing archive data volumes. Yet another advantage is the relatively low operating costs associated with the storage of data over a long period of time, because tapes are removable media and do not consume power all the time.

A disadvantage of magnetic tapes is that, due to their sequential characteristics, access times are low compared to optical media. Due to the relatively long loading and spooling process involved, access times to magnetic tapes are between 30 and 60 seconds. Furthermore, WORM tapes can only be used once and have to be disposed of when all data has expired. As with disk-based WORM storage systems, users and auditors today do not find WORM tapes generally acceptable for regulatory-compliant archiving. Therefore vendors of WORM tapes often engage an auditor or the like to carry out an audit of the regulatory compliance of their WORM tape systems.

Comparison and usage of WORM technologies

Table 8.2 summarises and compares the most important criteria for the WORM technologies described in the preceding section. The table shows that disk-based WORM storage is the most suitable option when data is to be archived for short retention periods and the user also requires fast access and high availability.

Optical WORM media is most suitable for small data quantities that are to be archived for long periods and where periodic access is needed. It offers medium access times and is therefore conditionally appropriate for online operations.

Due to their high capacity, WORM tapes are more suited to the long-term archiving of medium and large-sized amounts of data. Their sequential access nature means that magnetic tapes have longer access times, thereby making them a suitable option for infrequently accessed data.

Users will be afforded maximum flexibility if they use archive storage that supports all three WORM storage technologies combined with the means for automatic data migration from one technology to another based on the age of the data. This is the optimal way to utilise the special characteristics offered by the different WORM technologies and therefore cope with the changing needs that arise in the storage of archive data.

Additionally, these types of archive storage systems are not dependent on any specific storage technology, which is particularly an advantage when long data retention times are required. This allows a high degree of flexibility in changing a data storage location and, consequently, the particular WORM technology.

Table 8.2 A comparison of the three basic WORM technologies in respect of different criteria highlights the advantages and disadvantages of each one.

Criterion	Disk-based WORM storage	Optical WORM media	WORM tapes
WORM protection	Software	Hardware	Hardware
Scalability	High	Low	High
Regulatory compliance	Not accepted across the board	Generally accepted	Not accepted across the board
Procurement costs	Medium	Low	Medium
Operating costs	High	Low	Low
Access time	Fast	Medium	Slow
Data throughput	High	Low	High
High availability	Very good	Low	Good
Robustness of media	Minimal	High	Medium
Reusability	Yes	No	No

Compliance with regulatory requirements involves more than just considering the WORM technology used for archive storage. The entire archive system – from the archive application and the DMS through to archive storage – must provide functions that ensure that data is regulatory-compliant. These functions will be described in the following sections.

8.3.2   Data security

The law demands a guarantee of data security and data integrity for archive data (Section 8.2.2). In the context of archiving, data security means that the storage of and access to data must take place in a secure environment. This environment must ensure that any manipulation or deletion of data is impossible and only authorised users have access to the data. The data security of archive storage therefore relates to data and metadata access, to data transfer itself as well as to access over management interfaces. The standard techniques of data processing such as access control in combination with WORM techniques can be used to satisfy these requirements.

Data access refers to write and read operations. Metadata access includes the querying of metadata and enables the search for data based on the metadata. Protection from unauthorised access starts with basic things such as secured premises for all components of the entire archive system and isolated networks as well as authentication and encryption techniques and access control lists (ACLs).

The data traffic in archive systems must also be protected to prevent such occurrences as eavesdropping on the network. Standard techniques such as Secure Shell (SSH), Virtual Private Networks (VPN) and data encryption are used to provide this protection.

Lastly, access to the management interfaces of an archive system also needs protection. For example, a disk-based WORM storage system that suppresses changes to archived data is useless if the management interface is still configured with the vendor's standard password. This would enable any person with access to the LAN to reformat the disks, thereby destroying all the data. Therefore, the management interfaces for all components of an archive system must provide measures that prohibit unauthorised changes of and deletions to the data (Section 8.4.5).

8.3.3   Data integrity

Data integrity means that the data is preserved in its original state at all times during the archiving process and retention period. However, as storage media become older, single bits sometimes flip. In addition, system errors can cause changes to large amounts of data. This unintentional modification to archived data must be detected and logged automatically so that the integrity of the data can be assessed and restored when possible.

The data security discussed earlier is an important prerequisite for data integrity. In addition, proof of the integrity must be available during the period the archived data is stored. Standard techniques are available here too: checksums such as Cyclic Redundancy Check (CRC), cryptographic hashes such as Message Digest Algorithm 5 (MD5) and Secure Hash Algorithm (SHA), as well as digital signatures. Standard techniques such as RAID and error-correcting encoding (ECC) are also used to maintain and restore data integrity (Section 2.8).

Content Addressable Storage (CAS) is a special implementation for integrity checking that is also based on cryptographic hashes. For each data object that is stored by an application in a CAS system a hash or fingerprint is calculated and returned to the application. The application uses the hash subsequently as a reference or pointer to the data. When a data object is read, its hash is recalculated and compared to the hash stored as the address. This enables an efficient integrity check and also explains the term content address: the hash reflects the content of the data and is used as an address to reference the data object. The advantage of CAS is that the integrity checking for the data is embedded in the storage process. The disadvantage is that hash collisions can occur when two distinct data objects generate the same hash. These hash collisions must be resolved in order to prevent data loss. If a calculation of the hash is performed during the data storage process, data throughput is affected because of the additional effort required to calculate the hash. Another disadvantage is that if a hash method must be changed in a CAS system – for example from MD5 to SHA1024 – all content addresses have to be recalculated. A new hash must then be calculated for each data object. This can be a very time- and resource-consuming process and impact the entire archive system because the hashes must be updated in the CAS as well as in the applications that access the data objects via the hashes.

An integrity audit can be carried out on all layers of an archive system. For example, the application that generates the data or the DMS that manages the data can create, store and test checksums, hashes or digital signatures. The archive storage can also provide techniques (such as RAID and ECC) that detect faulty data and, if necessary, automatically reconstruct it. The advantage of multiple integrity auditing is that potential data errors are detected not only during data transfer between the DMS and archive storage but also during the retention period. On the other hand, there should be some control over multiple auditing for cost and time reasons.

An integrity audit of the data within archive storage should be run automatically. Continuous integrity audits that automatically run in the background only have a limited effect on actual archiving operations. An integrity audit that is executed during the read access to data is particularly ideal because the integrity of the data is verified at the moment of access. The associated performance decrease may well be accepted when the system can assure at any point of time that all data is valid. When integrity failures are detected, an error message should automatically be sent to the user or the operator.

When data integrity is restored – for example through ECC – it is important to ensure that the original state of the data has not been altered. If any doubt exists, the restoration should be abandoned and the data flagged as being faulty. It is also important that the audit and restoration of data integrity is logged using audit trails. The log must clearly show that the data corresponds to its original state.

Case study: data integrity in practice

In practice, multiple different methods of data integrity auditing and restoration are normally combined and implemented on different layers within an archive system. Figure 8.7 shows how this works using the typical architecture of disk-based archive storage. The system consists of a system controller that supports the interface to the DMS, which receives the data and metadata and forwards it for storage to the RAID controller. The system controller is also responsible for applying the retention times of the data accordingly and contains other functions such as data migration, backup and data deletion after expiration. The RAID controller converts the data to the particular RAID level used. The data is then stored on disks.

In this scenario the integrity of the data is protected on three levels. At the lowest level, current disks store data blocks with an ECC that enables the detection and reconstruction of the faulty bits stored on the disk.

In addition to data blocks, the RAID controller generates RAID stripes and parity blocks (Sections 2.4 and 2.5) and usually applies an additional ECC to the RAID stripes. If a RAID stripe on the RAID array is not readable or loses its integrity, the parity blocks and the ECC can be used to detect and correct this failure. For performance reasons RAID stripes and parity blocks are normally only checked if a fault occurs. However, other functions in the RAID controller, such as data scrubbing (Section 2.8), periodically check the RAID stripes and parity blocks.

The system controller, which contains the control software for the archiving functionality, generates checksums for the data that is sent from the DMS. The checksums are

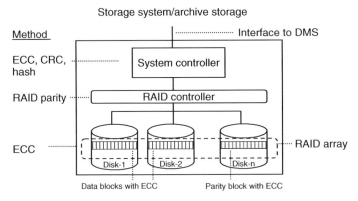

Figure 8.7   In practice, multiple components of a storage system support data integrity. Different techniques such as RAID, checksums and error-correcting codes (ECC) are used.

forwarded along with the data for storage to the RAID controller. Likewise, when the system controller reads the data, it can check the integrity of the data by recalculating the checksum and comparing it to the checksum that was stored initially. This allows the system controller to detect changes to the archived data independently of the RAID controller and the disk drive – for example, changes made through unauthorised manipulation on the storage media.

Integrity audits at multiple levels therefore protect against different types of errors that can then be most efficiently resolved at their respective levels. The extra effort involved in triple integrity audits therefore makes sense.

8.3.4   Proof of regulatory compliance

In the case of regulatory audits, evidence must be provided that no manipulation of the archived data has occurred. This is usually achieved through audit trails and may entail a complete history of all access to the data. Audit trails are usually kept in electronic form and record each access to archived data.

In addition to date and time, audit trails for data access should include the name of the application and of the user as well as the type and result of data access. Audit trail records are usually produced and stored by the DMS or the archive storage. They are often linked to monitoring systems that ensure a notification is sent to the system administrator when unauthorised access or access attempts have occurred. Audit trails may also be used to record all activities from the management interfaces.

In this context it is important to ensure that the audit trails cannot be altered, because a manipulated audit trail is not recognised as being regulatory-compliant. WORM technologies (Section 8.3.1) can be used to store audit trails in conjunction with methods for data security (Section 8.3.2).

For proof of regulatory compliance it is also important to create and maintain a complete set of documentation. This documentation must describe the end-to-end archival and retrieval processes as well as the measures assuring data security, data integrity and loss-free and continuous operation. Audit trails become part of this set of documentation that is used by the auditor to assess the archive system and data.

8.3.5   Deletion of data

The deletion of data upon expiration of the retention period is subject to legal requirements (Section 8.2.2). There are different ways in which data can be deleted in archive storage and this partly depends on the WORM storage technology used (Section 8.3.1). The deletion methods discussed apply to the deletion of the metadata, the deletion of the data itself and the destruction of the storage medium.

With the first method, deletion of the metadata including the references to data destroys the logical link to the data although the data itself remains on the storage medium. There is an advantage as well as a disadvantage to this method. On one hand, the method is very fast as only the references are being deleted and therefore only small amounts of data have to be processed. On the other hand, the data can still be reconstructed even after the references have been deleted. Therefore, certain legal requirements demand that data be completely destroyed on the storage medium, which means that this method cannot be applied. The deletion of references can be used with all WORM storage technologies. For disk-based WORM storage this option is the least expensive one because the storage capacity that was used by the deleted data can be reused. Other aforementioned WORM technologies do not allow the reuse of storage capacity.

With the second method, the data is deleted. The data is physically overwritten on the storage medium and the metadata including the reference to the data is also deleted. This ensures that the data cannot be reconstructed after it has been deleted. Also called data shredding, this method can only be used with disk-based WORM technology, which allows a medium to be overwritten (Section 8.3.1). This is an inexpensive variant as it allows the storage space released by the expired data to be reused. It is only applicable to disk-based WORM technologies or, more generally, to storage technologies that allow the overwriting of data.

With the third method, the storage medium on which the data is stored is destroyed – thereby definitively preventing any future reconstruction of the data. This method is especially suited to removable media (optical WORM storage and WORM tape) as this kind of media can be removed separately from archive storage and destroyed selectively. It is also appropriate for storage technologies that do not allow overwriting. With disk-based WORM storage it is not a good idea to destroy the disks. Not only would this be expensive and time-consuming, but it also means that more than one disk would have to be destroyed because with RAID the data is usually striped over multiple disks (Section 2.4). Irrespective of which WORM technology is used, with this method it is important to note that only those storage media on which all data has expired are destroyed. Overall this is a more expensive method than the first two mentioned, as storage media are physically destroyed and cannot be reused again.

Regardless of the method used, the deletion of data and metadata in an archive system must be synchronised over all layers (application, DMS, archive storage). This is the only way to ensure that the deletion process has been handled properly. Therefore, the interfaces between the layers must provide the appropriate deletion functions that work mutually together (Section 8.4).

It is an advantage if an archive storage system supports more than one deletion method, because, depending on the respective requirements, the most cost-effective method for deletion can then be selected. This approach allows different legal requirements for data deletion to be supported concurrently within one archive system.

8.3.6 Continuous operation

Continuous operation is something that is almost taken for granted with central applications such as archive systems. Certain legal requirements demand that the risk of data loss be kept to a minimum and that access to archived data is granted during the retention time (Section 8.2.2).

Continuous operation means that access to archived data is possible at all times during the retention time period. Sometimes the time span between a data request and the delivery of the data is also specified. Depending on the environment, this time span can range from a few seconds (for example, with interactive operations) up to several days (for audits).

Techniques for continuous operation have been dealt with in various sections of this book and will be covered again in detail in the following chapter (Chapter 9). Continuous operation must of course be guaranteed at all layers of the archive system (Section 8.1.4). Therefore, techniques such as clustering, RAID and data mirroring could be applied to all three layers of these systems (application, DMS, archive storage).

8.3.7 Loss-free operation

Loss-free operation means that no data or metadata will be lost. For an archive system this means that no data should be lost during the entire retention period, even if a major outage occurs. Considering the long lifespan of archive systems, techniques such as backup and data mirroring are essential.

The techniques used to back up archived data are the same as those for conventional backup (Chapter 7), the only difference being that backup copies of the archived data are created within the archive system (Figure 8.4). However, with the backup of archive data the regulatory compliance of the backup copies also has to be ensured. This can be achieved through the use of WORM storage technologies (Section 8.3.1) as the backup media. During the restore process there must be an assurance, and perhaps verification, that the copies of the data correspond to the original. Techniques to guarantee and restore data integrity (checksums, hashes, digital signatures and ECC, Section 8.3.3) can be used for this purpose. An archive storage system that integrates both automatic data backup and integrity auditing is optimal as it means that no additional backup software will then be required.

For data mirroring two compatible archive storage systems are normally coupled together and the data is mirrored between them (Figure 8.8). One of the two storage systems might be active (primary system) and the other one passive (standby system) or both systems can be active at the same time. Standard techniques such as synchronous and asynchronous data mirroring (Section 2.7.2) are used for the actual copying of the

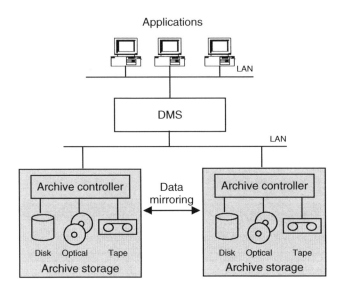

Figure 8.8 Two compatible archive storage systems are coupled together for data mirroring. The data is copied either synchronously or asynchronously from one archive storage system to the other one. The DMS can take over the control of the data mirroring as well, in which case the coupling between the archive storage systems is removed and the DMS is directly connected to the two archive storage systems.

data between the two systems. Similar to the backup of archive systems, there must be an assurance of regulatory compliance for both data copies. Additionally, automatic fail-over from system failure is an advantage because it reduces the manual effort involved as well as the fail-over time. The disadvantage of data mirroring performed by archive storage is that if the data received from the DMS is defective, the mirrored copy of the data will be defective as well.

Another mirroring principle is that of DMS mirroring. The DMS copies the data synchronously or asynchronously to two archive storage systems – similar to Figure 8.8 – the only difference being that the two archive storage systems are not coupled together. The DMS controls the data copying process and also the fail-over in the event one archive storage system fails.

With this type of mirroring the DMS has control over the data copies and automatically detects and corrects errors during the mirroring of the data. The capability of the DMS to fully recover an archive storage system in case all the data stored in this system has been completely lost due to a disaster is important, but not always a given. The benefit of this type of mirroring is that the DMS writes the data twice, and if one copy is defective the likelihood that the second copy is defective is low. The disadvantage of this method is that it is more complex to implement. In addition, for full redundancy and data protection two DMS systems are required, which also exchange the metadata and indices.

8.3.8   Data management: storage hierarchy and migration

Due to the long life span of archive data, special importance is given to the cost-efficient storage and management of the data. One important technique for this is Hierarchical Storage Management (HSM), which comprises multiple storage levels (tiered storage) and capabilities for automated data migration between these tiers. This technique helps greatly towards reducing storage costs because it allows data to be migrated between different storage technologies (Section 8.1.6).

Consequently, when large quantities of data are involved, archive storage must support tiered storage that incorporates different WORM storage technologies (Section 8.3.1). Figure 8.9 shows the storage hierarchy of an archive storage device using different storage techniques in conjunction with the life cycle of the data. Ideally, the data would be stored on the most cost-effective media based on access time and other operational requirements (Section 8.2.7).

With small and medium amounts of data it is wise to check carefully whether it is worth using multiple tiers of storage. With these data quantities it is often more feasible

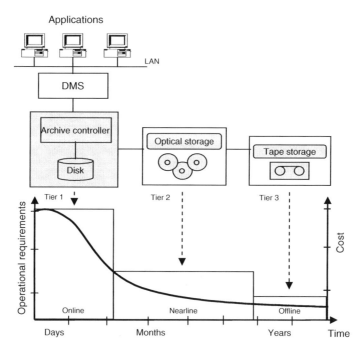

Figure 8.9   Different storage tiers in archive storage are aimed at meeting different operational requirements and cost objectives. The archive storage must store the data in the appropriate storage pools, and, when necessary, automatically migrate the data. This provides an optimal balance between requirements and cost.

to use only one or two different WORM technologies – for example, disks for the first shorter time period and magnetic tape for the second longer time period.

The management of a storage hierarchy must be automated through appropriate virtual-isation functions, because the users of archive systems do not want to concern themselves with the type of media on which their data is stored. Standard techniques of storage vir-tualisation can be used – such as data profiles, which allow a description of the access behaviour of different user groups of archive storage (Section 5.2.3).

Specific data control components in archive storage evaluate these data profiles and automatically send the archived data to the most cost-effective storage medium available. The data can be migrated in both directions – for example, from disk (tier 1) to tape (tier 3), but equally from tape to the disk. Again, the migration should have no effect on regulatory compliance and data integrity. This can be assured through audit trails that record the migration process and prove the data integrity (Sections 8.3.3 and 8.3.4).

The expression 'data migration' describes the process of moving data within a storage hierarchy as well as moving data from an old storage technology to a new one. In contrast to the storage hierarchy, data migration to new storage technologies is also relevant for small and medium quantities of data (Section 8.2.3).

Whereas applications and a DMS can be migrated comparatively easily from one server to another, the transfer of large amounts of archived data from one archive storage to another is more complicated. This again shows the advantages of archive systems that support different storage techniques along with data migration. In our experience, this particular problem is often overlooked during the procurement of archive systems and can result in the need for extra resources and budgets to accomplish data migration. It is much easier if archive storage can carry out the migration including audit trails without interaction with the DMS.

8.3.9   Component-neutral archiving

Today many archive systems store data separately from the metadata. Whereas the data is stored in the archive storage, the metadata remains in the DMS or in the application. Metadata can describe characteristics of the data, such as name, size, user, storage location, or the context of the data, such as retention times, access rights, access profiles or full-text index. The metadata also includes a link to the associated data that is stored in archive storage. Ultimately, the loss of the metadata can have just as disastrous an effect on an operation as the loss of the actual data itself.

Therefore, it is obvious that data and metadata should be protected together. The easiest way to implement this is by combining or encapsulating the data and the metadata, storing it together and then backing it up on a regular basis. Jeff Rothenberg presents a related approach in his report published in 1999 titled 'Avoiding Technological Quicksand: Find-ing a Viable Technical Foundation for Digital Preservation.' Rothenberg suggests storing data and metadata together as an encapsulated object that is technology-independent. The metadata is extended through the addition of information and instructions, thereby enabling the data to be read with any system and at any time. An emulation of the original software based on standards allows an interpretation and representation of the data based on the metadata.

Based on the encapsulation of data and metadata and the emulation, it would even be possible to read and interpret archived data many years or decades down the line – irrespective of which technology was used. This approach would therefore reduce the risk of obsolescence of systems and components (Section 8.2.3) and is therefore a valuable concept for archiving.

One weakness in this approach is that standards laying out the format of encapsulated data and metadata as well as the interfaces of the emulation have not yet caught on. Vendors generally have a difficult time agreeing on standards that are so far off in the future. Standardised interfaces, such as Java Content Repository (JCR, Section 8.4.2) and eXtensible Access Method (XAM, Section 8.4.4.), are suitable for supporting this approach as they support object-based data models that already contain metadata.

8.3.10   Selection of components and vendors

As archiving involves a long-term undertaking, it is especially important to be careful in the selection of the components of an archive system and the respective vendors. When a company buys components for its archive system, it automatically enters into a long relationship with the vendor. In addition to the components themselves, the vendor must also be able to provide maintenance, component refresh and other services during the course of the entire operating period.

It is therefore important to be aware of the strategic ambitions of the vendor – for example, on the basis of his commitment to his business and products, references from past and present customers and future plans. A vendor's reputation can also provide some insight into his strategies.

During the planning and implementation of an archive system it can be an advantage if all components come from one source, thus from one vendor. A careful choice of the right vendor reduces the risks involved.

Another option for reducing the risks associated with archiving (Section 8.2.5) consists of outsourcing IT operations or the archive system. This way most of the risk is transferred to the outsourcing vendor.

8.4   INTERFACES IN ARCHIVE SYSTEMS

In this section we add standardised interfaces to extend the reference architecture for archive systems (Section 8.1.4). The extended reference architecture recommends three basic interfaces (Figure 8.10): the interface between application and DMS (1), the

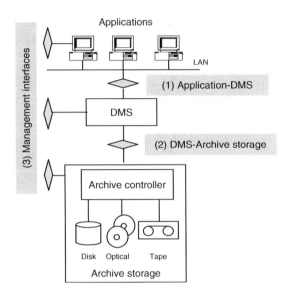

Figure 8.10 Three basic interfaces are located between the tiers of an archive system.

interface between DMS and archive storage (2) and the management interfaces of the archive system components (3).

Standardised and open interfaces form the basis for compatibility between the components from different vendors. In practice, due to the lack of standardisation, protocol converters are often necessary for integrating the components of different vendors (Figure 8.11). However, these protocol converters add to the complexity of an overall solution and make it more difficult to address essential requirements such as regulatory compliance and data protection. One positive exception is the standardised protocols such as Digital Imaging and Communications in Medicine (DICOM) and Health Level 7 (HL7) that are used as interface between applications and DMS in the medical-technical environment (Section 8.5.4).

The components of an archive system should therefore support standardised interfaces. Open standards promote the interchangeability of components and simplify the migration of data between them. This is especially important due to the long life of archive systems. Standardised interfaces also enhance adaptability and help in the integration of existing archive systems when companies are merged. Alternatively, they help to separate the corresponding archived data when parts of companies are split off (Section 8.2.6).

Interfaces are used to access and store data and are usually located between two or more components of the archive system. They virtualise data access because the accessing component usually does not know where the data is located. It is therefore important that interfaces are robust and changes made during the course of time do not challenge the reverse compatibility. This is the only way to guarantee that access to the data will still

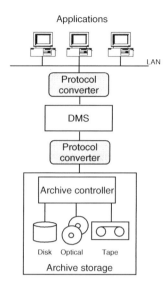

Figure 8.11 Due to the lack of standardised interfaces, protocol converters are required that are often installed as separate components. They are used at the interface between application and DMS as well as between DMS and archive storage.

be possible many years down the line. Besides being easy to implement, the interfaces in an archive system must also satisfy special archiving requirements. These include guaranteeing regulatory compliance, such as the requirements for the unalterability, integrity and security of the data (Section 8.3.2), and fulfilling operational requirements, such as throughput, access times and continuous operation (Section 8.3.6).

In the following discussion we will take a closer look at the three interfaces between application and DMS (Sections 8.4.1 and 8.4.2), DMS and archive storage (Sections 8.4.3 and 8.4.4) and the management interfaces (Section 8.4.5). We will also describe the interface that is used between multiple DMS systems (Section 8.4.6). We also present the actual status of the standardisation for all the interfaces and end by showing how the reference architecture for archive systems is extended through the addition of standardised interfaces (Section 8.4.7).

8.4.1 Interface between application and DMS

An operational data interface over which an application can communicate with the DMS is located between the two. Write and read operations are basic functions of this interface. Read access should include functionality that informs the user when access to data is taking longer than usual. This increases the transparency of data access if the data must be read from a slower medium within the storage hierarchy of the archive storage. The

DMS must therefore pass on certain information from the interface between DMS and archive storage to the interface between the DMS and the application.

Another important function of this interface is the exchange of metadata, such as the characteristics of data objects from the view of the application. This metadata is stored in the DMS and used later for data queries and searches. The DMS can also use the metadata to link up with the data of other applications.

A further important functionality of this interface concerns search functions. The search criteria are usually based on metadata – such as data characteristics and a full-text index – that is stored in the DMS. The search function should be embedded in the application to provide transparency for the user while the metadata stored in the DMS is being utilised.

This interface should also enable an application to forward retention times and rules for data to the DMS. For example, rules for retention times can be event-based with the occurrence of an event signalling either a shortening or a lengthening of the data retention time.

Deletion functions ensure that the deletion of data within an archive system is executed synchronously throughout all layers of the archive system (Section 8.3.5). This means that the data and the metadata in the application, in the DMS and in archive storage are deleted simultaneously. Deletion functions must of course satisfy the requirements for regulatory compliance, meaning that data and metadata cannot be deleted during the retention period.

Import and export functions that enable data to be extracted from or inserted into a DMS are necessary for the exchangeability of data with other archive systems. These functions enable a second application to read the data of a first application, essentially allowing applications and a DMS to be interchanged transparently.

Data security functions (authentication of users, access control and encryption of data and sessions) constitute another group of functions of this interface. This group also includes functions that allow audit trails to be generated and to record all data access (Section 8.3.4).

There are currently only a few archive solutions in which the application and the DMS support a common standardised interface, thereby ensuring the compatibility of both components. The medical field with its standards DICOM and HL7 is one example (Section 8.5.4). In other fields, protocol converters (Figure 8.11) are used or an application and the DMS are individually adapted by the vendor. In the next section we use the Java Content Repository (JCR) to present an approach for standardising the interface between application and DMS.

8.4.2 Java Content Repository (JCR)

The Java Content Repository (JCR) specifies the interface between an application and the DMS. JCR is an interface specification in the Java programming language.

JCR enables applications to access the DMS systems of different vendors. This largely reduces the dependency of applications on the DMS and means that they no longer have to be adapted to specific DMS systems. The expansion of an archive system through the addition of one or more DMS systems is then relatively simple, because the interface between application and DMS is not dependent on the DMS architecture. An existing application which supports JCR is therefore also able to communicate with multiple DMS systems simultaneously, which increases the scalability of archive systems and facilitates data exchange. Lastly, JCR also allows different DMS systems to be coupled together in a federated manner. It also facilitates the substitution of a DMS because, for example, an application can import data from one DMS and export it to another.

The key goal of this interface standard is to simplify the implementation of applications. At the same time, complex functions also have to be mapped that allow the data model of the DMS to be decoupled from the application. Both requirements can be met when JCR is implemented in two stages. During the first implementation stage, read, search and export operations are implemented in the application. This enables applications to search and read data in one or more DMS systems. The second stage contains write and import operations that allow an application to write and import data to a DMS. Both stages also take into account aspects such as data security (Section 8.3.2). Access control mechanisms are also available; data access is session-orientated.

The JCR object-based data model supports the use of metadata that is sent from the application together with the data to the DMS. This metadata can also be used later for search purposes as well as for controlling the life cycle of the data. JCR supports popular object types such as Files, WebDAV and XML, as well as database objects, and is therefore easy to integrate into current data models.

Version 1.0 of JCR is based on the standardised Java Specification Request (JSR)-170 that was approved in May 2005. The completion of JCR Version 2.0 which is based on JSR-283 is expected for the second half of 2009. So far JCR is only being supported by very few applications and DMS systems.

8.4.3   Interface between DMS and archive storage

A DMS communicates with archive storage over yet a different interface (Figure 8.10). This interface comprises basic write and read functions used by the DMS to write data into archive storage and to read it in from there. Read access should have functionality that allows the DMS to determine whether data access is taking longer than usual, for example, when the data is stored on a slower storage medium. It can then inform the application that the data must first be read in from a slower medium.

Some functions of this interface are similar to those of the interface between application and DMS. These functions will only be mentioned briefly here as they are covered in more detail in Section 8.4.1.:

- Exchange of metadata between DMS and archive storage
- Search functions based on data and metadata in DMS and archive storage

- Establishment of retention times and rules for regulatory-compliant storage of data in archive storage
- Deletion functions that allow explicit deletion of data in archive storage upon expiration of the retention period
- Import and export functions that allow the importing and exporting of data from archive storage
- Data security functions that protect data transfer and access and guarantee the integrity of data

Extended functions are required for the exchange of metadata between DMS and archive storage. This is because the DMS is aware of the context of the data that is suitable for controlling the life cycle and the archive storage has capabilities to control the life cycle (Section 8.3.8) but does not know the context of the data. For example, the DMS is aware of the type of data, how it is to be retained and how frequently the data is accessed over the life cycle. The interface between DMS and archive storage should therefore support functions that allow the DMS to notify archive storage of how the life cycle of the data should be controlled. For example, the DMS can instruct archive storage to store the data on a fast access storage medium for an initial time period and then transfer the data to a more cost-efficient medium later. Or it can instruct archive storage to transfer the data to a faster medium because of an anticipated increase in access frequency.

Due to the lack of standardisation, vendors of DMS and archive storage systems implement proprietary interfaces here. The use of protocol converters between DMS and archive storage becomes necessary as a result (Figure 8.11). These protocol converters are usually implemented into the DMS.

In a small number of cases the protocol converter is provided by archive storage – for example, as modified NAS gateways. A NAS gateway generally provides a file system protocol – such as NFS and CIFS – for communication between the DMS or the application and the archive storage. In contrast to conventional NAS gateways (Section 4.2.2), the modified NAS gateway for archive systems prevents data from being deleted or altered once it is stored.

These file system-orientated protocols were not created for archiving and therefore do not implement some of the important archiving functions. In particular, the protocols lack the functions that permit flexibility in establishing the length of retention times and life cycles for data, while at the same time guaranteeing regulatory compliance. The security of data access and assurance of data integrity are only reflected in a rudimentary form in these file system-orientated protocols. Due to the lack of standardised archiving interfaces, these protocols are used for archiving when they should not be.

The Storage Networking Industry Association (SNIA) has taken on this challenge and has worked out an interface standard between DMS and archive storage called eXtensible Access Method (XAM). Multiple DMS and archive storage system vendors have been involved in the creation and finalisation of the XAM standard. Version 1.0 of the XAM standard was approved by SNIA in July 2008. In contrast to the file system-orientated protocols discussed above, the range of functions offered by XAM is specifically tailored to the matters of archiving. XAM will be covered in more detail in the following section.

8.4.4   eXtensible Access Method (XAM)

The XAM specifies an open interface for communication between DMS and archive storage, with the aim of eliminating the tight coupling between the two. As a result, the DMS runs independently of the archive storage and therefore is able to use the archive storage systems of different vendors.

XAM specifies an object-based data model for data and metadata along with operations that can be applied to both data and metadata alike. XAM-specific metadata contains information about data and its life cycle. The encapsulation of data and metadata also permits storage of the data on a component-independent basis (Section 8.3.9).

For example, data-orientated metadata comprises annotations about the data generated by a DMS or an application. This can also include information about such things as data format, readability, indices and other characteristics. Data-orientated metadata therefore usually establishes a connection to the content of the data and to its context.

Life cycle-orientated metadata contains information about retention periods, operational requirements (access times, availability), regulatory-compliance requirements and controlling the life cycles for data. For example, the latter information includes instructions on whether and when data should be transferred to a different medium or to a different tier of storage. The archive storage is able to interpret the life cycle-orientated metadata and controls the data storage location in accordance with this metadata. XAM thus enables a DMS to send information about the life cycle of data objects to archive storage where this is then executed. This provides for cost-efficient data management (Section 8.3.8).

In XAM all data objects receive a globally unique name. This allows multiple DMS systems and multiple archive storage systems to coexist within one archive system. It also promotes the scalability of an archive system and helps to merge the data from different archive storage systems into a centralised archive storage – which, for example, becomes necessary when company mergers take place.

XAM simplifies the interchangeability of DMS and archive storage, even when they have been in operation for a long time and are full with archived data. For example, when a DMS stores data and metadata through the XAM interface into an archive storage system, a different DMS is then able to read and interpret this data again. Equally, XAM offers a starting point for the replacement of an archive storage system because, with XAM, data can be read out along with the metadata from an old archive storage system and migrated into a new one by the DMS.

Another advantage of XAM is that the DMS is not dependent on the storage technology used or on the storage location of the data and metadata within archive storage. This simplifies the implementation and maintenance of the DMS as no consideration has to be taken of the actual hardware in the archive storage system. A DMS vendor therefore no longer has to worry about developing and maintaining device drivers specifically for archive storage systems.

SNIA approved the first version of the XAM standard in July 2008. The standard includes a reference architecture and a Software Development Kit (SDK). In addition, XAM programming interfaces (application programming interface, APIs) in Java and C have been released. As with JCR, it still remains to be seen whether or how quickly XAM will succeed in the market.

8.4.5 Management interfaces

The management interfaces of archive system components are used for configuration, administration, monitoring and reporting. The management interfaces for archive systems must in addition support standard activities such as the analysis and handling of faults, performance and trend analyses and the generation of reports on configuration changes (Chapter 10). These interfaces must also satisfy regulatory requirements.

Archive systems require functions to produce complete audit trails for access to data and metadata (Section 8.3.4). This equally includes access to applications and DMS as well as access to the management interface. The availability of audit trails for the management interfaces is partly required due to regulatory stipulations.

The functions of the management interface must also guarantee regulatory compliance. It should not be possible for any changes or deletions to be made to data or metadata – neither in archive storage nor in the DMS nor in the application.

No specific standards currently exist for the management of archive systems. The vendors of components for archive systems implement general standards such as SNMP and SMI-S (Chapter 10). However, they must ensure that their implementations comply with legal requirements.

8.4.6 Interface between DMS systems

Along with the three basic interfaces for archive systems, the interface between multiple DMS systems is also important (Figure 8.12). This type of interface enables data exchange between multiple DMS systems, thereby moving them closer together. A company can operate multiple DMS systems simultaneously while at the same time centralising certain tasks such as search operations. With this integrated approach the user only has to connect to one DMS but can still search in the data of a different DMS and access it (Section 8.5.5).

The exchange of data and metadata is an important function of this interface because it allows the data from different DMS systems to be linked together. A central catalogue might be used to store the metadata from all other DMS systems.

Centralised search functions allow an application to search for data and information in all DMS systems without the need to send a separate search request to each individual DMS.

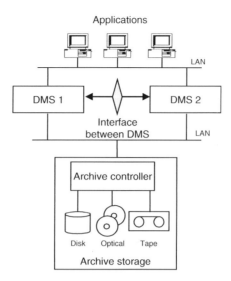

Figure 8.12 An interface between DMS systems enables data to be exchanged and searched between DMS systems of different kind.

Import and export functions also allow this interface to extract data and metadata from one DMS and to insert it into a different DMS. Therefore, data can be transferred from one DMS to another one, thereby allowing a migration of DMS systems.

The Association for Information and Image Management (AIIM) has made a brave effort by introducing Interoperable Enterprise Content Management (iECM) that standardises the interface between DMS systems. iECM specifies the services of the interface, the data model and the components of this type of integrated architecture that comprises multiple DMS systems. The future will decide whether iECM has a chance of success in the market.

8.4.7 Standardised interfaces for archive systems

Based on the considerations about interfaces in archive systems and the advantages of open standards discussed earlier, it is appropriate to apply standardised interfaces to the reference architectures for archive systems (Section 8.1.4). Figure 8.13 shows an integrated archive system with the interface standards introduced in this chapter: JCR, XAM and iECM.

The extended reference architecture illustrated in Figure 8.13 is an example of what to expect in the future. There is no doubt that open standards between components will help to ensure that the lifetime of archive systems can be extended for many decades. It is also obvious that standardised interfaces at the locations shown in Figure 8.13 are required

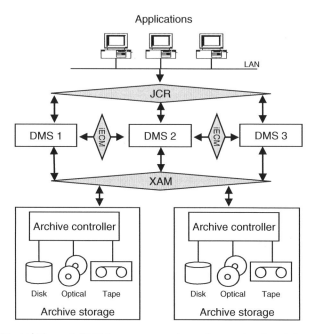

Figure 8.13 JCR, XAM and iECM are examples of standardised interfaces in archive systems. Standardised interfaces facilitate the replacement of components, enable data migration and provide long-term stability. This is very important due to the long life span of archive systems.

when they offer the scope of functions discussed in this chapter. The only question is whether the three standards JCR, XAM and iECM will gain acceptance in the market or whether it will be other standards that succeed. In any case, the recommendation to those planning or using an archive system is to check with vendors about their strategies for implementing standardised components in their archive systems.

8.5 ARCHIVE SOLUTIONS

An archive solution is interpreted as the implementation of the reference architecture (Section 8.1.4) for the archiving of data from specific data sources such as email or Enterprise Resource Planning (ERP). As the standards for archive systems are still at the development stage (Section 8.4.7), adjustments to the reference architecture are required.

In the section below we start by discussing archive solutions for the following data sources and environments: emails (Section 8.5.1), files (Section 8.5.2), ERP (Section 8.5.3) and hospitals (Section 8.5.4). We end by explaining centralised archive solutions that enable the integration of several data sources, DMS and archive storage systems (Section 8.5.5).

8.5.1   Archiving of emails

Today's business community could not survive without emails. Emails often contain business-relevant information, such as quotations, invoices and contracts, which for legal reasons must be archived (Section 8.2.2). The currently popular email systems such as IBM Lotus Domino and Microsoft Exchange must use archive systems to achieve the regulatory compliance required for the long-term storage of business-relevant information. The constant growth in data quantities in email systems is another motivation for archiving emails. Archiving old emails reduces the amount of data in the email server, and, consequently, the administration required. It also reduces the related protective efforts needed, such as data backup.

Conventional archiving solutions for emails transfer the data (emails and file attachments) from the storage area of an email server to a DMS or directly into archive storage. Email systems such as IBM Lotus Domino and Microsoft Exchange currently only have proprietary archiving interfaces, and, therefore, additional protocol converters (Figure 8.11) have to be used between the email server and the DMS and archive storage. Before we discuss two practical approaches to email archiving, we will first provide a short insight into the specific characteristics of email archiving.

Characteristics of email archiving

There are two basic concepts for email archiving: one is mailbox management and the other one is journaling. With mailbox management, emails are archived from the user's mailbox based on criterias. For example, one criterion might be the age of the email; another one might be the folder where the email is stored. The user can also specify what happens to the original email; in many cases the email is replaced by a reference that is also called stub-mail. This stub-mail is used for subsequent access to and searches for the archived email. The advantage of mailbox management is that it reduces the storage space needed in an email server. The emails and attachments are moved to an archive system and only a small stub-mail then resides in the email server. It also gives the user full control and can be combined with searches using full-text indices initiated from the mail-client. The fact that the user has full control over the archiving process is a disadvantage because it does not meet regulatory compliance objectives.

Journaling is based on the standard functions of email systems such as Lotus Domino and Exchange, where the email server creates a copy of each incoming and outgoing email in a journal. The journal is archived and consequently all emails are archived. This is an advantage because it ensures regulatory compliance. A separate search and discovery program is required to access the emails archived by the journaling function. With this program an administrator or an auditor can search for emails based on full-text indices in one or more mailboxes. This makes it easier for legal audits to be carried out. The disadvantage of the journal archive concept is that all emails are archived, including those that might not need to be archived. Furthermore, journaling does not reduce the storage space in the email server because all emails are copied to the archive. Other techniques in which the email data stream is intercepted and copied to an archive are available for email systems that do not support journaling.

Mailbox management and journaling can be combined to provide the advantages of both methods. With mailbox management the storage space in the mail server can be reduced while a user is given access to archived emails. With journaling all emails are archived, which complies with regulatory requirements. Of course, the combination of both methods requires more storage capacity in the archive system because emails and attachments might be archived twice. The single instance storage function can help to reduce the usage of storage capacity.

In many business sectors all emails and attachments are simply automatically archived using the journaling function – for instance, when the legal penalties for incomplete email archiving are especially high. The Sarbanes Oxley Act in the United States is an example. However, it is often a good idea to give some real thought as to which emails need to be archived and which do not. Business-relevant emails must be retained in compliance with regulatory requirements after they have been sent or received. However, today many spam emails as well as private emails with large attachments that are not business-relevant are also being sent through the system. Therefore, an automatic classification and analysis process that sorts out the relevant emails for archiving is desirable. There is still some need for innovation in this area.

Users tend to forward emails to their colleagues, thereby generating multiple identical copies of attachments that eat up storage space. Some archive solutions provide a single instance storage function that reduces the storage requirement for the archiving of attachments. The single instance storage function is equivalent to some data-deduplication approaches. This function detects when identical copies of an email body or file attachments exist and stores only one copy in archive storage. If other identical copies of the email body or file attachments are selected for archiving, this function detects that a copy already exists and generates only a reference to it in archive storage. The single instance storage function also works when multiple email servers exist because it is usually carried out by the DMS or the email archiving program.

Users of email systems need efficient search functions to find and access archived emails and file attachments. From a user perspective the search function should be implemented in the email client, so all searches in the archive are transparent. The search should be based on a full-text index, which enables searches to be carried out efficiently.

For auditing purposes it might be necessary to search for emails in multiple mailboxes or even in all mailboxes and to analyse the content. This process is also called discovery. The discovery function is usually implemented as a separate application program connected to the DMS and has tight access control. The search and discovery program allows mailbox-wide searches based on full-text indices and also enables an analysis of the content. It also allows selected emails to be extracted from the archive in a specific format for auditing purposes.

Among other things, regulatory compliance demands that archived emails cannot be altered or deleted during the specified retention period. This applies to the stub-mail in the email system as well as to the data and metadata stored in the DMS and archive storage. Regulatory compliance also requires that data is archived in its entirety. Mailbox management manifests some weaknesses in this regard because the stub-mail can usually be deleted by the mailbox user. Also, because the emails are selected for archiving based on specific criteria, the user may have the opportunity to delete or alter an email before it is archived. Therefore, only journaling offers regulatory compliance because it decouples the mailbox from the archive system. For achieving regulatory compliance additional organisational processes – for example, authentication methods and Access Control Lists (ACLs) (Section 8.3.2) – must be established for the email system and the archive.

In the following we describe two basic archive solutions for email archiving that are most often implemented: archive solutions without DMS (solution 1) and archive solutions with DMS (solution 2). Both solutions are applicable for mailbox management and journaling.

Solution 1 – Archive solution without DMS

In the first solution an email archiving program captures the emails from the email system and archives them directly into archive storage (Figure 8.14). The email archiving program

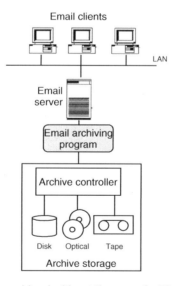

Figure 8.14 When emails are archived without the use of a DMS, an email archiving program fetches the data from an email server and stores it directly into archive storage. The email archiving program may also index the emails and attachments and store the indices in a local repository. Email data cannot be used for business process management and is not classified automatically.

therefore mediates between the incompatible interfaces of the email system and archive storage and acts as a protocol converter. The email archiving program may offer additional functions, such as indexing of emails and attachments, search and discovery functions and single instance storage.

The main advantage of this solution is its simplicity. The disadvantage of this solution is that the archived email data cannot be used for business process management which makes email archiving a stand-alone system. Also no automatic classification of emails is provided.

Solution 2 – Archive solution with DMS

The second solution for archiving emails uses a DMS to provide business process management and an automated classification of emails (Figure 8.15). A protocol converter captures the emails from the email server and transfers them into a DMS that indexes the email data and stores it in archive storage. The indices – for example, full-text indices – enable efficient search and discovery. The DMS also allows email data to be mapped to business processes – for example, by linking emails to the data of other business systems connected to the DMS (such as ERP systems). Therefore, complete business processes can be sampled, starting with email data related to contract negotiations, continuing with contracts and invoices created in an Enterprise Resource Planning (ERP) system

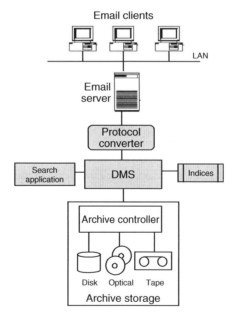

Figure 8.15   When emails are archived using a DMS, a protocol converter captures the email data from an email server and transfers it to a DMS. The DMS links the emails to data of other business processes and stores it in the archive storage.

all the way up to complaints related to the trade business from customer relationship management (CRM) systems.

The main advantage of this solution is that the DMS allows archived emails to be brought into a business context. Thus email archiving is no longer a stand-alone system. This solution is also better suited to guaranteeing regulatory compliance, because the DMS decouples the email system from the archive storage. This solution is therefore often used with journaling, because journaling enables regulatory compliance. It is also a more scalable solution because it enables a multitude of applications – for example, email, ERP and CRM systems – to be integrated into a single central archive (Section 8.5.5). One disadvantage is the complexity of the solution in that it requires a tight integration of DMS with email systems and archive storage. Due to the lack of standardised interfaces between DMS and archive storage, DMS systems only support a small proportion of the archive storage systems available in the market today. With XAM (Section 8.4.4) this might change.

Comparison of the two solutions

The first solution does not include a DMS and is therefore simpler and less expensive to implement than the one that uses a DMS. This solution does not allow emails to be linked to business processes. It is used mainly for mailbox management with the primary objective of reducing storage capacity in an email system.

The second solution uses a DMS and therefore allows email data to be linked to business processes. It essentially integrates email archiving into the overall context of a business and is more scalable. This solution also enables regulatory compliance because the DMS decouples the email system from the archive storage. On the downside this approach is more complex to implement and to maintain. This approach is used mainly in large environments where different types of data are to be archived and linked together.

8.5.2 Archiving of files

Companies use files to store information such as contracts, correspondence letters, product specifications, drawings and marketing presentations. Many types of files generated in companies and enterprises are therefore subject to legal requirements that make it necessary for files to be archived (Section 8.2.2). The enormous data growth in file systems and the resulting effort required for management and backup are other important reasons for archiving files if their content will not change again.

Files are normally stored in a file system that is located either locally on an applications computer or remotely over the network (NAS storage, Chapter 4). The interface between application and file system implements standards such as NFS and CIFS. Before we explain three different file archiving solutions, we will explore some special characteristics related to file archiving.

Characteristics of file archiving

Selecting which files to archive is a real challenge as only those files that will not change again should be archived. Therefore, file systems provide metadata that indicates the date and time of the last read and write operation. A protocol converter between applications computer and archive can use this file system metadata to decide which files need to be archived based on age, context and file type. For example, a protocol converter can analyse the last write and read access date, directory and file names as well as file types in order to select files for archiving based on underlying rules. An archiving rule could state, for example, that only PDF and JPG files that have not changed in four weeks are to be archived. Another rule could stipulate that video films (MPEG) and music files (MP3) will not be archived in principle as they do not contain any business-related content.

With the archiving of files, usually only one version of a file – the last and final version – is archived. In most cases, files should not be archived until a business trans-action or a project has been completed and the corresponding files will not be changed again. Sometimes the archive solution for files will require multiple versions of a file to be managed. The implementation of this requirement is normally subject to the functionality of the application and the DMS.

Among other things, regulatory compliance requires that data cannot be altered or deleted before expiration of the specified retention period. The deletion of archived data in a file system should only be an option if access to the archived data is still possible afterwards. The same applies to references to archived files (such as stub-files) because in many implementations the deletion of a stub-file destroys the link to the archived file and makes it impossible to access it. The deletion of files or stub-files in a local file system can be prevented through organisational measures such as authentication and ACLs.

The efficiency of the search functions is an important distinguishing feature of archiving solutions for files. In addition to file content, the search criteria includes meta information such as file name, directory name, file size, date of creation, last read and write access or a full-text index. In describing the different solutions available, we will consider searches for archived files in detail.

In the following we present three alternative solutions for file archiving that differ from one another in terms of complexity and their guarantee of regulatory compliance. The first two solutions do not deploy a DMS. With the first one, the file system interface is integrated into the archive storage system (solution 1); with the second one, a protocol converter operates the file system interface for archiving (solution 2). The third alternative utilises the services of a DMS in conjunction with a protocol converter (solution 3).

Solution 1 – Archive storage with file system interface

Some archive storage systems make their storage capacity available through a file system interface over which applications (network clients) can send their files directly into archive storage (Figure 8.16). This means that standardised file systems such as NFS and CIFS form the interface here between applications and archive storage. The archive storage

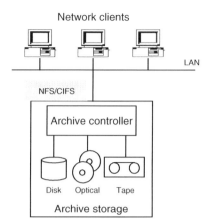

Figure 8.16 Archive storage makes its storage capacity available over a modified file system interface. Applications store the files in the archive storage system in the same way as on a file server, the difference being that the files cannot be altered or prematurely deleted once they are stored in the archive storage.

system functions as a file server and ensures that once data has been stored it cannot be changed again or deleted before expiration of the retention period, thereby guaranteeing regulatory compliance.

Standard file system interfaces do not offer any archiving-specific functions. As a result, in these archive solutions the existing functions of the file system tend to be misused for the exchange of archiving-specific metadata (Section 8.4.3). For example, the file attributes 'last access date' or 'read only' are used in order to specify the retention period of a file and to prevent deletion. The application in this case must know exactly which file attributes will be "mis"used by the underlying archive storage system to determine archiving related metadata such as the required retention period. Thus some level of integration of the application with the archive storage is required.

The search for files can be based on the standard functions included in current operating systems. Many operating systems provide tools for a simple file search, such as the Unix commands 'find' and 'grep' and the search functions in Microsoft Windows Explorer. These functions allow searches based on attributes and content of the files and are partially based on the metadata of the file systems, such as file name, directory, size and date of last change. If the operating system functions are not adequate, then the application has to store additional metadata, such as a full-text index. A separate search tool like Google Desktop is another inexpensive option for implementing efficient searches.

The main advantage of this solution is its simplicity. Many applications support standardised file system interfaces and can therefore use archive storage with only small adjustments. Furthermore, the application is independent of the archive storage because a standardised interface decouples the application from the storage technology. This means that the archive storage system can easily be replaced by another system supporting the same interface. The search for archived files can easily be implemented with the help of file systems tools or additional programs. Regulatory compliance can be guaranteed with this solution as the data is stored directly into archive storage and cannot be altered or deleted there until the retention period has expired.

The main disadvantage of this solution is that the file system protocols that are used were not designed for archiving. There is also a problem with the integration of certain applications as it is not possible for data to be added to an existing file in the file system. This means that archiving-specific functions have to be implemented in the application to adopt the proprietary extensions of the file system interface. This produces dependencies between the applications and archive storage. Another disadvantage is that the files cannot be linked with data from other sources and business processes.

Solution 2 – Archive storage using protocol converters

With the second archive solution for files, the archive storage offers no file system interface (Figure 8.17). Instead the applications (network clients) store their files in local or remote file systems and an additional protocol converter automatically transfers the files from the respective file system into archive storage. Depending on the implementation, the protocol converter may copy the file to archive storage, in which case a copy of the file remains in the file system. The protocol converter may also remove the file from the file system and optionally replace it with a reference (stub-file) in the file system. The selection of files

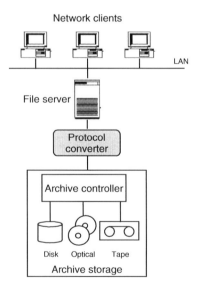

Figure 8.17   Files are archived by a protocol converter that captures the files from a file system and stores them directly into archive storage. The protocol converter may leave a reference (stub-file) in the file system allowing transparent access. The archived data is neither indexed nor linked to other data.

to be transferred to the archive storage can be based on criteria such as the age of the file, the last access date and the folder location of the file. The archive storage prevents any deletion or alteration of data during the retention period.

There are two implementations of the protocol converter that we are going to explain: one is the archiving client and the other one is the Hierarchical Storage Management (HSM) client of a network backup system (Chapter 7). Both implementations are similar in the way in which they select files for archiving; however, they function very differently when accessing archived files from a user perspective.

The HSM client can be configured to leave a reference – also called stub-file – which is serving as a link to the archived file in the file system after the file has been archived. This frees up storage space in the file system while the user or application continues to have transparent access to the archived file via the stub-file. When the user or the application accesses a stub-file, the HSM client intercepts this access, transfers the file content from archive storage into the file system and then finally grants access to the application. This enables the application to work with the archived files without any further modifications.

This transparency during access to archived files is an important advantage of HSM clients. A disadvantage of this implementation is that the stub-file in the file system can be deleted – in which case an application can no longer access the file in archive storage. The file can also be changed in the file system with the HSM client usually replacing the previous version in the archive storage. This is in conflict with regulatory compliance and can be prevented through additional organisational measures, such as authentication and ACLs in the file server. Another option is for the HSM client to implement methods that ensure regulatory compliance. For example, to prevent changes to archived files, the HSM client does not replace a version of a file once it has been archived. Alternatively, the HSM client in conjunction with archive storage can implement versioning whereby all versions of a file are retained. To prevent deletions, the HSM client can recreate a deleted stub-file as long as the retention period of the associated file has not expired.

Searches for files that have been archived by the HSM client can be implemented through standard file system functions and utilise the metadata stored with the stub-file. However, if the stub-file is opened, e.g., when the 'grep' command is used, the HSM client retrieves the file from the archive storage. Depending on the storage medium, this may take some time. Therefore, bulk searches based on the content for archived files are not recommended.

The HSM client usually does not work with all types of file systems because it requires an additional interface to the file system. The implementation of the HSM client with the file system is usually based on the Data Management API (DMAPI), which is not provided by all file systems.

In contrast to the HSM client implementation, an archiving client does not establish a stub-file to the archived file in the file system during the archiving process. The archiving client can be configured either to leave a copy of the archived file in the file system or to remove the archived file from the file system. It manages the reference to the file in a separate metadata structure. Access to archived files is therefore only handled explicitly through the archiving client. When a file is removed from the file system the associated storage space is freed up; however, access to the archived files is not transparent to the user. This is a key disadvantage of archiving clients compared to HSM clients. Furthermore, searches for files require using the archiving client and are usually limited to file attributes, because the archiving client usually does not perform full-text indexing. On the positive side, the use of an archiving client makes it easier to prevent the deletion of archived data and thus enables regulatory compliance as no stub-file pointing to the archived files can be deleted from the file system.

The main advantage of this kind of archiving solution, which includes the two implementations, is again its simplicity. It is used when the archive storage does not offer its own file system interface. Another plus is its easy integration into existing network backup systems as these often have their own clients for archiving and HSM (Section 7.2). The key disadvantages are the limited search capability offered and the fact that files cannot be linked to other data for business process management because there is no DMS involved.

Solution 3 – Archive solutions with DMS

In comparison with the two solutions discussed above, the third archiving solution for files additionally uses a DMS (Figure 8.18) and therefore helps to improve the search

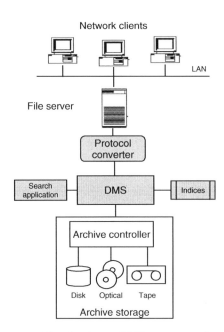

Figure 8.18   When files are archived using a DMS, a protocol converter captures the data from a file system and transfers it to a DMS. The DMS indexes and links the data and then stores it in the archive storage.

capabilities and guarantee regulatory compliance. It also provides the basic foundation to allow business process management for archived files. In practice, a protocol converter is frequently needed between the file system and the DMS. The protocol converter captures the files from the file system, transfers them to the DMS and, depending on the implementation, leaves a stub-file in the file system. In many implementations this reference is a hyperlink that only provides semi-transparent access to the user or the application. The DMS indexes the files and may be configured to produce a full-text index for known and unencrypted file types in order to provide full-text search capabilities. It can also be configured to link files with data from other sources, such as an ERP system, thereby enabling business process management. The files are then stored by the DMS into the archive storage that then prevents any deletion or alteration during the course of the retention period.

This solution is more complex and requires more time and personnel for operation than the other two archiving solutions due to the addition of a DMS. As a result, this solution tends to be mainly considered for large environments with multiple sources of data that must be linked together. This solution is also very appropriate when particularly efficient searches and discoveries are needed as the DMS can index the archived files. Usually an extra search and discovery program is connected to the DMS and in some instances this program can be accessed via a thin client or a web browser from the network clients.

This solution makes it easier to safeguard regulatory compliance because the DMS decouples the file system from the archive storage while providing search and discovery capabilities. Although it is still possible to delete a stub-file in the file system, the data continues to be accessible through the DMS. In addition, files can also be retrieved and extracted from the archive using the search and discovery program – which is important for legal audits. As in the solutions above, certain measures are required to protect the files that are not yet archived.

Comparison of different solutions

Due to the use of standardised interfaces in network file systems, the first solution is simple and inexpensive. However, the standard file system interfaces (NFS, CIFS) were not made for archiving and, as a result, some important functions are missing. In addition, minimal adaptations of the applications to this modified interface are necessary. On the other hand, this solution allows the use of system tools for searching files even based on their content. Regulatory compliance is also safeguarded because data is directly stored into the archive storage. This solution is mainly used when files can be archived directly into a file system and there is no need to link files with other kinds of data for business process management.

The second solution is somewhat more complicated than the first one because it also requires an additional protocol converter in the form of an archive client or an HSM client. The search for data is usually limited to file attributes and cannot always be executed directly from the file system. There is no total guarantee of regulatory compliance because in many practical implementations today the stub-file can be easily deleted. This solution is an inexpensive one and is used when the main goal of file archiving is to free up space in the file systems.

The third solution is the most complex because it also requires a DMS. On the other hand, it offers efficient search and discovery mechanisms and a linking of files with data from other systems. Regulatory compliance is ensured because the DMS decouples the file system from the archive storage. At the same time the application may retain transparent access to the data; in some cases, however, this is only a hyperlink. This solution is best suited for larger environments where business process management and regulatory compliance are the main objectives.

8.5.3 Archiving of ERP systems

Enterprise Resource Planning (ERP) systems such as SAP help companies to use and manage existing resources (capital, equipment and personnel) efficiently. These systems contain large quantities of data, such as invoices, correspondence and logistical information, that must comply with the legal requirements for archiving. The huge growth in data quantities in these systems is another reason for transferring old information from production ERP systems to archive storage and releasing storage space in the production system.

Characteristics of archiving ERP data

ERP systems mostly use commercial databases to store and to link data. In this respect the archiving solutions of ERP systems described here are closely connected to the archiving of databases. Typically ERP systems provide proprietary interfaces for archiving such as ArchiveLink for SAP. Protocol converters establish the connection between the ERP system and the DMS or the archive storage. The ERP system can therefore control the archiving process and seamlessly gain access to the archived data.

ERP systems manage a variety of different data types, ranging from data sets in relational databases all the way to print lists as well as outgoing documents such as letters and incoming documents such as scanned invoices.

The archiving and search methods vary according to data type with the ERP system usually controlling the archiving process. The data sets in a relational database are often archived when they have reached a certain age, whereas incoming and outgoing documents are usually archived immediately. In both cases the ERP system activates the protocol converter that moves the data into the archive. During this ingesting process it produces its own metadata that enables the search in the ERP system.

Data in ERP systems is often subject to strict legal conditions. ERP systems and protocol converters must therefore prevent any deletion of archived data during the prescribed retention period. This applies to archived data as well as to the references remaining in the ERP system. Organisational measures such as authentication and ACLs are essential.

Searches for archived data are usually directed towards business processes. The metadata stored in the ERP system supports these searches. If the search functions of the ERP system are not adequate, they can then be supplemented with a DMS that allows an extensive indexing of the data and permits search and discovery.

In practice, we consider two main solutions for the archiving of ERP data: archiving solutions without DMS (solution 1) and archiving solutions with DMS (solution 2).

Solution 1 – Archiving solution without DMS

The first archiving solution for ERP systems does not deploy a DMS (Figure 8.19). The main purpose of this solution is to relieve the burden on the production database of the ERP system. Little importance is given either to providing extended search capabilities or linking to data from other systems. A protocol converter is again required and connects the ERP system directly to archive storage. It translates the archiving interface of the ERP system into the protocol of the archive storage. The archiving process is usually triggered by the ERP system and the protocol converter usually deposits a reference to the archived data that is used later for access.

This solution is marked by its simplicity. However, its shortcomings are in the search capabilities and the linking with data from other applications and data sources. Because search is the basic foundation for finding data after long periods of time, access to data may be limited due to the lack of a full-text index. This solution may also lack some aspects that enable regulatory compliance because the ERP system is solely responsible for preventing deletions and modifications of the archived data. One would expect an ERP system to safeguard these requirements, but it should be kept in mind that an ERP system is not an archiving system that satisfies regulatory compliance. Thus additional organisational

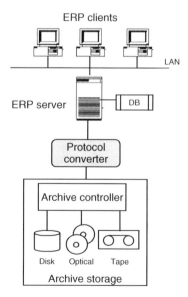

Figure 8.19   When ERP data is archived without the use of a DMS, an ERP system activates a protocol converter that captures the data from the ERP system and stores the data directly into archive storage. The archived data is neither indexed nor linked to other data.

methods may be required. This solution is usually applied when the main objective is to free up storage capacity in the ERP system by transferring old data into the archive.

Solution 2 – Archiving solution with DMS

The second archiving solution for ERP systems additionally uses a DMS (Figure 8.20). The protocol converter in this case works with the DMS and not directly with the archive storage. The DMS can index the data as well as link it to data from other applications before storing it in archive storage. ERP systems often supply proprietary interfaces such as Document Finder from SAP so that the metadata in the DMS can be evaluated together with the metadata in the ERP system during a search.

The key advantage of this solution is the efficient search and discovery capability enabled by the DMS. The DMS can index all the archived data and provide convenient full-text searches. This is not only advantageous for the user of an ERP system but also guarantees regulatory compliance because the DMS decouples the ERP from the archive storage and still provides access to the data via the search and discovery function. Another positive aspect is the possibility of linking data from an ERP system with data from other data sources, such as emails to enable business process management. All in all this is a more complex solution than the first one. However, this aspect is gladly accepted

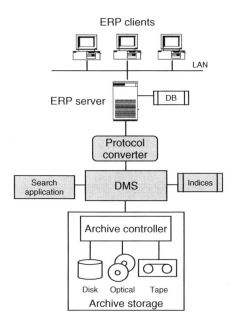

Figure 8.20 When ERP data is archived using a DMS, the protocol converter captures the data from the ERP system and forwards it to the DMS. The DMS indexes and links the data and stores it in the archive storage.

by companies that are already operating archive systems for other applications, such as email, and want to integrate an ERP system into a central archive system (Section 8.5.5). Furthermore, this solution offers flexibility for expansion even if no archiving of emails or other data is being carried out at the time.

Comparison of the solutions

The first solution operates without a DMS and is therefore simpler and less expensive to implement. The search for data is limited because no full-text index exists. This solution also does not allow ERP data to be linked to data from other systems. Additional operational measures that prevent users from deleting references to archived data are required in the ERP system to ensure regulatory compliance. This solution is mostly used when the main goal is to release the burden on the ERP server.

The second solution uses a DMS and thus allows data to be indexed and linked. It also offers an efficient search and discovery function that can partly be integrated with the ERP system. The requirements of regulatory compliance are easier to fulfil because the management of the archived data is the responsibility of the DMS, which decouples the ERP system from the archive storage. This solution is used when efficient search functions and regulatory compliance are the overriding requirements. It is also an expandable solution and can be used in the archiving of other data.

8.5.4 Archiving in hospitals

Hospitals generate large quantities of data that are legally required to be retained for long periods of time. In Germany, for example, X-rays sometimes have to be kept for 30 years or longer. The size of X-ray films and the long retention times required make them an inevitable candidate for storage in an archive system. State-of-the-art diagnostic procedures, such as three-dimensional films taken of the heart during a stress electrocardiogram (ECG), are a joy for every sales person selling archive storage systems.

Characteristics of archiving in hospitals

A Picture Archiving and Communication System (PACS) collects, indexes and archives digital imaging data from modalities such as radiology, nuclear medicine and pathology. To exchange images and data, the PACS usually communicates with the modalities over a standardised interface called Digital Imaging and Communications in Medicine (DICOM). In the reference architecture for archive systems PACS takes on the role of a DMS. A PACS normally also includes a database for metadata derived from the DICOM data, such as image type and size, patient name and date of image, as well as for metadata from other systems, such as a Radiology Information System (RIS).

Hospitals have a great deal of other data and data sources in addition to radiology films. This includes the family history data of patients, test results, medical procedures performed and billing. Hospital Information Systems (HIS) manage this data centrally, thereby using it to support the medical and administrative processes. A HIS generally communicates with the modalities using Health Level 7 (HL 7). Similar to a PACS, a HIS functions as a DMS that indexes and links data.

HIS and PACS store data in an archive storage system. PACS vendors usually test the interaction of their products with the archive storage of different vendors. Certification of an archive storage system by a PACS vendor ensures that the systems are compatible and will function for many years.

Data can be searched over the PACS or HIS. In many practical implementations today PACS and HIS are separate systems, making it difficult to coordinate a search of PACS and HIS data over both systems. Some initial approaches towards integration share a common DMS so that data from PACS and HIS can be linked together and therefore managed centrally.

In the following sections we will be explaining two basic archiving solutions for hospitals. With the first one, individual PACS and HIS archive their data directly into an archive storage system (solution 1). With the second one, a DMS integrates the information from several PACS and HIS and then stores the data in the archive storage (solution 2).

Solution 1 – Archiving solution without DMS

Figure 8.21 shows a typical archiving solution with PACS. The PACS functions like a DMS and the different modalities represent the applications. The communication between

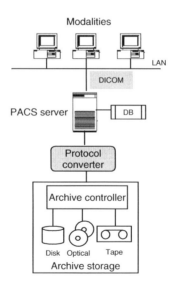

Figure 8.21 With the archiving of PACS data, the PACS functions as a DMS. It indexes and the links the data and stores it in the archive storage. A comparable solution can be implemented for a HIS.

the PACS and the modalities takes place over the standardised DICOM protocol. The communication between the PACS and the archive storage takes place via the interface of the archive storage. In many instances, network file system protocols (Section 4.2) are used for this communication. An archiving solution can equally be implemented with HIS, the difference being that the modalities are other ones and the standardised HL7 protocol is used between the modalities and the HIS.

The key benefit of this PACS solution is that it is easy to implement. Searches can be conducted over the PACS. Due to the lack of integration with HIS, cross-system searches and data linking are not possible.

Solution 2 – Archiving solutions with combined DMS

In the second archiving solution for hospitals PACS and HIS are integrated over a combined DMS (Figure 8.22). This simplifies the search operations and allows data to be linked because a central archiving system now exists for all the data of the hospital.

The communication between PACS and HIS to the applications (modalities) continues unchanged over the standardised interfaces DICOM and HL7. The same interfaces might also be used to connect to the DMS: DICOM for PACS and HL7 for HIS. The DMS stores the metadata in its own database and forwards the data that is being archived

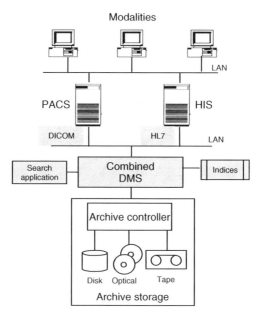

Figure 8.22   The archiving of PACS and HIS data can be integrated over a combined DMS. This DMS links the data from both systems and permits the search for data from both systems based on additional indexes.

to archive storage. The combined DMS can create additional indexes such as a full-text index for textual information and link the data from PACS and HIS together. This enables a comprehensive search across the data from different modalities. In addition, it allows patient records such as X-rays, medical results, historical data and invoices to be all linked together. Despite the integration, PACS and HIS can continue to be operated independently of one another.

The main advantage of this solution is that the data from PACS and HIS can be linked bringing all patient data together. A central search can be undertaken from the DMS for the data of both systems. This integrated archiving solution is more complicated to implement than the first solution presented earlier. Nevertheless, when multiple PACS and HIS are involved, it still pays to consolidate them centrally using a combined DMS.

Comparison of solutions

The first solution is simple and inexpensive. However, it is a separate solution that can be implemented either for PACS or HIS. This solution is mainly found in small specialised practices where only one of the systems is operated.

The second one is more complicated because it requires a DMS. On the plus side it provides an efficient and comprehensive search for the data of both PACS and HIS. It also allows data from a PACS to be linked to data from a HIS and vice versa. This solution is appropriate for medium-sized and large hospitals where, in addition to radiological data, other patient-related data needs to be archived and linked together.

8.5.5   Central archives

The integration of all data sources of a hospital into a central archive system as shown in the previous section can also be used effectively for other business sectors and data sources. The separate solutions for emails (Section 8.5.1), files (Section 8.5.2) and ERP systems (Section 8.5.3) can equally be merged into a central archive. This enables an integral mapping of data to business processes. For example, an email concerning a contract can now be linked to the corresponding delivery notes and invoices from an ERP system. All the data of a company is linked together and thus maps all the information relating to a business process.

Characteristics of central archives

A central archive solution is a very complex entity alone because of its size and long life. The use of standardised interfaces such as JCR, XAM and iECM is essential (Section 8.4.7). This is the only way in which requirements such as the replacement of old components including data migration and scalability can be met.

In practice, there are three obvious approaches to central archiving today. What they all have in common is that they integrate more than one data source and application. A simple approach only establishes one central DMS that supports multiple applications (solution 1). The other two approaches support a number of applications, multiple DMS and multiple archive storage systems and are therefore more scalable. These two approaches differ from one another in the integration of the DMS: depending on the architecture, a federated DMS creates a metadata catalogue for all DMS (solution 2) or the integrated DMS systems communicate directly with one another (solution 3).

Solution 1 – Central archive with one central DMS

The simplest approach to a central archive uses one central DMS and one central archive storage (Figure 8.23). The central DMS takes the data and metadata from the protocol converters of the different applications (data sources), indexes them, stores the metadata in a separate database and then stores the data in the archive storage. The protocol converters are the same as described in the separate solutions (Section 8.5.1 to 8.5.4), although some protocol converters can support multiple applications simultaneously.

The main advantage of this solution is that the data from different sources can be logically linked together and thus enables business process management. This solution

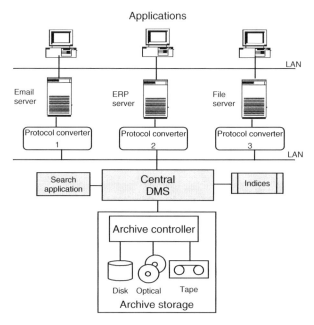

Figure 8.23 With centralised archiving, a central DMS collects data from different sources and stores it in an archive storage system. The DMS additionally links the data from different sources and enables a cross-application search and process management.

also allows an efficient search of the archived data of different applications. On the down side, only one DMS is available, which restricts scalability. This solution is mainly used when multiple different data sources with small amounts of data exist and one DMS is sufficient.

Solution 2 – Central archive with federated DMS

Multiple DMS systems in combination with one or more archive storage systems can be used to increase the scalability of a central archive. One approach to integrating multiple DMS is to use a higher-ranking DMS that bundles the services of the multiple individual DMS (Figure 8.24). The higher-ranking DMS – also called federated DMS – normally provides its own index database in which the indices of all data of the individual DMS is stored (central catalogue). Using these indices, an application can transparently access the information of all DMS systems that are connected to the federated DMS. The individual DMS systems remain hidden from the application as they only communicate with the federated DMS.

With this solution the applications may only have an interface to the federated DMS. This DMS allows access to all other DMS. The federated DMS also allows the data of the other DMS systems to be linked and enables a cross-DMS search to data because it stores a centralised index (centralised catalogue). Theoretically, this solution also permits the migration of data from one DMS to another. The disadvantage of this solution is that

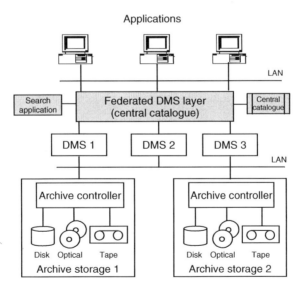

Figure 8.24 With centralised archiving using a federated DMS, this DMS collects data from the other DMS systems and creates a central catalogue. The applications may only communicate with the federated DMS and use it to access data in the other DMS systems.

it is more complex and expensive to implement and maintain. Also, it is not arbitrarily scalable because the federated DMS can only communicate with those DMS systems that support the interface to the federated DMS. Uniform interface standards such as iECM (Section 8.4.6) could be providing this kind of scalability in the future. Solution 2 is therefore an option when a company already has multiple DMS systems in place and there is a need to integrate them.

Solution 3 – Central archive with integrated DMS

An alternative approach to integrating multiple DMS systems requires a standardised interface between the different DMS in order to exchange data and metadata (Figure 8.12). The applications work with one or more DMS. If necessary, the individual DMS systems forward requests to other DMS systems so that from the user's view the cluster of multiple DMS represents one large DMS. A DMS can also link its data with the data of other DMS. We have already looked closely at this approach in Section 8.4.6. It is based on a common interface between DMS systems – such as the standardised interface iECM.

The benefit of this solution is that it is scalable so that any number of DMS systems can communicate with one another. Another advantage is that the applications continue to communicate with the same DMS as before and, therefore, no changes are required to the application. This solution also allows data to be migrated from one DMS to another DMS. However, only those DMS that support the appropriate standardised interface can be integrated. This solution is mainly selected when more than one DMS already exists or a single DMS is no longer adequate.

Comparison of solutions

The first solution is the simplest one and supports different applications simultaneously. The disadvantage is that it is not highly scalable because it is restricted to one DMS. This solution is mainly used when a single DMS is sufficient to cope with the amounts of data being archived.

The second solution allows multiple DMS systems to be federated and enables a cross-DMS search and linking of the data. The applications only communicate with the federated DMS. This solution also allows data from one DMS to be migrated to another one. On the negative side, this solution is very complex and expensive. Furthermore, it is restricted only to those DMS that are supported by the federated DMS. This solution is appropriate when a company already has multiple DMS and there is a need for them to be integrated.

Like solution 2, solution 3 integrates a number of DMS systems so that a cross-DMS search and linking of data is possible. The applications communicate with one or multiple DMS, which, if necessary, forward requests to other DMS systems. In contrast to solution 2, there is no central catalogue. Therefore, each DMS is responsible for the management of its own data. This solution requires an integrative interface between the different DMS systems – something that is still in its infancy. This solution is suitable for enterprises with more than one DMS system providing the interfaces for communication with other DMS systems.

8.6 OPERATIONAL AND ORGANISATIONAL ASPECTS

Archive systems have an elementary and long-term significance for a company. The appropriate operational and organisational measures must therefore be available to support the selection and operation of such systems. This is the only way in which the requirements (Section 8.2) for an archive system derived from basic needs and necessities can be fulfilled.

Archiving must be viewed as a strategy that comprises not only the archive system itself but also other resources, such as staffing, computers, networks, processes for operation and, last but not least, provisions for risk management. The strategy should reflect company-wide guidelines regarding the architecture and the operation of archive systems. The constancy of a strategy is also important: The chosen path should not be changed quickly again as archive data has considerable longevity. A great deal of time and cost can be saved if a strategy is well thought out at the beginning. This is not only true for the company itself but also for the vendor providing the infrastructure required for archiving.

Risk management must be an essential component of the archiving strategy. This is the only way to avoid the risks of today or to reduce their impact (Section 8.2.5). The risk management for archive systems must also test and establish measures and processes such as emergency plans and data migration to new technologies.

Due to the long life of archive data, the use of standardised interfaces is highly recommended. They facilitate the migration of data and systems as well as the exchangeability of the components of archive systems. Furthermore, standardised interfaces that have been widely accepted by a large number of vendors have a longer life span.

It is important to refer once again to protective measures such as access-controlled premises and the regular checking of access authorisation to premises, networks, systems and archived data. This is the only way of fully guaranteeing the regulatory compliance for an entire archive system. In addition, an early detection of errors and a precautionary elimination of possible fault sources are essential for the 'vitality' of an archive system. Internal audits should be undertaken regularly to ensure that all requirements are being fulfilled and irregularities do not come to light until an audit is carried out by external authorised auditors or the tax authorities. This also means that the processes related to regulatory-compliant archiving must be documented and periodically checked and revised because this is one way to perform an in-house check for regulatory compliance.

In our experience, it is extremely critical if an organisational unit of a company that operates an archive system is additionally chartered to perform other non-archiving related work such as backup and other administrative tasks. The daily pressure to cope with IT system failures is often so high that the strategic planning for archiving is literally neglected to a 'punishable' degree. Therefore, medium and large companies should set up a central organisational unit that deals exclusively with archiving and establish company-wide guidelines for it. This should not only relate to a specific kind of data requiring archiving, such as emails, but also include all kinds of data that must be archived. This organisation that is chartered for company-wide archiving will then be able to combine and develop the knowledge and experience within a small group of responsible people. This will also promote the type of constancy that is urgently needed because of the long life span of archiving retention periods.

8.7 SUMMARY AND OUTLOOK

The digital and regulatory-compliant archiving of data in accordance with existing operational and legal requirements has already affected the IT environments of companies in many sectors (such as medicine, finance, industry and production). Certain requirements are specified through laws, standards and other conditions and are partly subject to the control of authorised auditors. Other requirements are defined by users or evolve from certain conditions and needs. One of the biggest challenges that arises in connection with digital archiving is cost and this is something that has to be considered throughout an entire archiving period. The length of an archiving period substantially differentiates archiving from other applications in data processing. Many of the requirements and conditions mentioned above can be addressed through the strategic planning of archive systems and the use of suitable technologies. Archiving solutions today are very specifically tailored to data sources. However, this can change when standardised interfaces become more common in archive systems. For example, an archive solution that can equally be used for different data sources might be available in the future. The use of standardised interfaces can also reduce the risks that arise in conjunction with the introduction of new technologies or with economic risks. This is because standardised interfaces allow an easy replacement of old archive system components and an easy integration of new components.

This chapter began with a definition of digital archiving, the presentation of a reference architecture for archive systems and a delineation between archiving and backup as well as Information Lifecycle Management (ILM). We presented two important reasons for archiving on the basis of legal regulations and the reduction of loads on productive systems. Due to the long life of archive systems, special measures must be taken to assure the durability of these systems. This includes the adaptability and exchangeability of the components as well as the adaptability and scalability of the entire archive system. Lastly, we came to the conclusion that a company should view an archive system as a strategic investment and give it the appropriate attention. For the implementation of archive systems we first made a comparison between disk-based WORM storage, optical WORM media and magnetic WORM tape. This showed that archive storage systems that support multiple storage technologies and Hierarchical Storage Management (HSM) represent a future-proof investment. The discussion also covered the implementation of requirements such as data security, data integrity, proof of regulatory compliance, data deletion, continuous and loss-free operation as well as the use of a storage hierarchy for archived data. Two other important aspects in the implementation of archive systems are the selection of the vendor of the components and the interfaces in the system. With JCR, XAM and iECM we presented three important interface standards for archive systems. Finally, the characteristics of the different archiving solutions for emails, files, ERP data as well as PACS and HIS data were discussed. The chapter ended with an impression of the benefits of central archiving solutions as well as a look at the operational and organisational aspects of archiving.

With this chapter on archiving we have continued our study of basic applications and basic architectures for storage networks. The continuous and loss-free operation of IT systems plays an important role for all basic applications. In the next chapter we use Business Continuity to delve into this subject more deeply.

9

Business Continuity

In the past it was mainly human error and natural catastrophes, such as storms, earthquakes and flooding, that posed a threat to the operation of IT systems. In recent years, however, other but not less dangerous risks have been added to the list. The attacks of September 11, 2001, and the more recent denial-of-service attacks point to the complexity of possible threats to the IT infrastructures of companies. Added factors are the ever growing amount of data, new and often stricter regulatory requirements and increasingly more complex IT environments. The risk of making expensive errors increases in this environment.

Continuous access to data is a deciding factor in the continuing existence or failure of an enterprise. It is therefore essential that business operations be maintained on a continuous and loss-free basis even in crisis situations. This is where appropriately designed storage solutions have a role to play. The considerations, operational requirements and technologies needed all fall under the area of 'business continuity.' Business continuity encompasses the technical as well as the organisational strategies for guaranteeing continuous and loss-free business operations. Within the framework of this book we will only be able to address some key organisational aspects. We will concentrate on the technical considerations of business continuity.

The aim of this chapter is to discuss the special requirements of business continuity and to present a detailed discussion of specific storage technologies and solutions. The strategies and solutions described should help in the understanding and implementation of the requirements for storage technologies that exist in a large-scale business continuity plan. During the discussion we will be referring to basic technologies and solutions that have already been explained in previous chapters.

We start this chapter by motivating the background reasons and the necessity for continuous and loss-free business operations (Section 9.1). In so doing we identify some of

Storage Networks Explained: Basics and Application of Fibre Channel SAN, NAS, iSCSI, InfiniBand and FCoE, Second Edition
U. Troppens R. Erkens W. Müller-Friedt N. Haustein R. Wolafka   © 2009 John Wiley & Sons, Ltd

the risks that can affect IT operations and, by extension, a company's continuous business processes. We use risk management to determine the steps needed to create a business continuity plan. We then explain the different strategies of business continuity (Section 9.2) and introduce parameters for the purpose of comparing these strategies (Section 9.3). Using this as a basis, we discuss a seven-tier model for business continuity that is useful in analysing the economics of different business continuity solutions (Section 9.4). We end the chapter by presenting some selected technical solutions and positioning them within the seven-tier model (Section 9.5).

9.1 GENERAL CONDITIONS

To lead into the subject, we start this section by taking a detailed look at the general conditions of business continuity. After establishing some of the terms (Section 9.1.1) and the target audience of business continuity (Section 9.1.2), we present a classification of potential risks (Section 9.1.3) and IT outages (Section 9.1.4). We then take a look at these IT failures in the context of business processes (Section 9.1.5) and describe the dependencies that exist with the recovery of IT operations (Section 9.1.6). We follow this by extending the web architecture introduced in Section 6.4.2 with the addition of the aspect of business continuity (Section 9.1.7) and then consider the cost optimisation of such solutions (Section 9.1.8). The chapter ends with an explanation of the connection between business continuity and risk management (Section 9.1.9) and a description of how a business continuity plan is created from a design perspective (Section 9.1.10).

9.1.1 Terminology

In the course of this chapter we frequently use the labels 'business continuity program' and 'business continuity plan.' A business continuity program comprises and formulates the entire strategy and long-term vision of a company in its effort to guarantee continuous and loss-free operation. The implementation of such a program usually takes place iteratively and, depending on size and complexity, extends over a longer period of time. This ensures that the implementation has a controlled effect on an existing IT operation and, as a result, its quality is constantly being improved. A business continuity program is therefore a strategic baseline against which actual operating decisions and plans can be orientated.

A business continuity plan, on the other hand, describes in detail which actions and tactics are to be executed during and after a crisis. This type of plan comprises the technologies, personnel and processes needed in order to resume IT operations as quickly as possible after an unplanned outage. This all happens on the basis of detailed, reliable and tested procedures. A business continuity plan also includes all the non-IT-related elements and functions that are necessary for restarting operations. Examples include operating facilities such as offices and call centre telephone systems.

9.1.2 Target audience

Companies involved in areas such as stock broking, airline safety and patient data and Internet companies such as Amazon and eBay are the ones apt to be mentioned in connection with business continuity. However, business continuity is not only an important issue for large companies. Many small and medium-sized firms are now finding it increasingly important to have the ability to run continuous business operations. Today these companies are making deliveries to customers all over the world or are closely integrated into the production processes of large companies, such as car manufacturers, through just-in-time production and contractually fixed short delivery times. Continuous business operations are especially crucial to the existence of companies that market their goods and services exclusively on the Internet.

9.1.3 Classification of risks

Business continuity strategies must incorporate measures for planned and unplanned outages. The literature on business continuity includes endlessly long lists of possible risks to IT systems that can result in unplanned outages. These can be roughly divided into the following categories: human error (user or administrator), faults in components (applications, middleware, operating systems, hardware) and occurrences in the environment of IT systems (Table 9.1).

Table 9.1 The risk factors for IT operations can be divided into the areas of human error, IT components and environment.

Risk factor	Examples
User	Unintentional (and intentional) deletion or tampering of data
Administrators	Unintentional stopping of important applications
Applications and middleware	Program crash, memory leak
Network infrastructure	Interruption to network connections
Operating system	System crash, problem with device drivers
Hardware	Disk failure, failure of power supplies
Operational environment	Power outage, fire, sabotage, plane crash, terrorist attack
Natural catastrophes	Flood, earthquake

Alone between the attacks on the World Trade Centre and the printing of this book in 2009 numerous other events have had an adverse effect on business operations. Component faults, overloading and electricity poles downed by snow have led to widespread power outages in the USA and in Europe. Internet traffic in Asia has been badly affected twice due to damaged underwater cables. Last but not least, there were natural catastrophes in Germany (flooding), Indonesia (Tsunami) and New Orleans, USA (hurricanes).

Added to these unplanned and usually sudden and unexpected outages are the planned and often unavoidable interruptions, such as those due to maintenance work and changes to IT infrastructure. These are necessary so that applications and systems can be maintained at an efficient level and problems can be eliminated. Examples include the replacement of defective or obsolete hardware components, the installation of additional RAM into computers and the rollout of new software versions.

9.1.4 Classification of outages

Ultimately, both planned and unplanned disruptions affect the availability of data and services. To provide a more detailed description of the effects of outages in IT operations, we make a distinction between 'minor outages' and 'major outages'. This distinction helps to provide a clearer differentiation between the requirements for business continuity and position the appropriate technical solutions.

Minor outages, such as the failure of a component within a system (host bus adapter, hard disk) or even the breakdown of a complete system (server, disk subsystem), have only a minor or medium impact on IT operations. Outages of this kind statistically occur somewhat frequently and usually affect only one or a few components within the IT environment. High-availability solutions such as RAID (Section 2.4) and server clustering (Section 6.3.2) offer protection against minor failures by eliminating or reducing their impact.

On the other hand, major outages (catastrophes), such as the total breakdown of a data centre caused by fire, have a far-reaching impact on the IT operation of a company. The entire IT operation of a data centre comes to a halt. Major outages are also put into the category of catastrophes. Disaster recovery solutions like remote mirroring (Section 2.7.2) are aimed at protecting against this type of outage and minimising its effects.

For an IT operation a disaster is not always equated only to the loss or damage of expensive equipment and infrastructure. In this context a disaster is instead each crucial, far-reaching catastrophic event for an IT operation. For example, it can be a catastrophe for a company if business-critical applications crash as a result of the incorporation of software patches. Therefore, a disaster does not necessarily have to be an event that causes the destruction of physical assets but instead is defined through the business-related damage caused by an event.

9.1.5 IT failures in the context of business processes

A failure in IT operations can have an immediate catastrophic impact on a company. In certain business areas this kind of interruption can result very quickly in losses ranging in millions. In the financial sector, in automated production operations and in the telecommunications area losses are calculated at millions of dollars per hour. The amount of a loss increases the longer an IT operation is impaired, making a speedy resumption of operations essential for the survival of a company. Irrespective of the size of a company, even a short outage in IT operations can jeopardise a company's existence. If a company cannot provide its services as contractually agreed due to an IT failure, the claims for compensation from its clients alone can mean financial ruin – for instance, if orders cannot be placed in time in the securities market.

Beyond the direct damages incurred, this kind of IT outage usually also produces indirect longer-term losses. In addition to financial losses, a company can face forfeiture of market share, reduced productivity and competitiveness, damage to its business reputation and a drop in customer loyalty. Repercussions of this kind are more difficult to measure than direct damage.

The effects of outages and technical countermeasures such as high-availability and disaster recovery solutions should always be assessed in a business context. Each failure of a server is of course initially a disruption to IT operations. However, if, as part of a high-availability solution, the affected IT services and applications are automatically taken over after a short period of time by another server, or if less critical applications are being run on the server, this outage is either not important or is of little importance to the course of a company's business.

Every company must take an individual approach towards determining potential risks and the repercussions for its IT operation. Depending on the sector involved, companies are subject to more or less strict legal requirements (Section 8.2.2) so costs for the loss of data and services can vary substantially from company to company. Business continuity solutions are therefore specifically tailored to the needs of each company. For example, it could be more important for a company to resume its IT operation as quickly as possible instead of being concerned about data consistency. However, another company may be able to tolerate a slower resumption of its operation because it requires a maximum level of data consistency and must under all circumstances avoid any loss of data. Yet other companies absolutely require continuous IT operations while at the same time avoiding any loss of data.

9.1.6 Resumption of business processes

As with the evaluation of IT outages, the restart of IT systems after a fault has occurred should be viewed in conjunction with the entire business process involved. The restart time of the business processes is not only dependent on data availability but also comprises the following four phases (Figure 9.1):

- Resumption of data availability
- Resumption of IT infrastructure
- Restoration of operational processes
- Restoration of business processes

A business continuity strategy must include these four phases of resumption in its planning. Delays in the resumption of any one of these phases will affect the overall time required for the restart.

The first thing that has to be restored after a catastrophe is the availability of the data and the components needed to implement this. This is a prerequisite for the subsequent restart phases and is realised, for example, by providing the data on storage systems in a backup data centre. Building on this, all other components in the required IT infrastructure, such as the storage networks, must also be recovered so as to make the data available to other systems and services. The first two phases then enable a resumption of the operational processes of a company. This includes the starting of network services and database servers, which depends on the availability of data and infrastructure. These measures are a necessary condition so that those business processes needed for the normal running of a company can finally be resumed.

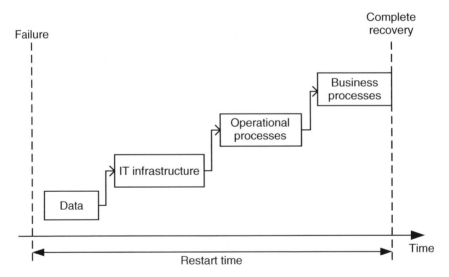

Figure 9.1 The restart time for IT systems until a complete recovery consists of several phases. Delays in any one of the phases will affect the overall time needed for the restart.

The requirements for restarting business processes often can only be achieved through a switch over of operations to a backup data centre. For example, after the attacks of September 11 the US government divided all the banks into three categories, depending on their importance to the US economy, and issued strict guidelines for the resumption of business activities. For their core processes – data, IT and operations – banks in the highest category of importance must have the means to switch over automatically to standby systems that will enable them to adhere to strict requirements for the switch over time of the entire business operation.

9.1.7 Business continuity for the web architecture

For business continuity the entire infrastructure of a company must be considered. This includes the technologies from the different levels of its IT systems (storage system, storage network, servers, networks, operating system, applications) as well as business processes and operating facilities. The web architecture (Section 6.4.2) is a good starting point for identifying all the components that should be protected through the strategies of business continuity (Figure 9.2).

In the planning and implementation of a business continuity plan the entire infrastructure must be considered as a whole. A coordinated configuration of all the components is essential for a reliable overall solution. For example, it is not sufficient to replicate all business-critical data in a backup data centre unless the corresponding servers and applications that can access this data are available there and, if necessary, can resume business operations.

9.1.8 Cost optimisation for business continuity

There are many possibilities for shortening the time needed to restart IT systems. The business continuity strategy of a company must find an economically viable balance between the business value of the data and services for the company and the cost required to protect an operation against the loss of such data and services. It may not be a good idea if the money invested in protecting data would be more than the effects of the loss of this data in a crisis situation.

The huge amount of data and numerous applications in an IT landscape usually have varying degrees of importance to business operations. Consequently, it is economically feasible to implement different types of storage solutions with different business continuity parameters (Section 9.3). These graduated qualities of service for storage are also referred to as storage hierarchies (Section 8.3.8). Users of the storage are charged according to the quality of service selected. Therefore, a best-possible price-performance ratio is established between the business value of the data and the operating costs of the data.

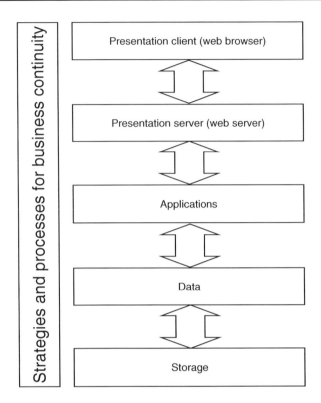

Figure 9.2 The strategies and processes for business continuity must encompass all the levels of an IT system.

It should be noted that the importance of data and, consequently, its value to a company can change over the course of time. These kinds of considerations fall under the area of Information Lifecycle Management (ILM). The requirements of business continuity and ILM must be coordinated with one another so that a cost-optimal overall solution can be achieved.

The optimisation of restart time and cost is marked by two opposing aspects (Figure 9.3): On the one hand, the shorter the restart time in a crisis, the more expensive the storage solution is. On the other hand, the longer a system failure lasts, the greater the financial loss is to a company.

Therefore, the optimisation of the restart time and of the cost curve is an important aspect to be considered in the planning and selection of a business continuity solution. An optimal solution is embedded in the area where the cost curve of the solution intersects with the cost curve of the failure or outage. This area is also referred to as the optimal cost-time window for a business continuity solution.

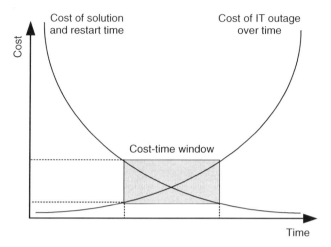

Cost of solution
and restart time

Cost of IT outage
over time

Cost

Cost-time window

Time

Figure 9.3 The optimal business continuity solution is found in the area of the cost-time window. This window shows the cost of a solution in an economic ratio to the costs incurred in the event of an IT outage.

9.1.9 Risk analysis and risk management

The general methods of risk management are helpful in optimally implementing a business continuity plan. Risk management sets the costs that arise through the occurrence of risks in relationship to the costs involved in avoiding a risk and the costs of reducing the effects of a risk.

A business continuity plan is used to protect against and minimise failures that occur as a result of business risks. It is therefore recommended that proven processes and methods of risk management are taken into account when a business continuity plan is being created and implemented. In simplified terms, risk management in the context of business continuity encompasses the processes of analysis, avoidance and control of business risks. These processes work together in a continuous cycle (Figure 9.4).

The creation of a business continuity strategy begins with a risk analysis of business-critical processes and systems. The goal of this analysis is to identify and evaluate the potential risks and consequences that could affect business processes, data and IT infrastructure. The main steps are:

1. Identification of risks and an assessment of the probability of an occurrence and the extent of loss as a result of such an occurrence.

2. Definition of the measures and actions required in order to avoid or reduce the extent of loss and the probability of occurrence.

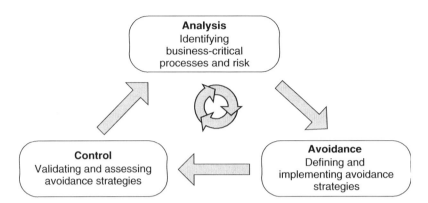

Figure 9.4   Risk management consists of the processes analysis, avoidance and control. These processes are repeated in a cycle, thereby enabling a continuous validation of and improvement to the business continuity plan.

After a categorisation has been made, the probability of occurrence and the potential extent of loss are defined in step 1 for each identified risk (Section 9.1.3). The extent of a loss depends on the business value of the data and services and this can vary for different areas of a company. Based on the different requirements that exist, certain potential risks are also intentionally not considered within the framework of the business continuity strategy in this step. Consequently, no measures to protect against the consequences of these risks are introduced for these occurrences.

In step 2 a decision is made on which strategies to follow to avoid or minimise the risks identified in step 1. The aim is to define measures that reduce the extent of loss or the probability of occurrence or to eliminate them altogether.

Avoidance strategies ensuring continuous and loss-free IT operations are then derived from the results of the risk analysis. Initially these strategies should be considered separately without any reference to specific products and vendors and relate exclusively to methods that offer protection against the consequences of the risks. This will then enable the defined business continuity strategies to be realised using concrete technologies and the appropriate operational processes. The main priority in the context of business continuity is the implementation of technologies, solutions and operational processes. In this phase, in particular, the costs incurred for risk avoidance must be weighed against those that arise as a result of the unprotected consequences of risks (Section 9.1.8). It may be necessary to seek a more cost-effective strategy, to dispense with a business continuity strategy altogether or, if possible, to reduce the likelihood of an occurrence or the extent of a possible loss to an acceptable level.

Finally, controlling risks involves validating the effectiveness and the success of avoidance measures – for example, through tests. This process must be run continuously as this is the only way in which the effectiveness of the measures can be tested. One consequence of this process could be that new risks are identified, which initiates the process of analysis. Another consequence could be that the avoidance measures are too weak or not effective at all. In this case, the process of analysis is initiated again.

9.1.10   Creation of a business continuity plan

A business continuity plan describes the strategies of a company for continuous and loss-free business operations. It encompasses all planning and implementation activities for preventing and minimising identified risks. This relates to personnel, technologies and business processes.

The following steps are necessary for creating a business continuity plan:

1. *Analysis of business-critical processes, services and data*
   The first step is an analysis and identification of the processes, services and data that are most important for the course of business. Which ones are essential under all circumstances? This also includes a definition of the requirements for these processes, services and data, such as the level of availability demanded. The result of this analysis provides the basis for the next steps in creating a business continuity plan.

2. *Analysis of business-critical IT systems*
   In this step an analysis determines which IT systems provide and support the business-critical processes, services and data identified in step 1. What now has to be determined is which direct and indirect losses are likely to occur in the event of a system outage. This is where a classification of the IT systems and applications is helpful: for example, according to 'absolutely critical,' 'critical,' 'less critical' and 'not critical.' This information and classification are required for the risk analysis in the next step.

3. *Analysis and evaluation of possible risks for these systems (risk analysis)*
   As the business-critical systems were identified in the preceding step, an analysis and an assessment of the possible risks to these systems can now be undertaken. In this step answers must be found to the following questions: Which risks exist for the identified systems? How high is the probability of these risks and what is the potential impact on business operations? The potential risks that can be covered by a business continuity strategy are identified as well as those that will deliberately not be considered, such as a simultaneous power outage in both the primary and the backup data centres.

4. *Identification of required parameters and definition of requirements for technology and infrastructure*
   In this step the requirements for technology and infrastructure are identified and quantified. How long should it be before IT operations are running again (Recovery Time Objective, RTO) (Section 9.3.4) and what is the maximum data loss tolerated in a disaster situation (Recovery Point Objective, RPO) (Section 9.3.4)? What is the maximum time allowed for an outage of IT operations (availability) (Section 9.3.1)? Specific requirements can be assigned to each category based on the classification of IT systems in step 2. Service Level Agreement (SLA) (Section 9.4.1) can then be formulated

from these concrete requirements. Based on these requirements, a selection of suitable technologies and infrastructure can be provided in the next step.

5. *Selection of suitable technologies and infrastructure in accordance with the defined requirements in the different categories*
   The requirements for a business continuity solution established in the preceding step now enable a selection of technologies and infrastructure for the implementation of these requirements. In accordance with the defined requirements, the seven-tier model for business continuity (Section 9.4) and the division into different solution segments (Section 9.5.2) offer a good starting point and are useful in the selection of a strategy.

6. *Implementation and continuous testing of a solution*
   This step involves the implementation of the designed solutions and infrastructure. This also includes a test of the entire implementation at regular – for example, half-yearly – intervals. The test should check whether the infrastructure including the backup data centre is really fulfilling the defined requirements. The question is: Can the productive IT operation actually be resumed at a different location in the required amount of time?

7. *Management of the solution*
   This refers to the validation and, if necessary, updating of the business continuity plan. This step includes the following questions: Is the business continuity plan still fulfilling the requirements? Are new business-critical IT systems and applications added since the previous assessment? Has there been a change in the assessment of the business processes and risks? This for example can be done twice a year or whenever a big change occurs in the infrastructure or in requirements – such as when there is a reduction in the restart time.

A business continuity plan must include all business-critical resources. In addition to IT systems, this includes the personnel of a company, facilities (for example, data centres) and the operational infrastructure (such as electricity, telephone systems). Definitions of the corresponding processes and procedures for disaster recovery and restart after a catastrophe also constitute part of this plan.

9.2 STRATEGIES OF BUSINESS CONTINUITY

A business continuity program uses different strategies (Figure 9.5) depending on the specific requirements of business processes for continuous and loss-free operation. These requirements can be divided into high availability, continuous operations and disaster recovery.

Later we will see that technologies and products are usually optimised for one of these areas. A business continuity plan must therefore combine a number of different technologies and products in order to cover the entire range of requirements. It is not

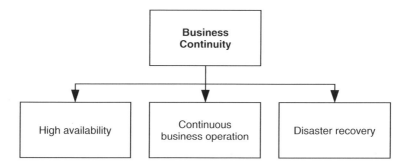

Figure 9.5 Business continuity strategies are divided into high availability, continuous business operations and disaster recovery.

only the technical solutions that have a determining influence on these strategies but also the operational areas of a company.

In this section we will be presenting a detailed discussion on high availability (Section 9.2.1), disaster recovery (Section 9.2.2) and strategies for continuous business operations (Section 9.2.3).

9.2.1 High availability

High availability describes a system's ability to maintain IT operations despite faults in individual components or subsystems. A high availability solution tolerates or corrects faults, thereby reducing the outage times of an IT system. High availability solutions also protect IT operations from component failures (minor outages). The main task of these solutions is to safeguard continuous access to data and IT services. Hence, the goal of these solutions is to guarantee continuous operation.

For the most part this happens automatically and is achieved through the use of fault-tolerant components for hardware and software. When component failure occurs, additional monitoring and management modules ensure that operations continue with the remaining components. Highly available systems must avoid single points of failure so that a fault-tolerant overall system can be maintained.

In the course of this book, we have presented a number of high availability strategies: RAID to protect against failures of individual or multiple disks (Section 2.4), redundant I/O buses against failures of I/O paths (Section 6.3.1), server clusters against the failures of servers or the applications running on them (Section 6.3.2) and data mirroring with volume manager mirroring and disk subsystem-based remote mirroring against failures of storage systems (Section 6.3.3). Lastly, with reference to the protection provided against the outage of a data centre, we have shown how these technologies can be combined together (Section 6.3.5).

9.2.2 Disaster recovery

What is meant by disaster recovery is protection against unplanned outages in IT operations due to widespread failures and catastrophes. Disaster recovery solutions enable the restart of IT services after a major outage and, like high availability solutions, ensure that the length of the outage times of an IT system is reduced. The main function of these solutions is to restore access to data and IT services as quickly as possible and, if possible, with the latest data state. Disaster recovery solutions therefore ensure loss-free resumption of business operations.

Backup data centres that are located some distance away are frequently used for this purpose because catastrophes can extend over a large geographical area. The necessity for this geographical separation was clearly illustrated by the power outage on the East coast of the USA on 14 August 2003. Data is mirrored between the data centres so that operations can be continued in the backup facility when a failure occurs in the primary data centre. This is achieved through replication techniques and a reliable and precisely defined restart of IT operations on supplementary systems. In practice, sometimes even more than two data centres are used for disaster recovery.

The consistency of data in backup data centres is especially important for a smooth switch over of IT operations to the backup facility. With synchronous remote mirroring (Section 2.7.2) and consistency groups (Section 2.7.3) we have presented two technologies that ensure the consistency of mirrored data. We will be referring back to these technologies during the course of this chapter.

9.2.3 Continuous business operation

Continuous business operation means finding a way for data backup and maintenance in an IT infrastructure without affecting the availability of applications, IT services and data. In the process all components should remain intact whilst providing an acceptable level of performance. It is also important that the impact of administrative activities on production operations is as minimal as possible.

For example, in the past network backup meant scheduling a backup window during which applications and data would not be available for hours because a consistent backup copy of the data was being created. Applications, such as databases, were either halted or put into a special mode in which temporarily no further changes could be made to the data. This affected running business operations but was necessary in order to guarantee the consistency of the data being backed up. If a fault occurred, the data could be restored to its state at the time of the last backup.

Numerous techniques are now available for carrying out network backup with no or little impact on running operations (Section 7.8). For example, we discussed how through the use of log files and archived log files databases can be backed up in parallel with running operations and the effects on business operations kept to a minimum (Section 7.10).

9.3 PARAMETERS OF BUSINESS CONTINUITY

High availability, disaster recovery and continuous business operations are aspects that only give a rough description of the requirements and strategies for business continuity. In this section we introduce parameters that more precisely characterise the requirements, strategies and products for business continuity, thereby enabling an effective evaluation and comparison.

We begin with availability (Section 9.3.1) and contrast different forms of availability using the parameters Mean Time between Failure (MTBF), Mean Time to Repair (MTTR) and Mean Time to Failure (MTTF) (Section 9.3.2). We then show how the availability of composite systems is calculated from the availability of individual components (Section 9.3.3). Two other important parameters are Recovery Time Objective (RTO) and Recovery Point Objective (RPO) (Section 9.3.4): These indicate how quickly operations must be recovered (RTO) and the maximum amount of data that is allowed to be lost (RPO) when an outage occurs. We conclude with another parameter Network Recovery Objective (NRO) (Section 9.3.5).

9.3.1 Availability

Availability is an important parameter for high availability solutions. It specifies how long components, applications and services should be functioning properly over a certain period of time. The general rule of thumb is that the higher the availability, and thus the stability of a solution, the more expensive it is.

Availability is indicated in absolute values (for example, 715 out of 720 hours during the previous month) or in percentage terms (99.3%). The availability is calculated from the operating time and the outage time as follows:

$$\text{Availability} = \text{operating time}/(\text{operating time} + \text{outage time})$$

N × M IT operation is frequently mentioned in connection with availability; the N means N hours per day and M means M days of the week. The now very common description is 24 × 7, which means round-the-clock IT operations and, due to the increase in the networking of business processes beyond company boundaries, is becoming more and more important. For many IT services shorter service times are totally sufficient – for example, from Monday to Saturday from 6 in the morning until 8 in the evening. When availability is indicated in percentage terms, it is important to note that this relates to the required service time – such as 99.99% of a 14 × 6 operation.

Product specifications and marketing documents, in particular, state the availability of components or solutions for 24 × 7 operations in percentage terms. It is very helpful to convert these figures to an annual absolute outage time (Table 9.2). For example, with a requirement of 99.999% availability per year a system is only allowed to fail a total of 5.26 minutes. Considering the boot times of large Unix servers after a restart, it is easy to

Table 9.2 High availability demands that outage times are short. The table shows the outage times for a 24 × 7 operation.

Availability	Outage times per year
98%	7.30 days
99%	3.65 days
99.5%	1.83 days
99.9%	8.76 h
99.99%	52.6 min
99.999%	5.26 min
99.9999%	31.5 s

see that the frequently mentioned 'five nines' represents a high requirement that is only achievable through the use of redundant components for hardware and software and with automatic switch over.

9.3.2 Characterisation of availability (MTBF, MTTR and MTTF)

Availability on its own only delivers a conclusion about the ratio between outage time and service time – for example, the total outage time over a specific period of time. Therefore, other information is necessary to arrive at the frequency of outages and the duration of each individual disruption.

The parameters Mean Time between Failure (MTBF), Mean Time to Recovery (MTTR) and Mean Time to Failure (MTTF) close this gap exactly (Figure 9.6). The MTBF is the average time between two successive failures of a specific component or a specific system. MTTR refers to the period of time before a component or a system is recovered after a failure. The MTTF gives the average period of time between the recovery of a component or a system and a new outage.

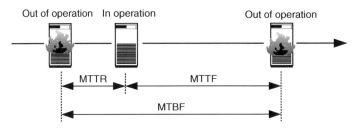

Figure 9.6 The parameters MTTR, MTTF and MTBF describe the course of an outage and thus characterise different styles of availability.

Based on these new parameters, availability can be calculated as follows:

$$\text{Availability} = \text{MTTF}/(\text{MTTF} + \text{MTTR})$$

This formula shows that availability increases substantially as the MTTR is reduced, i.e., with a faster restart of IT operations. This can be achieved with preinstalled and preconfigured emergency equipment, such as hot-standby computers, and short support routes.

9.3.3 Calculation of overall availability

The availability of the individual components has to be considered when the overall availability of a system is calculated because of the importance of how these components interact with one another. Whether components are coupled together sequentially or in parallel has a major influence on overall availability. RAID arrays with multiple disks and Fibre Channel fabrics with multiple switches are examples of composite systems.

RAID 0 (Section 2.5.1) and RAID 1 (Section 2.5.2) are good examples of how overall availability is dependent on the coupling. With RAID 0 (striping) data is written onto multiple disks. If even only one disk fails, all the data is lost (sequential coupling). With RAID 0 using two disks, each with an availability of 90%, the data on the RAID 0 array only has an availability of 81% (Figure 9.7).

However, if the data is mirrored on two disks through RAID 1 (mirroring), the availability of the data on the RAID 1 array increases to 99% (Figure 9.8). The data is still available even if a disk fails (parallel coupling).

With RAID 0, twice as much data can be stored than with RAID 1. However, with RAID 1 the availability of the data is higher than with RAID 0. This example shows that the protection of data comes at a cost and therefore a compromise has to be found between cost and benefit (cost-time window, Figure 9.3).

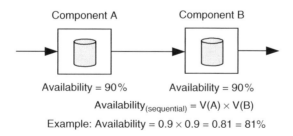

Figure 9.7 Sequentially-coupled components reduce the overall availability of a system.

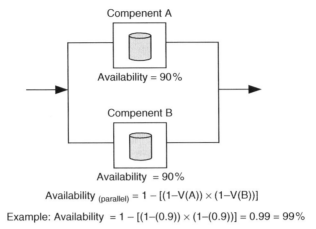

Availability $_{(parallel)}$ = 1 − [(1−V(A)) × (1−V(B))]

Example: Availability = 1 − [(1−(0.9)) × (1−(0.9))] = 0.99 = 99%

Figure 9.8   Parallel-coupled components increase the overall availability of a system.

9.3.4   Characterisation of failures (RTO and RPO)

Recovery Time Objective (RTO) and Recovery Point Objective (RPO) are two important parameters for characterising failures and the business continuity strategies that protect against them. The RTO indicates the maximum allowable length of time for resuming IT operations after a failure (Figure 9.9). As a rule, this is given in minutes. For example, an RTO of 60 means that IT operations must be restarted within 60 minutes. The maximum restart time in this case is therefore 60 minutes.

High availability solutions such as server clustering strive for as low an RTO as possible. This means that a crisis or failure situation will result in a relatively short interruption to IT operations. The RTO requirements are established according to the importance of the applications and the data. For business-critical areas a very short restart time of a

Figure 9.9   The Recovery Time Objective (RTO) specifies the maximum length of time for the restart of a business continuity solution.

few minutes or even seconds may be required. For less important applications a restart time of several hours or even days may suffice. Each company must – as part of the business continuity plan (Section 9.10) – determine the length of time an IT failure can be tolerated in each of its areas.

In this scenario a good compromise should be found between restart time and IT costs (cost-time window, Figure 9.3). Depending on the complexity and the size of an IT infrastructure, the realisation of a very low RTO is usually associated with considerable costs. The shorter the restart time, the more expensive and complicated is the technical solution. A cost saving can be effected in the implementation of high availability solutions if products and technologies that deliver a sufficient RTO (Section 9.1.8) are used instead of technically best solutions.

The RPO indicates the maximum level of data loss that can be tolerated in a crisis situation (Figure 9.10). Explained in a different way, the RPO determines the period of time that recovered data is not in an up-to-date state. An RPO>0 thus means that a certain amount of data loss is tolerated in a crisis. The RPO can be decreased, for example, if data is backed up at shorter intervals. The same rule applies here: the lower the RPO, the more expensive and complicated the storage solution is.

In our experience, RTO and RPO are always being confused in practice, which is why we want to explain these two parameters again using an example. A company backs up all its data from a disk subsystem to tape every night at 24:00 hours. One day due to the breakdown of an air-conditioning system at 12:00 hours, a fire occurs in the disk subsystem and results in a total loss. As part of the procedure to restart the IT operation, the data is restored from tape to a second disk subsystem. It takes a total of 8 hours to restore the data from tape to the second disk subsystem and to execute other steps, such as configuration of the virtual hard disks and zoning.

In this scenario the recovery time (RTO) amounts to 8 hours. Based on the last data backup the recovery time is 'minus 12 hours,' which means that an RPO of 12 hours has been achieved. Had the fire occurred later in the evening, with the same RTO all the changes made the entire day would have been lost and the RPO would have been considerably higher.

Figure 9.10 The Recovery Point Objective (RPO) specifies the maximum tolerated level of data loss in a business continuity solution.

9.3.5   Network Recovery Objective (NRO)

The NRO is another parameter for business continuity. The NRO indicates the maximum allowable time for the recovery of the operation of a data network (LAN and WAN) after an outage. It is an important aspect of an overall business continuity program as it must take into account all the components and services of the IT infrastructure (Section 9.1.7). It only makes sense for data and applications to be restored quickly after an outage if users can also reach them over LANs, WANs or the Internet.

9.4   QUALITY OF SERVICE FOR BUSINESS CONTINUITY

As with many architectural and design decisions, enormous costs can also be saved with business continuity if a solution that is good enough to support the business processes is implemented rather than one that is technically the best (Section 9.1.8). The considerations discussed in this section should be helpful for working out the requirements needed to find a solution that falls into the cost-time window for recovery (Figure 9.3) when a business continuity plan is implemented.

In this context we also take a brief look at Service Level Agreements (SLAs) in respect of business continuity (Section 9.4.1). We then distinguish between the varying requirements for high availability and disaster recovery (Section 9.4.2) and, using the seven-tier model for business continuity, present a method for categorising business continuity requirements (Section 9.4.3). The discussion ends with a detailed look at the individual tiers of the model (Sections 9.4.4 to 9.4.11).

9.4.1   Service Level Agreements (SLAs)

The different applications in an IT operation vary in their importance to the support of business processes and consequently have different RTOs and RPOs. The requirements for a specific application can occasionally change, even during the course of a day. For example, high availability is demanded for many applications during business hours (RTO close to zero), but outside this time period a longer recovery time is tolerated (RTO clearly greater than zero). Therefore, maintenance work – even on standby systems – often has to be performed at night or on weekends.

A cost-efficient business continuity plan is one in which the RTOs and RPOs of the corresponding storage solution are commensurate to the value and the importance of the respective data for a company. The parameters introduced in Section 9.3 are suitable for specifying Service Level Agreements (SLAs) for IT services between provider and client. SLAs describe in detail and in a quantifiable form the requirements a customer of IT services demands of an IT service provider and therefore indirectly determine the design of the respective technical solutions. SLAs usually constitute part of a contract between IT provider and customer and can also be negotiated between different departments within a company. Non-compliance of contractually agreed SLAs usually results in penalties for the provider.

The parameters introduced in Section 9.3 enable the requirements for business continuity to be precisely described, quantified and monitored. A provider is therefore able to offer services at different service levels and to charge his clients for these services accordingly. The client, on the other hand, is in a position to control the agreed quality criteria and thus pays for a solution that matches his requirements.

9.4.2 High availability versus disaster recovery

'How is it possible that I am losing my data even though I am running RAID 5?' is a question that is always asked with amazement when a complete RAID array is lost. The terms high availability and disaster recovery are often used as equivalents or synonymously although both pursue different goals and therefore have to be supported with different technologies. RAID 5 is a high availability solution for individual disk failures but not a disaster recovery solution for multiple failures. Even if two disks fail simultaneously in a RAID 5 system, the goal of RAID 5 has been achieved – namely, by offering a high availability solution to the failure of individual components. If, in addition, protection is needed against the failure of the entire RAID 5 array, then a disaster recovery solution such as mirroring RAID 5 arrays to a second disk subsystem is required.

'Why am I losing my data even though I am mirroring it with a volume manager in two data centres?' This is another question that is always being asked with the same amazement when a current copy of data does not exist in the backup data centre after the loss of the primary data centre. The requirements for high availability and disaster recovery have not been understood correctly with this solution either. Data mirroring using the majority of volume managers usually provides protection against the failure of an entire RAID array. If the mirror fails in the backup centre, then operations are continued with the copy in the primary data centre. If subsequently the entire primary data centre has an outage, then the backup data centre will not have a current copy of the data. Nevertheless, the purpose of classical data mirroring with the volume manager would have been fulfilled. If protection is wanted against loss of the primary data centre, then the mirroring must be carried out synchronously. However, synchronous mirroring contradicts the concept of high availability.

High availability solutions guarantee continuous access to data and IT services even in the event of component failure. The requirements for high availability solutions are quantified through an RTO. Solutions of this type have an RTO close to zero and therefore must be automated. Consequently, it is not the point of high availability solutions to store data synchronously at two sites so that, if necessary, IT operations can be restarted. The priority is undisturbed access to data rather than a restart in the backup data centre after a catastrophe.

In comparison, disaster recovery solutions provide for consistent data mirroring at two different locations, thereby guaranteeing a restart in the backup data centre after an outage in the primary data centre. An RPO quantifies the requirements for how current the data is in the backup centre. For important applications an RPO of zero is required. This means that data may only be written if immediate mirroring to the backup data centre is possible (Section 2.7.2). If any doubt exists, it is preferable to terminate a write process than to have an obsolete data set in the backup data centre. However, this aborting of the write process is a contradiction of the requirements for high availability. Disaster recovery solutions are therefore only conditionally suited to realising high availability requirements.

In summary, it is important to emphasise that high availability and disaster recovery are quantified through different parameters and consequently should be considered independently of one another. Using high availability technologies for disaster recovery is just as unwise as using disaster recovery technologies for high availability. Table 9.3 compares the parameters of high availability and disaster recovery. The table clarifies the different requirements that exist for each one.

High availability (RTO near zero) as well as rapid disaster recovery (RPO close to zero) is required for important applications. These applications require a combining of the techniques for high availability with those for disaster recovery. We will be presenting this type of solution later in Section 9.5.6. As shown in Table 9.3, weaker requirements for RTO and RPO are adequate for many applications. Using the seven-tier model for business continuity, we will now present approaches to determining these requirements and therefore finding a cost-optimal solution (Figure 9.3).

Table 9.3 High availability and disaster recovery have different RTO and RPO requirements.

Business Continuity solution	RPO	RTO
High availability	None	Close to 0
Disaster recovery (synchronous mirroring)	Close to 0	None
Disaster recovery (asynchronous mirroring)	>0	None

9.4.3 The seven-tier model

The requirement grades for high availability are relatively easy to distinguish: fully automated high availability solutions, automated switching after activation by a person and deliberately no provisions for high availability. In contrast, the requirement spectrum for disaster recovery is considerably more diverse. Therefore, in 1992 the SHARE user group in the USA in collaboration with IBM created a seven-tier model for business continuity requirements (Figure 9.11). The objective of this model is the classification and quantification of technologies for business continuity for the purposes of positioning of them in

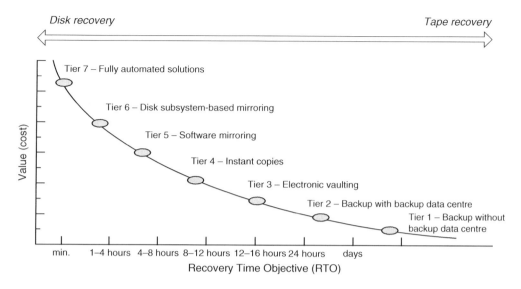

Figure 9.11   Seven-tier model for business continuity.

respect of the requirements of business processes. In the foreground is a comparison of the concepts and technologies on an abstract level, providing a comparison of requirements and costs but not of the individual products of certain vendors. In the course of this section we will take a close look at the different steps of the model.

Above all, the model emphasises the different RTOs of the various steps. In the course of time these have been adapted to comply with technical progress. As a by-product, typical RPOs have also evolved for the different tiers. Table 9.4 presents a compilation of typical parameters for the different tiers of the model for the year 2009. The values in the

Table 9.4   The different tiers of business continuity can be characterised through RTO and RPO. At the lower layers time spans up to several days are tolerated. At the higher layers the time spans shrink to an area of seconds or even to zero.

Tier	Typical RTO	Typical RPO
7	In seconds	Zero or several seconds
6	In minutes	Zero or several seconds
5	1–4 h	>15 min
4	4–8 h	>4 h
3	8–12 h	>4 h
2	1 or more days	Several hours to days
1	Several days	Several days
0	None	None

table should only be viewed as a rough guide. The values achieved in a concrete solution depend on the respective implementation and can therefore deviate from the information in the table.

9.4.4   Tier 0: no data backup

No plans or techniques to maintain or restart IT operations in the event of a catastrophe are available for tier 0. No data is backed up, there is no contingency plan and no special storage technologies are deployed. The recovery time for these environments cannot be assessed. There is not any indication yet whether data can even be restored at all if a failure occurs. It is therefore obvious that a solution at tier 0 is the least expensive one but does not offer any protection of data whatsoever. Nevertheless, it still makes sense within the framework of risk analysis to identify which applications, data and services will not be protected (Section 9.1.9) to avoid the unnecessary cost of countermeasures.

9.4.5   Tier 1: data backup without a backup data centre

Solutions at tier 1 back up data on tapes that are transported at a later time from the primary data centre to a different location (vaulting). If a catastrophe occurs in the primary data centre, a copy of the data is still available at the other location. In this scenario the backup is usually from the preceding day or even from a much earlier point in time (very high RPO). No additional hardware is available that would enable a quick restoration of the data in a catastrophe. After an outage the replacement hardware first has to be procured and installed and the backup tapes then transported to this location before the data can be restored (very high RTO).

9.4.6   Tier 2: data backup with backup data centre

With the solutions at tier 2, the data is backed up on tape just as with tier 1 and the tapes are transported to a different location. In contrast to tier 1, with tier 2 the complete infrastructure (servers, storage, networks, WAN connections, office premises and so forth) is available in a backup data centre. This setup enables data to be restored from the tapes and the services and applications to be started again in the event of a catastrophe. The restart of the operation after a catastrophe is performed manually. Therefore, a potential data loss of several hours or days is still tolerated in tier 2 (very high RPO). However, the required restart time is easier to plan and can be predicted more accurately (high RTO).

9.4.7   Tier 3: electronic vaulting

In contrast with tier 2, with solutions at tier 3 the data is electronically vaulted rather than manually. For example, a network backup system can back up data over LAN and WAN in a remote backup data centre. Backups produced in this way are usually more up to date than manually transferred tapes. Therefore, in a catastrophe the potential for data loss is lower than with the solutions at tiers 1 and 2 (medium RPO). In addition, these data backups can be imported again faster and more easily because the data is available over the network (medium RTO).

9.4.8   Tier 4: instant copies

From tier 4 onwards disks instead of magnetic tapes are used for data backup. Instant copies from intelligent disk subsystems and snapshots from modern file systems are particularly utilised for this purpose. These techniques make it considerably easier and faster to create backups than is the case with copies on magnetic tapes or over a WAN. This allows the frequency of the backups to be increased and, consequently, reduces potential data loss from outages (low RPO). Equally, the restart time sinks because the backups can be imported quickly again after a failure (low RTO). However, it should be noted with solutions at tier 4 that data and security copies are stored in close proximity to one another. Therefore, protection against certain risks like fire and lengthy power outages is lost compared to solutions at tiers 1 to 3.

9.4.9   Tier 5: software mirroring

Solutions at tier 5 require that the data states in the primary data centre and in the backup data centre are consistent and at the same level of update. This is achieved when data is mirrored through the application – for example, through the replication mechanisms of commercial database systems. In contrast to tier 4, the data is again stored at different locations. A very minimal to non-existent data loss is tolerated (very low RPO). The mechanisms required depend solely on the applications used. The mirroring is not supported or even guaranteed by functions in the storage network. The restoration of the data from these mirrors is usually integrated in the applications so that an operation can quickly switch to the copies (very low RTO).

9.4.10   Tier 6: disk subsystem-based mirroring

Solutions at tier 6 offer the highest level of data consistency. These solutions are applied when only very little data loss or none whatsoever is tolerated in a crisis situation (minimal

RPO) and access to data must be restored as quickly as possible (minimal RTO). Unlike the techniques at tier 5, these solutions do not depend on the applications themselves providing functions for data consistency. This is implemented instead through functions in the storage network: for example, through synchronous remote mirroring between two disk subsystems located in different data centres.

9.4.11 Tier 7: fully automated solutions

Solutions at tier 7 comprise all the components from tier 6. In addition, this solution is fully automated so that IT operations can be switched without human intervention from the primary data centre to a backup data centre. This results in the best possible restart time (RTO near zero). These solutions also do not normally tolerate any data loss (RPO = 0). Tier 7 offers the best possible, albeit most expensive, solution for the restoration of data and data access after a catastrophe. In order to meet the high requirements for RTO and RPO, the solutions at tier 7 are usually customer-specific solutions with high standards.

9.5 BUSINESS CONTINUITY SOLUTIONS

This section deals with the implementation of business continuity requirements. In our discussions we refer back to a number of techniques that were presented in the preceding chapters of this book and show which RPOs and RTOs they support.

We first start with the basic techniques for high availability and disaster recovery (Section 9.5.1). We then simplify the seven-tier model for business continuity into the three solution segments: backup and restore, rapid data recovery and continuous availability (Section 9.5.2); and present selected solutions for each of the three segments (Sections 9.5.3 to 9.5.6).

9.5.1 Basic techniques

High availability techniques form the basis for continuous IT operation. The main aim is to prevent the failure of individual components and systems through the use of redundant components. When a failure occurs, an automatic transfer of data and services ensures that IT operations will continue without disruption. In Section 6.3 we compiled numerous techniques that protect against the failure of hard disks, I/O buses, servers and storage devices. The goal of all these techniques is to realise an RTO close to zero in the event that the mentioned components fail.

Replication techniques for copying data to remote locations form the basis for loss-free IT operation. The goal is to restart IT operations with a consistent state of these data copies after an outage. In this book we have presented numerous techniques for disaster recovery. In the course of our discussion we will be referring back to these techniques and positioning them in the seven-tier model to highlight their different RPOs.

9.5.2 Solution segments of the seven-tier model

In our further consideration of solutions for loss-free operation we simplify the seven-tier model into three solution segments (Figure 9.12). Tiers 1 to 3 are essentially based on magnetic tape storage solutions with traditional network backup systems. Solutions for tiers 4 to 6 are based on the replication techniques in storage systems as well as on applications and are used to provide rapid data recovery. The solutions at level 7 emphasise continuous and loss-free IT operation as a means of protecting against all possible forms of outages.

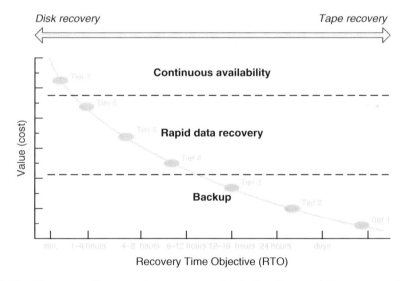

Figure 9.12 The seven tiers can be summarised into the three solution segments: backup and restore, rapid data recovery and continuous availability. This simplifies the positioning of requirements, techniques and solutions.

9.5.3 Backup and restore

Classical network backup systems (Chapter 7) are the standard tool for all three tiers of the solution segment backup and restore. These systems enable data to be backed up on tape (tiers 1 to 3) and can even supervise and control the transfer of tapes to a backup data centre (Section 7.3.4). Depending on the configuration, they also allow backup over LAN and WAN onto backup media in a backup data centre, thereby fulfilling the requirements of tier 3. Data backup and restore using instant copy, snapshots, remote mirroring and replication are increasingly being integrated into network backup systems, thus making these systems also suitable for dealing with the requirements of the solution segment rapid data recovery (Section 9.5.4) in the future.

9.5.4 Rapid data recovery using copies

Instant copies of disk subsystems (Section 2.7.1) and snapshots of modern file systems (Section 4.1.3) enable fast backup and restoration of data. Both techniques allow virtual copying of large quantities of data in a few seconds and allow access to the data copy within a very short time. An assurance of the consistency of data that has been copied this way is an important aspect of backup using instant copies and snapshots in guaranteeing a smooth recovery (Section 7.8.4). The characteristics of instant copies and snapshots in respect of business continuity are very similar, so we will use instant copies as being representative of both techniques in the discussions that follow.

Positioning of instant copy

Instant copies copy data at specific points in time. In a failure situation, data can be restored to the point of the last backup. All changes made between the last backup and the failure will be lost. Today the normal practice is to repeat backups at intervals of from one or several hours to one day (RPO in the area of hours to one day). Advanced features of instant copy, such as Space Efficient Instant Copy and Incremental Instant Copy, reduce the time needed for instant copies (Section 7.8.4) and therefore permit an increase in the frequency of the backups. This reduces the time span for potential data loss (RPO of one hour or less).

Instant copies are available immediately after copying via an alternate access path (Section 7.8.4). If the application can continue operation over this access path when a failure occurs, the restart time is reduced (low RTO). This is particularly an advantage in file systems: Users tend to edit a file so frequently that it cannot be used anymore or they unintentionally delete the wrong file. Modern file systems use special subdirectories to provide an older state of the file system produced per snapshot. These subdirectories immediately allow users to restore an older version of a file (RTO = 0). If operation of the application cannot be continued over the access path of the copy, then the data must first be mirrored back from the copy to the original location. A large number of disk subsystems shorten this process through a reversal of the instant copy (Section 7.8.4). The quality of the reversal of the instant copy varies considerably between different products, thus requiring a careful analysis of the achievable RTO (RTO product-specific).

Solution variant: continuous data protection (CDP)

A weakness in using instant copy for backup is that quite a bit of time might have passed between an outage and the last backup. Unless other measures are taken, data will therefore often be lost (RPO>0). In the database environment protection is provided through the backup of all log files that the database system has created since the previous backup. The forward recovery procedure of the database imports these log files so that the most recent state of the database can be created before the applications are restarted (Section 7.10.2).

Continuous Data Protection (CDP) transfers the log mechanism of databases to file systems and disk subsystems whereby it simultaneously records all changes to a data set immediately in a log file. This enables each version of the data from the recent past to be restored and not only the state since the last instant copy (RPO = 0 plus versioning). CDP is a new but very promising technology, and the assumption is that its use will increase over the coming years.

Solution variant: disk-to-disk-to-tape (D2D2T)

Another problem of rapid data recovery with instant copy is the proximity and close coupling of the original and the copy. If the disk subsystem suffers a loss – for example, through a reformatting of the wrong physical disk or as a result of a fire – both the original and the instant copy are affected. Unless other measures are taken, instant copies and CDP would offer protection from user errors but no protection from administrator error or even physical damage. For this kind of protection the instant copy must be decoupled and separated from the original.

One approach to resolving this problem is disk-to-disk-to-tape (D2D2T) backup. D2D2T solutions first back up data on a disk and then make a second copy from there to tape. This principle has been applied for a long time in classical network data backup systems where data is first backed up on a hard disk and then copied a second time on tape within the network backup system (Section 7.3.4). With the newer forms of D2D2T, as with instant copy and CDP, data is first copied within the storage system and then copied on tape through network backup (Figure 9.13). The assumption is that in the future these new versions of D2D2T will be integrated efficiently into network backup systems.

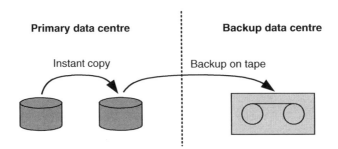

Primary data centre **Backup data centre**

Instant copy Backup on tape

Figure 9.13 A D2D2T solution first copies data via instant copy and then copies it from there a second time using network backup.

9.5.5 Rapid data recovery using mirrors

Remote mirroring between disk subsystems (Section 2.7.2), mirroring using volume managers (Section 4.1.4) and the replication of file systems and applications allow business operations to be resumed in a backup data centre with up-to-date data. A synchronous mirroring of data at both locations provides the basis for a loss-free restart (Section 9.4.2).

Positioning of remote mirroring

Remote mirroring copies data continuously between the primary data centre and the backup data centre. Synchronous remote mirroring (Section 2.7.2) ensures that the data in both data centres is in the same state of being up to date (RPO = 0). If the data of an application is distributed over multiple virtual disks or even over multiple disk subsystems, the mirroring of all disks must be synchronised using a consistency group to ensure that the data in the backup data centre remains consistent (Section 2.7.3). If consistency groups are not used, there can be no guarantee that the applications will restart with the data in the backup data centre (RPO uncertain). If the data is mirrored using asynchronous remote mirroring, then it will be copied with a delay to the backup facility (RPO configuration-dependent). If here too the data is distributed over multiple virtual disks or even over multiple disk subsystems, then the mirroring of all disks must be synchronised over a consistency group that also maintains the write order consistency (Section 2.7.3). Despite the delay in transmission, the data in the backup data centre is therefore always consistent. Without these measures the consistency cannot be guaranteed (RPO uncertain).

Information about recovery time can not necessarily be derived from the use of remote mirroring (RTO independent). However, the effort involved in using remote mirroring is so high that it is often only used for applications that require a very fast restart. Therefore, with remote mirroring it is common to have standby computers on hand for fast restarts in backup data centres (low to medium RTO). The switch to the copies is often automated using cluster solutions (minimal RTO).

Solution variant: mirroring over three locations

For secure protection from widespread outages data must be kept in the same state over long distances. However, this correlates with a huge reduction in performance (Section 2.7.2). Synchronous remote mirroring keeps data in the same state (RPO = 0) at the cost of latency in the write operations. The latency increases the further apart the two disk subsystems are located. Asynchronous remote mirroring eliminates the restrictions of long distances and even when distances are great has no influence on latency. On the other hand, depending on the configuration, the mirrors can diverge considerably from one another (low to medium RPO).

One approach to resolving this situation is a solution that combines synchronous and asynchronous remote mirroring (Section 2.7.2 and Figure 9.14). Depending on the write performance required of an application, the distance for synchronous remote mirroring is between a few kilometres and 100 kilometres. It is important here to consider the actual cable length between disk subsystems; this can be considerably longer than a true flight path or a road link. Usually, this link is designed with in-built redundancy (standby connection). Therefore it is important to ask the provider about the length, latency and throughput for the standby connection. The performance of standby connections is sometimes worse in multiples than for active connections and the data can no longer be mirrored fast enough. If a fault occurs at the provider side, this can lead to an outage of the primary data centre. These faults with providers should therefore definitely be planned

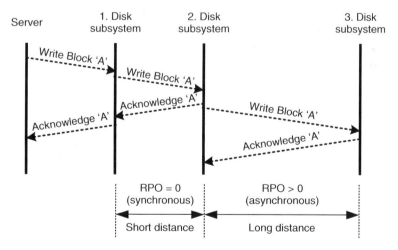

Figure 9.14   This again shows the combination of synchronous and asynchronous remote mirroring (Figure 2.23), but expanded with the addition of the parameters RPO and distance.

for when testing such a solution. No distance restrictions exist for asynchronous remote mirroring. In actuality, mirroring is even extended to other continents.

Solution variant: mirroring with data backup

Data mirroring over three locations satisfies high requirements for availability and disaster recovery. However, this requires three data centres and WAN capacity, which does not come cheaply. Therefore, the third mirror is often left out. The assumption here is this: If a primary data centre and a backup data centre fail at the same time, the fault is so major that business operations have come to a standstill due to other reasons and a third copy would only be "a drop in the ocean."

Backing up data mirrors on tape is a good compromise between mirroring between two locations and mirroring between three locations. In this case, the data is mirrored through synchronous remote mirroring in both data centres (RPO = 0). In addition, an incremental instant copy is regularly generated in the backup centre in order to create a consistent backup of the data from it on tape (Figure 9.15). The tapes are then transferred to a third location. This approach reduces the costs for a third copy of the data at the trade-off of the RTO and RPO.

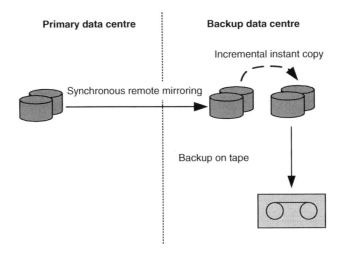

Figure 9.15 As an alternative to mirroring over three locations, here data is mirrored through synchronous remote mirroring and then a copy for the third data centre is generated via incremental instant copy and backup of the remote mirror.

Positioning of volume manager mirroring

Depending on the implementation, volume managers support synchronous and asynchronous mirroring. The volume manager normally helps to guarantee continuous operation when a disk subsystem fails. The mirrors are run synchronously when both copies are available. If one of the mirrors fails, the volume manager continues to work on the remaining one and an automatic switch over is made to asynchronous mirroring. The two mirrors can therefore diverge (RPO uncertain) in which case no disaster recovery exists, even if both mirrors are separated by a fire protection wall (Figure 6.20). A volume manager does not become a disaster recovery solution until it works synchronously (RPO close to zero). Individual products should be looked at very closely to determine whether the RPO is equal to zero or whether it is only close to zero.

An asynchronous volume manager is a high availability solution (minimal RTO). Redundant I/O paths and cluster solutions can be used to increase data availability even further (minimal RTO, Figure 6.20). However, if synchronous mirroring is used for disaster recovery, then the high availability is lost (RTO uncertain).

A volume manager cannot be run synchronously and asynchronously at the same time, which means that it can be used either for high availability or for disaster recovery. This important difference is unfortunately often overlooked in practice, and a volume manager is falsely considered simultaneously as a disaster recovery and a high availability solution over two locations. This mistake results in unanticipated data loss (Section 9.4.2).

Positioning of replication

Replication is a copying technique of file systems and applications such as email systems. It synchronises data between client and server or between two servers at regular intervals. The mirroring is therefore asynchronous and both data copies can diverge (low to medium RPO). Depending on the configuration, the time between synchronisation varies from a few minutes up to one day.

File systems and applications with replication techniques are frequently equipped with high availability techniques. If a client cannot reach a file server or an applications server, it automatically reroutes the request to another one. For the user the request only takes a few moments longer (low RTO).

9.5.6 Continuous availability

In this section we finally come to the solution segment continuous availability and, consequently, the top tier of the seven-tier model. The goal of this tier is the merging of continuous operation for high availability (RTO close to zero) and loss-free operation with disaster recovery (RPO = 0).

Solution variant: volume manager mirroring with remote mirroring

The preceding section clarified that individual copying techniques are operated either synchronously or asynchronously. This means that they are used either for disaster recovery or for high availability. However, important applications have a requirement for both high availability and disaster recovery at the same time.

Two mirrors must be combined together to guarantee continuous business operation (Figure 9.16). In the standard solution the data is first mirrored asynchronously with a volume manager and then also mirrored through synchronous remote mirroring to a backup data centre. Therefore, in total four copies of the data are stored on four different disk subsystems. In configurations of this type the I/O paths between servers and disk subsystems (Section 6.3.2) and the connections between disk subsystems are always designed with redundancy; therefore, we will not be considering the failures of individual paths.

Asynchronous mirroring in the volume manager guarantees continuous operation when a disk subsystem fails (RTO close to zero). In addition, per synchronous remote mirroring each mirror is mirrored in the backup data centre, which means that at least one of the two mirrors in the backup facility is in the current data state (RPO = 0). If one of the four disk subsystems or the remote mirroring fails in one of the two mirrors, operations continue to run and the data continues to be mirrored synchronously between the two

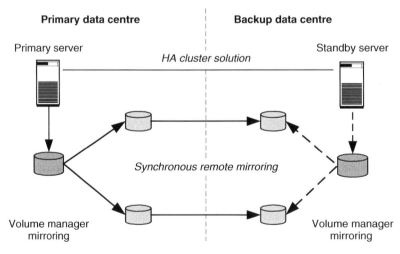

Figure 9.16   The combination of volume manager mirroring and synchronous remote mirroring offers high availability and disaster recovery.

data centres. This guarantees that operations are continuous and loss-free even during a failure (RPO = 0 and RTO almost zero during failure).

Solution variant: no volume manager mirroring

The combination of volume manager mirroring and remote mirroring offers some good properties in respect of high availability and disaster recovery. However, the cost of this solution is enormous: Firstly, four disk subsystems have to be operated with a corresponding expansion of the storage network. Secondly, the data for each mirror is sent separately to the backup data centre, which means a high bandwidth is required.

In practice, the mirror in the volume manager is therefore often dispensed with (Figure 9.17). In this case, only one copy of the data is kept in each data centre and the solution manages with only two disk subsystems and less bandwidth between both data centres.

Synchronised remote mirroring ensures that the data states are the same in both data centres (RPO = 0). The interaction between remote mirroring and the cluster software replaces the task of asynchronous mirroring with a volume manager. If the disk subsystem in the backup data centre or the remote mirroring fails, operations are continued in the primary data centre without remote mirroring (low RTO). If the disk subsystem in the primary data centre fails, the cluster software reverses the direction of the remote mirror and transfers the application from the primary data centre to the backup data centre (low RTO). The data is no longer mirrored after the first failure and therefore operations are interrupted when a second failure occurs (RTO uncertain) and data may even be lost (RPO uncertain). The deterioration of the parameters RTO and RPO during a failure is the key difference between this solution and the combination of volume manager mirroring and remote mirroring.

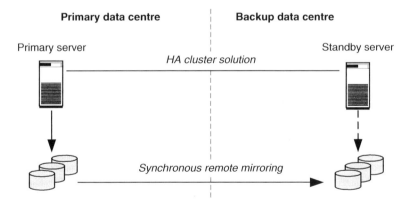

Figure 9.17 Synchronous remote mirroring offers better disaster recovery than the volume manager mirroring shown in Figure 6.20. On the other hand, the volume manager offers higher availability.

Solution variant: extension through instant copies

The two solution variants presented offer good protection against many types of failures. If the data of an application is distributed over multiple virtual disks and multiple disk subsystems, then the consistency groups of these solution variants should guarantee that an operation can always be continued with the data in the backup data centre (Section 2.7.3). However, in practice, the data set in a backup data centre is not always consistent – despite consistency groups. When a real catastrophe occurs, the entire operation does not break down at the same time at the exact same millisecond. Instead the effects of the catastrophe spread gradually until the whole operation comes to a standstill (rolling disaster). As a result, some applications will already have been disrupted while other ones are still running and the changes to them are being mirrored in the backup data centre. This means that the data in the backup facility is no longer consistent in the sense of applications logic.

As a means of protecting against a rolling disaster, remote mirroring is combined with incremental instant copies in the backup centre (Figure 9.18). In this case write access to the data is halted briefly at regular intervals to ensure that the data on the disk subsystems is consistent (Section 7.8.4). Through synchronous remote mirroring the data in the backup data centre is also consistent. An incremental instant copy of the entire data set is produced there and this copy is also consistent. Now write access is permitted in the primary data centre again and data is mirrored anew to the backup data centre. If a rolling disaster occurs now, an attempt is first made to continue the operation with the data of the remote mirror. If it turns out that this data set is inconsistent, then the operation is continued using the incremental instant copy. This copy will definitely be consistent, albeit to the cost of the RPO.

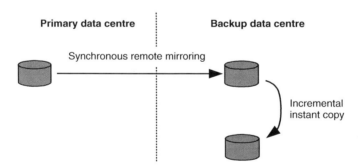

Figure 9.18   The combination of remote mirroring and incremental instant copy protects against a rolling disaster.

Solution variant: extension through double instant copies

Every so often an operation has to be switched over to a backup data centre because of maintenance work. However, if the mechanisms for incremental instant copies are only

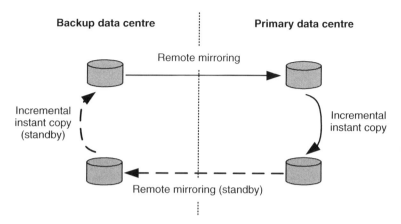

Figure 9.19 The cycle of remote mirroring and instant copy protects against a rolling disaster when operations are switched to a backup facility.

provided in the backup data centre, the operation has to be switched back to the primary data centre once the maintenance work has been completed. This produces an unnecessary further interruption to the IT operation.

Alternatively, both data centres can be set up symmetrically so that the roles "primary data centre" and "backup data centre" are interchangeable between both data centres after a switch over. This involves establishing an incremental instant copy for remote mirroring in both data centres (Figure 9.19) so that the mechanism is available in both directions.

Fully automated integrated solutions

The solution variants presented in this section show that the solutions at tier 7 must integrate several techniques together. Data is mirrored multiple times through a volume manager and remote mirroring and, depending on the requirements, is also copied using instant copy. The deployment of other high availability techniques, such as redundant I/O paths and RAID, and the use of consistency groups is a given.

These techniques must be integrated into a cluster solution that monitors the status of the different components and mirrors and initiates appropriate countermeasures when a failure occurs. This involves the coordination of numerous processes. For example, access to data must be restored when an operation is transferred to a backup data centre. Depending on the configuration, the direction of the remote mirroring has to be reversed before the applications in the backup data centre access the data and therefore can be started there. Due to the large number of components and configuration changes that have to be dealt with in the right sequence and under the pressure of time, it is essential that these processes are fully automated. Human intervention would take too long and would be too error prone.

In Section 6.3 we presented a number of high availability solutions to protect against failures of individual components and in Section 6.3.5 we showed how these solutions can be combined for the purposes of designing an important highly available database (Figure 6.22). In the example given there, the log files of the database were imported into the backup data centre with a 2-hour delay (RPO equal to 2 hours). If a loss-free operation should be guaranteed for the high availability solution in Figure 6.22 (RPO = 0), then the mirroring between the two data centres must be converted to synchronous remote mirroring and consistency groups. As an alternative to solutions of this type, Figure 9.20 shows an implementation that combines the techniques of high availability and disaster recovery in order to do justice to the high requirements of RTO and RPO.

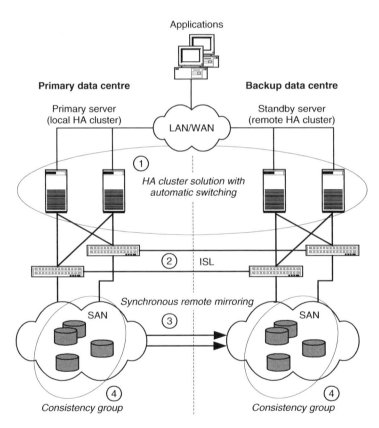

Figure 9.20   The example combines techniques for high availability and disaster recovery. The cluster between the computer centres protects against server failures (1). The storage network that is designed with in-built redundancy tolerates component failures, such as switch failures and breaks in cables (2). The synchronous remote mirroring (3) and consistency groups (4) provide disaster recovery. For increasing availability the data can also be mirrored over a volume manager (not shown in the example).

9.6   SWITCH OF OPERATIONAL LOCATION

Many of the business continuity solutions presented in this chapter are based on a switch of the operation from one location to another. A location switch involves numerous configuration changes that must be coordinated and executed without error. Depending on the restart time required, an integrated automation of these configuration changes is essential. Consideration should also be given to automation even if the requirement is for medium restart times. A failure that forces an switch of the operational location can occur at any time and without warning. A facility usually only has reduced staffing in the evenings, on weekends and during holiday periods. These people normally have less training than regular staff. If they have to analyse the current status of the systems and execute all configuration changes manually and under the pressure of time, it is likely that errors will occur.

Some companies require that the initiation of a location switch must be confirmed by an operator. This delays the start-up in the backup data centre and should therefore be well thought out. Even if a business continuity plan permits this kind of release through human intervention, the location switch itself should be largely automated. A fully automated solution is not always required. Thoroughly tested scripts that can be executed manually should be considered as a minimum.

It is not only emergencies that initiate an urgent switch of the operational location. Maintenance work is often needed in a data centre and this too requires a switch in location. In these cases the switch of location can be planned in advance and initiated outside of peak hours. Operations in the primary data centre are shut down in a coordinated way and then switched over with consistent data to the backup data centre. A high level of automation should be strived for here to reduce the number of administrator errors.

Every switch of location is generally associated with risks. Even if it is totally automated, each location switch is a special situation in which many configuration changes are executed in a short period of time. Even minor occurrences, such as a short fault in a LAN or a delay in shutting down an application, can be misinterpreted by people as well as by automated solutions. Therefore, the decision for a switch in location – whether it is to be initiated manually or is to be automated – should be well thought out and based on criteria of the business continuity plan that can be clearly measured.

Business continuity requirements for emergency operations are an important aspect of a location switch. Depending on the disruption, primary data centres and backup data centres will not be usable for a long period of time. In these situations the requirements for RTO and RPO are therefore disregarded. A complete business continuity strategy must also take into account phases with longer-term emergency operation.

9.7   ORGANISATIONAL ASPECTS

Business continuity solutions are complex creations that require large numbers of components to be coupled closely together. In medium-sized environments several ten or hundred virtual hard disks need to be mirrored or copied onto different disk subsystems at different locations. This requires a careful configuration of the techniques presented in this chapter so that the RTOs and RPOs established in a business continuity plan can actually be achieved.

It is therefore important to validate and test different failure scenarios on a regular basis. This begins with a review of the current business continuity plan and ends with a test of the identified recovery procedures. These actions check whether all emergency plans and procedures are complete and can be carried out by personnel. In addition, personnel receive training in the execution of a restart so that the necessary actions will be well rehearsed if a serious situation arises. These tests determine whether the requirements stipulated in a business continuity plan can actually be achieved. Actual situations have shown only too often that an orderly restart was not possible in an emergency. These types of tests find errors in a controlled environment before it is too late and they are not discovered until a situation becomes serious.

The qualifications and experience of staff are also very important during planning, implementation and tests and not only in a crisis. For example, IT personnel must have an absolutely clear perception of the point at which a company is facing a catastrophe and when to initiate disaster recovery. This requires that the appropriate people are familiar with and always understand the business continuity plan. Business continuity solutions can be extensively automated. If the automation does not take effect, it is essential that trained personnel intervene to control the situation. Even the best conceived business continuity solution is doomed to failure if the processes are not adequate and there is a lack of trained personnel to resume IT operations in a crisis situation.

The planning, implementation and testing of business continuity solutions must be underpinned by respective resources (personnel and material). A business continuity program does not consist only of technical content. The organisational aspects of a company also play a major role for business continuity, if not the main role. The necessity for business continuity must be firmly anchored in a company's strategy and the management must fully support these strategies. Only then it is ensured that the topic of business continuity will be awarded the necessary importance and allocated the appropriate resources at all levels.

9.8 SUMMARY

In today's markets it is important for many companies to guarantee the continuous running of their business processes. This requires continuous and loss-free operation and, consequently, a rapid recovery of data and IT services in the event of a failure. This chapter presented the general conditions of business continuity and showed that it is possible to fall back on the proven methods of risk management in the creation of a cost-efficient business continuity plan. In this connection the context of business processes should always be considered in an assessment of risks, failure and their consequences.

Another focus of the chapter was the presentation of parameters for quantifying the requirements for business continuity solutions, with special importance given to Recovery Time Objective (RTO) and Recovery Point Objective (RPO). In addition, the seven-tier model for business continuity was explained and typical RTO and RPO requirements provided for each tier.

We have used these resources to highlight the differences between the two business continuity strategies, which are high availability and disaster recovery, and derived solutions for different requirements. In so doing we made use of a number of techniques that we had presented earlier in the book and contrasted them in respect of their RTO and RPO. The basic techniques for high availability are elimination of single points of failure and automatic switching to a backup facility in a failure situation. The basic techniques for disaster recovery are data copying with the variants of instant copies, remote mirroring, volume manager mirroring, replication techniques and network backup. For continuous business operations a combination of the techniques for high availability and disaster recovery is necessary. Lastly, we showed that the techniques must be supported through appropriate organisational measures and qualified personnel to ensure that IT systems are designed and operated in a way that meets a company's requirements for the continuity of its business processes.

Today business continuity is an important issue for the competitiveness of companies and will be even more important in the future. Service Level Agreements (SLAs) in the form of contracts and, increasingly, regulations are more frequently requiring continuous access to data and applications. Business continuity solutions must always be tailored to the requirements of the respective enterprise. The concepts presented in this chapter have pointed out possibilities for developing a better understanding of the requirements for the different areas of business continuity and making an effective selection of the appropriate solutions.

This chapter ends the discussions on the use of storage networks. In the two chapters that follow we deal with the management of storage networks.

10

Management of Storage Networks

In the course of this book we have dealt with the different techniques that are used in storage networks and the benefits that can be derived from them. As we did so it became clear that storage networks are complex architectures, the management of which imposes stringent demands on administrators. In this chapter we therefore want to look at the management of storage networks.

We start by considering the requirements for management systems in storage networks from different angles, arriving at five basic services that an integrated management system must provide an administrator over a central management console (Section 10.1). The components of storage networks offer a variety of different interfaces so that these can be integrated into management systems and monitored and controlled there (Section 10.2). We then take a detailed look at standardised interfaces for in-band management (Section 10.3) and out-band management (Section 10.4). We close the chapter with a discussion about the operational aspects of storage network management (Section 10.5).

10.1 REQUIREMENTS

The requirements for a management system for storage networks can be divided into user-related (Section 10.1.1), component-related (Section 10.1.2) and architecture-related (Section 10.1.3) requirements. This produces the necessity for *one* integrated management tool for storage networks (Section 10.1.4) that provides five basic services from a central management console for all components in the storage network (Section 10.1.5).

Storage Networks Explained: Basics and Application of Fibre Channel SAN, NAS, iSCSI, InfiniBand and FCoE, Second Edition
U. Troppens R. Erkens W. Müller-Friedt N. Haustein R. Wolafka © 2009 John Wiley & Sons, Ltd

10.1.1   User-related requirements

The management of storage networks is of different significance to various technical fields. For example, the classical network administrator is interested in the question of how the data should be transported and how it is possible to ensure that the transport functions correctly. Further aspects for him are the transmission capacity of the transport medium, redundancy of the data paths or the support for and operation of numerous protocols (Fibre Channel FCP, iSCSI, NFS, CIFS, etc.). In short: to a network administrator it is important how the data travels from A to B and not what happens to it when it finally arrives at its destination.

This is where the field of interest of a storage administrator begins. He is more interested in the organisation and storage of the data when it has arrived at its destination. He is concerned with the allocation of LUNs to the servers (LUN mapping) of intelligent storage systems or the RAID levels used. A storage administrator therefore assumes that the data has already arrived intact at point B and concerns himself with aspects of storage. The data transport in itself has no importance to him.

An industrial economist, on the other hand, assumes that A, B and the route between them function correctly and concerns himself with the question of how long it takes for the individual devices to depreciate or when an investment in new hardware and software must be made.

A balanced management system must ultimately live up to all these different requirements equally. It should cover the complete bandwidth from the start of the conceptual phase through the implementation of the storage network to its daily operation. Therefore, right from the conception of the storage network, appropriate measures should be put in place to subsequently make management easier in daily operation.

10.1.2   Component-related requirements

A good way of taking into account all aspects of such a management system for a storage network is to orientate ourselves with the requirements that the individual components of the storage network will impose upon a management system. These components include:

- *Applications*
  These include all software that processes data in a storage network.
- *Data*
  Data is the term used for all information that is processed by the applications, transported over the network and stored on storage resources.
- *Resources*
  The resources include all the hardware that is required for the storage and the transport of the data and the operation of applications.
- *Network*
  The term network is used to mean the connections between the individual resources.

Diverse requirements can now be formulated for these individual components with regard to monitoring, availability, performance or scalability. Some of these are requirements such as monitoring that occur during the daily operation of a storage network, others are requirements such as availability that must be taken into account as early as the implementation phase of a storage network. For reasons of readability we do not want to investigate the individual requirements in more detail at this point. In Appendix B you will find a detailed elaboration of these requirements in the form of a checklist. We now wish to turn our attention to the possibilities that a management system can offer in daily operation.

10.1.3 Architectural requirements

In the conventional server-centric IT architecture it is assumed that a server has directly connected storage. From the perspective of system management, therefore, there are two units to manage: the server on the one hand and the storage on the other. The connection between server and storage does not represent a unit to be managed. It is primarily a question of how and where the data is stored and not how it is moved.

The transition to storage-centric IT architecture – i.e. the introduction of storage networks – has greatly changed the requirements of system management. In a storage network the storage is no longer local to the servers but can instead be located in different buildings or even different parts of the city. Thereby servers and storage devices are decoupled by multiple virtualisation layers, which makes it even more difficult to understand the assignment of storage capacity to servers. Such IT environments need suitable management systems to answers questions like, which application will be impacted by a two hour maintenance of a certain disk subsystem.

In the network between the servers and storage devices, numerous devices (host bus adapters, hubs, switches, gateways) are used, which can each affect the data flow. In a storage network there are thus many more units to manage than in a server-centric IT architecture. Now administrators have to think not only about their data on the storage devices, but also about how the data travels from the servers to the storage devices. The question, which applications are impacted by the maintenance of a switch, is difficult to answer as well, if respective management systems are not available.

10.1.4 One central management system

As explained in the previous section, a storage network raises numerous questions for and imposes many requirements upon a management system. Many software manufacturers tackle individual aspects of this problem and offer various management systems that address the various problem areas. Some management systems concern themselves more with the commercial aspects, whereas others tend to concentrate upon administrative interests. Still other management systems specialise in certain components of the storage network such as applications, resources, data or the network itself. This results in numerous different system management products being used in a complex and heterogeneous storage network. The lack of a comprehensive management system increases the complexity of and the costs involved in system management. Analysts therefore assume that in addition to the simple costs for hardware and software of a storage network, up to ten times these costs will have to be spent upon its management.

What is then needed is an integrated management system that can be used from a central location to manage, if possible, all aspects of the storage network. This management system must encompass all the components of a storage network. This ranges from block-based storage, such as disk subsystems and tape libraries, and network components, such as cables and switches, all the way to file systems and databases and the applications which run on them. Therefore, the goal is to develop a central management system in which all components are integrated and can be managed from a central location.

10.1.5   Five basic services

To assist with the daily tasks involved, an integrated management system for storage networks must provide the administrator with five basic services over a central management console.

The discovery component detects the applications and resources used in the storage network automatically. It collects information about the properties, the current configuration and the status of resources. The status comprises performance and error statistics. Finally, it correlates and evaluates all gathered information and supplies the data for the representation of the network topology.

The monitoring component compares continuously the current state of applications and resources with their target state. In the event of an application crash or the failure of a resource, it must take appropriate measures to raise the alert based upon the severity of the error that has occurred. The monitoring components performs error isolation by trying to find the actual cause of the fault in the event of the failure of part of the storage network.

The central configuration component significantly simplifies the configuration of all components. For instance, the zoning of a switch and the LUN masking of a disk subsystem for the setup of a new server can be configured centrally where in the past the usage of isolated tools was required. Only a central management system can help the administrator to coordinate and validate the single steps. Furthermore it is desired to simulated the effects of potential configuration changes in advance before the real changes are executed.

The analysis component collects continuously current performance statistics, error statistics and configuration parameters and stores them in a data warehouse. These historic data enables trend analysis to determine capacity limits in advance to plan necessary expansions on time. This supports operational as well as economic conclusions.

An further aspect is the spotting of error-prone components and the detection of single point of failures.

The data management component covers all aspects regarding the data such as performance, backup, archiving and migration and controls the efficient utilization and availability of data and resources. The administrator can define policies to control the placement and the flow of the data automatically. Data management was already presented in context of storage virtualization (Chapter 5).

The management system needs mechanisms and interfaces for all resources in the storage network, to provide previously presented requirements and to integrate all management tasks in a *single* central management tool. These interfaces will be presented in the next sections.

10.2 CHARACTERISATION OF MANAGEMENT INTERFACES

In the literature and in the standards for storage network management, servers and storage devices are frequently described as end devices (Figure 10.1). They are connected together over connection devices such as switches, hubs and gateways. Connection devices and end devices are also referred to as nodes because they appear to form the nodes of a network, which is formed from cables. End devices are then appropriately referred to as end nodes.

The interfaces for the management of storage networks are differentiated as in-band interfaces (Section 10.2.1) and out-band interfaces (Section 10.2.2). Standardised (Section 10.2.3) as well as proprietary (Section 10.2.4) interfaces are used for both types. This section closes with the maxim that all these interfaces have their purpose (Section 10.2.5).

10.2.1 In-band interfaces

The interfaces for the management of end-point devices and connection devices are differentiated into in-band and out-band interfaces. All devices of a storage network have an in-band interface anyway. Devices are connected to the storage network via the in-band interface and data transfer takes place through this interface. In addition, management functions for discovery, monitoring and configuration of connection devices and end-point devices are made available on this interface. These are generally realised in the form of components of the current protocol of the in-band interface. Thus, for example, in a Fibre Channel SAN, the Fibre Channel protocol makes the appropriate in-band management functions available. The use of these services for the management of storage networks is then called in-band management.

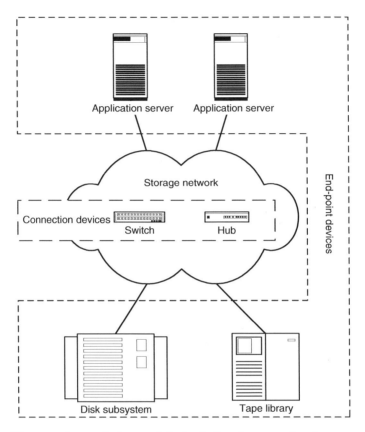

Figure 10.1   There are two main types of device in the storage network: connection devices and end-point devices. Devices are also called nodes since they seemingly form the nodes of the network.

10.2.2 Out-band interfaces

Most connection devices and complex end-point devices possess one or more further interfaces in addition to the in-band interface. These are not directly connected to the storage network, but are available on a second, separate channel. In general, these are LAN connections and serial cables. This channel is not intended for data transport, but is provided exclusively for management purposes. This interface is therefore called out-of-band or out-band for short. Management functions are made available over this additional interface using a suitable protocol. Thus Fibre Channel devices generally have an additional LAN interface and frequently possess a serial port in addition to their

Fibre Channel ports to the storage network. The use of the management services that are provided by means of the out-band interface is called out-band management.

10.2.3 Standardised interfaces

The standardisation for in-band management is found at two levels. The management interfaces for Fibre Channel, TCP/IP and InfiniBand are defined on the in-band transport levels. In Section 10.6.1 we will discuss in detail the management interface of the transport levels of the Fibre Channel protocol.

Primarily SCSI variants such as Fibre Channel FCP and iSCSI are used as an upper layer protocol. SCSI has its own mechanisms for requesting device and status information: the so-called SCSI Enclosure Services (SES). In addition to the management functions on transport levels a management system can also operate these upper layer protocol operations in order to identify an end device and request status information.

Special protocols such as Simple Network Management Protocol (SNMP) (Section 10.7.1) and Web-Based Enterprise Management (WBEM) with Common Information Model (CIM) (Section 10.7.2) as well as Storage Management Initiative Specification (SMI-S) (Section 10.7.3) are used for the out-band management.

Standardisation organisations, such as Storage Networking Industry Organisation (SNIA), Internet Engineering Task Force (IETF) and Fibre Channel Industry Association (FCIA), are working on the standardisation of management interfaces. The goal of this standardisation is to develop a uniform interface that can be used to address all the components of a storage network. For the suppliers of management systems this will reduce the complexity and cost involved in integrating components into these systems. It will enable customers to purchase storage components and management systems for storage networks from different vendors, thereby providing them with a greater choice in the procurement of new components.

10.2.4 Proprietary interfaces

Non-standardised interfaces are also referred to as being proprietary. They are usually vendor-specific and sometimes even device-specific. The standardisation usually lags behind the state of technology. Therefore, proprietary interfaces have the advantage that they offer services beyond those of the standard and thus enable more intensive and more device-specific management intervention for certain devices.

However, a big investment in terms of development and testing is made in propriety interfaces to enable them to be integrated into central management systems, because a corresponding software module has to be developed for each proprietary interface. But this effort and expense does not deter many manufacturers from supplying device support that is based more closely on the requirements of central management systems than standardised interfaces are able to provide. If a management system does not support proprietary interfaces, customers will not have the flexibility to combine established management systems with any number of other components.

Proprietary interfaces are differentiated as application programming interfaces (APIs), Telnet and Secure Shell (SSH) based interfaces and element managers. A great number of devices have an API over which special management functions can be invoked. These are usually out-band, but can also be in-band, implementations.

Many devices can also be configured out-band via Telnet. Although Telnet itself is not a proprietary mechanism, it is subject to the same problems as an API regarding connection to a central management system. For this reason we will count it amongst the proprietary mechanisms.

An element manager is a device-specific management interface. It is frequently found in the form of a graphical user interface (GUI) on a further device or in the form of a web user interface (WUI) implemented over a web server integrated in the device itself. Since the communication between element manager and device generally takes place via a separate channel next to the data channel, element managers are classified amongst the out-band management interfaces.

Element managers have largely the same disadvantages in a large heterogeneous storage network as the proprietary APIs. However, element managers can be more easily integrated into a central management system than can an API. To achieve this, the central management system only needs to know and call up the appropriate start routines of the element manager. WUIs are started by means of the Internet browser. To call up a GUI this must be installed upon the computer on which the management system runs. In that way element managers to a certain degree form the device-specific level in the software architecture of the management system.

10.2.5 Conclusion

There are advantages and disadvantages to all these interfaces. It depends mainly on which interfaces are used by the devices in a storage network and how extensive they are. In some storage network environments the use of an additional out-band interface may not be possible, such as for security reasons or because of a restriction in the number of LAN connections to a storage network servicing different computer centres. It is an advantage if a management system can use all existing interfaces and is able to correlate all the information from the interfaces.

10.3 IN-BAND MANAGEMENT

In-band management runs over the same interface as the one that connects devices to the storage network and over which normal data transfer takes place. This interface is thus available to every end device node and every connection node within the storage network. The management functions are implemented as services that are provided by the protocol in question via the nodes.

In-band services can be divided into the following two groups. Operational services serve to fulfil the actual tasks of the storage network such as making the connection and data transfer. Management-specific services supply the functions for discovery, monitoring and the configuration of devices. However, not only the management-specific services are of interest from a management point of view. The operational services can also be used for system management.

In order to be able to use in-band services, a so-called management agent is normally needed that is installed in the form of software upon a server connected to the storage network. This agent communicates with the local host bus adapter over an API in order to call up appropriate in-band management functions from an in-band management service (Figure 10.2).

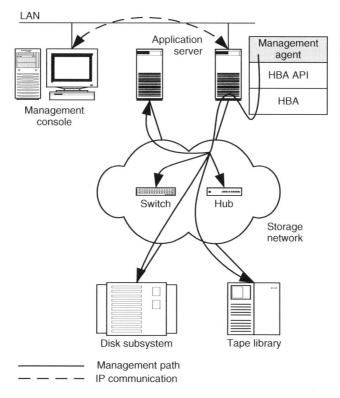

Figure 10.2   In-band management runs through the same interface that connects devices to the storage network and via which the normal data transfer takes place. A management agent accesses the in-band management services via the HBA API.

For Fibre Channel, the SNIA has already released the Fibre Channel Common HBA API, which gives a management agent easy and cross-platform access to in-band management services. The computer on which this management agent runs is also called the management agent.

For a central management system, this either means that it acts as such a management agent itself and must be connected to the storage network or that within the storage network there must be computers upon which – in addition to the actual applications – a management agent software is installed, since the in-band management services otherwise could not be used. If the management system uses such a decentral agent, then a communication between the central management console and the management agent must additionally be created so that the management information can travel from the agents to the central console (Figure 10.2). Normally this takes place over a LAN connection.

Typically, the management agent is also used for more services than merely to provide access to the management services of the in-band protocol. Some possibilities are the collection of information about the operating system, about the file systems or about the applications of the server. This information can then also be called up via the central console of the management system.

10.3.1   In-band management in Fibre Channel SAN

The Fibre Channel Methodologies for Interconnects (FC-MI) and Fibre Channel Generic Services 4 (FC-GS-4) standards defined by the American National Standards Institute (ANSI) form the basis for the in-band management in the Fibre Channel SAN. The FC-MI standard describes general methods to guarantee interoperability between various devices. In particular, this defines the prerequisites that a device must fulfil for in-band management. The FC-GS-4 standard defines management services that are made available over the so-called Common Transport Interface of the Fibre Channel protocol.

Services for management

There are two Fibre Channel services that are important to Fibre Channel SAN management: the directory service and the management service. Each service defines one or more so-called servers. In general, these servers – split into individual components – are implemented in distributed form via the individual connection nodes of a Fibre Channel SAN but are available as one single logical unit. If an individual component cannot answer a management query, then the query is forwarded to a different server component on a different node. This implementation is comparable to the Domain Name Services (DNS) that we know from IP networks.

The Fibre Channel standard defines, amongst other services, a name server, a configuration server and a zone server. These servers are of interest for the management of storage networks. These services make it possible for a management system to recognise and configure the devices, the topology and the zones of a Fibre Channel network.

The name server is defined by the directory service. It is an example of an operational service. Its benefit for a management system is that it reads out connection information and the Fibre Channel specific properties of a port (node name, port type).

The configuration server belongs to the class of management-specific services. It is provided by the management service. It allows a management system to detect the topology of a Fibre Channel SAN.

The zone server performs both an operational and an administrative task. It permits the zones of a Fibre Channel SAN fabric to be configured (operational) and detected (management-specific).

Discovery

The configuration server is used to identify devices in the Fibre Channel SAN and to recognise the topology. The so-called function Request Node Identification Data (RNID) is also available to the management agent via its host bus adapter API, which it can use to request identification information from a device in the Fibre Channel SAN. The function Request Topology INformation (RTIN) allows information to be called up about connected devices.

Suitable chaining of these two functions finally permits a management system to discover the entire topology of the Fibre Channel SAN and to identify all devices and properties. If, for example, a device is also reachable out-band via a LAN connection, then its IP address can be requested in-band in the form of a so-called management address. This can then be used by the software for subsequent out-band management.

Monitoring

Since in-band access always facilitates communication with each node in a Fibre Channel SAN, it is simple to also request link and port state information. Performance data can also be determined in this manner. For example, a management agent can send a request to a node in the Fibre Channel SAN so that this transmits its counters for error, retry and traffic. With the aid of this information, the performance and usage profile of the Fibre Channel SAN can be derived. This type of monitoring requires no additional management entity on the nodes in question and also requires no out-band access to them.

The FC-GS-4 standard also defined extended functions that make it possible to call up state information and error statistics of other nodes. Two commands that realise the collection of port statistics are: Read Port Status Block (RPS) and Read Link Status Block (RLS).

Messages

In addition to the passive management functions described above, the Fibre Channel protocol also possesses active mechanisms such as the sending of messages, so-called events. Events are sent via the storage network in order to notify the other nodes of status changes of an individual node or a link.

Thus, for example, in the occurrence of the failure of a link at a switch, a so-called Registered State Change Notification (RSCN) is sent as an event to all nodes that have registered for this service. This event can be received by a registered management agent and then transmitted to the management system.

The zoning problem

The identification and monitoring of a node in the Fibre Channel SAN usually fail if this is located in a different zone than the management agent since in this situation direct access is no longer permitted. This problem can be rectified by the setting up of special management zones, the placing of a management agent in several zones or the placing of further management agents.

The Fibre Channel protocol, however, has an even more elegant method of solving this problem. It defines services that permit the collection or entering of information. A management agent is capable of requesting the necessary information about these services, since these are not affected by the zoning problem.

The fabric configuration servers provides a so-called platform registration services. End nodes, such as a servers or a storage devices, can use this service to register information about itself in the configuration server. Management agents can retrieve this information from the configuration server later on. The end-point node information is a new proposal for the Fibre Channel protocol, which was entered under the name of fabric device management interface. This services allows the fabric itself to collect information about the properties and state statistics of ports centrally, thus management agents can retrieve these data without being limited by zoning.

10.4   OUT-BAND MANAGEMENT

Out-band management goes through a different interface than the interface used by data traffic. In Fibre Channel SANs, for example, most devices have a separate IP interface for connection to the LAN, over which they offer management functions (Figure 10.3).

For out-band management an IP connection must exist between the computer of the central management system and the device to be managed. For security reasons it can be a good idea to set up a separate LAN for the management of the storage network in addition to the conventional LAN for the data transfer.

For a long time, the Simple Network Management Protocol (SNMP) was most frequently used for Out-Band management (Section 10.4.7). In addition there are more recent developments such as the Common Information Model (CIM) and the Web Based Enterprise Management (WBEM), which can be used instead of SNMP (Section 10.4.2). Finally, Storage Management Initiative Specification (SMI-S) represents a further

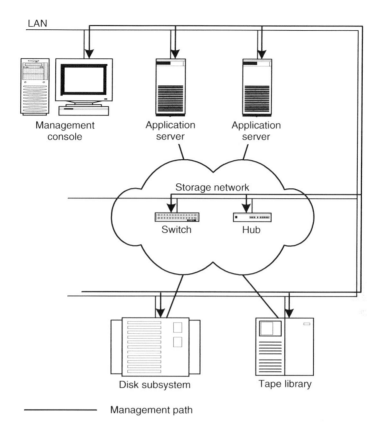

Figure 10.3   Out-band management goes through a different interface than the interface used by the data traffic. In order to operate out-band management from a central management system the management consoles and the device to be managed generally have to be able to make contact via an interface provided by IP.

development of WBEM and CIM that is specially tailored to the management of storage networks (Section 10.4.3). Furthermore, there are other protocols such as Common Management Information Protocol (CMIP) and Desktop Management Interface (DMI) that specialise in server monitoring (Section 10.4.4).

10.4.1   The Simple Network Management Protocol (SNMP)

The first version of the Simple Network Management Protocol (SNMP) was ratified in 1988 by the IETF and was originally a standard for the management of IP networks. Although there are, even now, protocols for this field that can be better adapted to the devices to be managed, SNMP is still the most frequently used protocol due to its simple architecture. Perhaps this is also the reason why SNMP has gained such great importance in the field of storage networks.

SNMP architecture

In SNMP jargon, a management application is called Network Management System (NMS) (Figure 10.4). This could be a central management system which utilises SNMP among other protocols. However, even the Syslog-Daemon of a Unix system can be used as an NMS to monitor the status of devices via SNMP. All devices are referred to as Managed Device. Each Managed Device runs a SNMP Agent, which enables the NMS to retrieve the status of the Managed Device and to configure it.

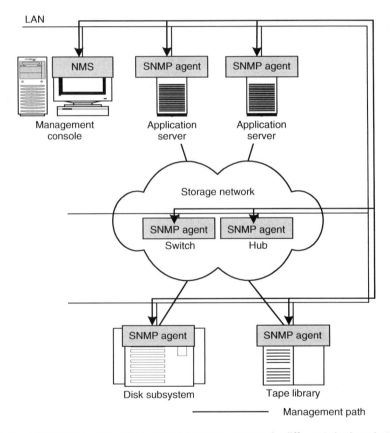

Figure 10.4 An SNMP agent can be installed upon extremely different devices in the storage network. This allows MIB information to be read via IP from a network management system (NMS).

SNMP models devices as Managed Objects. At the end, these are variables which represent the status of a device. Scalar objects define precisely one object instance. Tabular objects bring together several related object instances in the form of a so-called MIB table.

SNMP organises Managed Objects in so-called Management Information Bases (MIB). If the NMS knows the MIB of the device to be managed, then it can interrogate or change individual MIB objects by appropriate requests to the SNMP agent. The information regarding the MIB in question is loaded into the NMS in advance by means of a so-called MIB file. Since an MIB can also exist as precisely one Managed Object, Managed Objects are also called MIB objects or even just MIB. In this manner a Managed Object is identified with its MIB.

All the MIBs on the market can be divided into two groups. Standard MIBs cover general management functions of certain device classes. In addition to that vendors can develop private or so-called enterprise MIBs as proprietary MIB. Management functions can thus be offered that are specially tailored to individual devices and extend beyond the functions of the standard MIBs.

There are two important standard MIBs for the management of a Fibre Channel SAN. The Fabric Element MIB developed by the SNIA is specialised for Fibre Channel switches and supplies detailed information on port states and port statistics. Likewise, connection information can be read over this.

The Fibre Channel Management MIB was developed by the Fibre Alliance. It can be used to request connection information, information on the device configuration or the status of a device. Access to the fabric name server and thus the collection of topology information is also possible.

In order to differentiate between the individual managed objects there is an MIB hierarchy with a tree structure (Figure 10.5). The various standardisation organisations form the top level of the tree. From there, the tree branches to the individual standards of this organisation and then to the actual objects, which form the leaves of the hierarchy tree. In this manner an individual MIB object can be clearly defined by means of its position within the MIB hierarchy. In addition, each managed object is given a unique identification number, the so-called object identifier. The object identifier is a sequence of digits that are separated by points. Each individual digit stands for a branch in the MIB tree and each point for a junction. The full object identifier describes the route from the root to the MIB object in question. For example, all MIB objects defined by the IBM Corporation hang under the branch 1.3.6.1.4.1.2 or in words iso.org.dod.internet.private.enterprises.ibm (Figure 10.5). Thus all object identifiers of the MIB objects that have been defined by IBM Corporation begin with this sequence of numbers.

SNMP defines four operations for the monitoring and configuration of managed devices. The Get request is used by the NMS in order to request the values of one or more MIB object instances from an agent. The GetNext request allows the NMS to request the next value of an object instance within an MIB table from an agent after a prior Get request. The Set request allows the NMS to set the value of an object instance. Finally, the Trap operation allows the SNMP agent to inform the NMS independently about value changes of object instances.

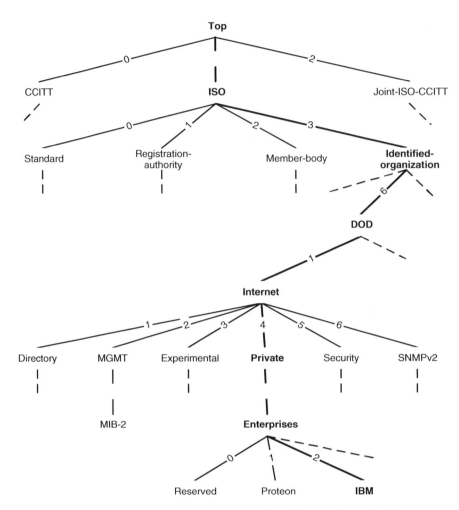

Figure 10.5 The MIBs are kept in a MIB hierarchy. The section of this tree structure represented shows the path through the tree to the MIB object that has been defined by the IBM Corporation. The actual MIB objects – the leaves of the tree – are not shown.

SNMP has no secure authentication options. Only so-called community names are issued. Each NMS and each SNMP agent is allocated such a community name. The allocation of community names creates individual administrative domains. Two communication partners (an NMS and an SNMP agent) may only talk to each other if they have the same community name. The most frequently used community name is 'public'.

If, for example, an NMS makes a Set request of an SNMP agent, then it sends its community name with it. If the community name of the NMS corresponds with that of the SNMP agent, then this performs the Set operation. Otherwise it is rejected. Thus anyone who knows the community name can make changes to the values of an object instance. This is one reason why many providers of SNMP-capable devices avoid the implementation of Set operations on their SNMP agent, because community names only represent a weak form of authentication. In addition, they are transmitted over the network unencrypted.

Management of storage networks with SNMP

SNMP covers the entire range of devices of a storage network. An SNMP agent can be installed upon the various devices such as servers, storage devices or connection devices. Particular MIBs such as the previously mentioned Fabric Element MIB and the Fibre Channel Management MIB ensure elementary interoperability between NMS and SNMP Agent.

A management system can address the SNMP agent on a connection device in order to interrogate the properties of a device and to obtain information from it about the connected devices. It thus also gets to know the immediate neighbours and with that information can continue scanning the end devices insofar as all devices lying on this route support SNMP. In this manner a management system finally obtains the topology of a storage network.

SNMP also supports the management system in the monitoring of the storage network. The SNMP agents of the end nodes can be addressed in order to ask for device-specific status information. Corresponding error and performance statistics can thus be requested from the SNMP agent of a connection node, for example, from a Fibre Channel switch.

Due to the Trap operation SNMP is also familiar with the concept of messages. In SNMP jargon these are called traps. In this manner an SNMP agent on a device in the storage network can send the management system information via IP if, for example, the status has changed. To achieve this only the IP address of the so-called trap recipient has to be registered on the SNMP agent. In our case, the trap recipient would be the management system.

In contrast to the Registered State Change Notification (RSCN), in the in-band management of a Fibre Channel SAN, in which all registered nodes are informed about changes to a device by means of a message, an SNMP trap only reaches the trap recipients registered in the SNMP agent. In addition, the connection-free User Datagram Protocol (UDP) is used for sending a trap, which does not guarantee the message delivery to the desired recipient.

SNMP also offers the option of changing the configuration of devices. If the device MIB is known, this can be performed by changing the value of the MIB variables on a managed device by means of the Set operation.

10.4.2 CIM and WBEM

Today, numerous techniques and protocols are used for system management that, when taken together, are very difficult to integrate into a single, central management system.

Therefore, numerous tools are currently used for system management that all address only a subsection of the system management.

Web Based Enterprise Management (WBEM) is an initiative by the Distributed Management Task Force (DMTF), the aim of which is to make possible the management of the entire IT infrastructure of a company (Figure 10.6). WBEM uses web techniques such as XML and HTTP to access and represent management information. Furthermore, it defines interfaces for integrating conventional techniques such as SNMP.

WBEM defines three columns that standardise the interfaces between resources and management tools (Figure 10.7). The Common Information Model (CIM) defines an

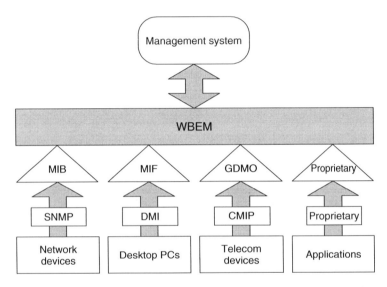

Figure 10.6 The objective of Web-Based Enterprise Management (WBEM) is to develop integrated tools that can manage the entire IT infrastructure of a company.

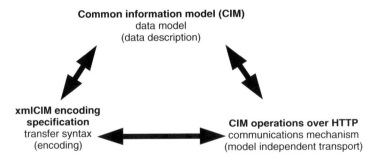

Figure 10.7 The three columns of WBEM standardise the modelling language (CIM), the transfer syntax (xmlCIM) and the transport mechanism (CIM operations over HTTP).

object-oriented model that can describe all aspects of system management. It is left up to the components participating in a WBEM environment how they realise this model, for example in C++ or in Java. The only important thing is that they provide the semantics of the model, i.e. provide the defined classes and objects plus the corresponding methods outwards to other components.

The xmlCIM Encoding Specification describes the transfer syntax in a WBEM environment. It thus defines precisely the XML formats in which method calls of the CIM objects and the corresponding returned results are encoded and transmitted. As a result, it is possible for two components to communicate with each other in the WBEM architecture, regardless of how they locally implement the CIM classes and CIM objects.

Finally, CIM operations over HTTP provide the transport mechanism in a WBEM environment that makes it possible for two components to send messages encoded in xmlCIM back and forth. This makes it possible to call up methods of CIM objects that are located on a different component.

Common Information Model (CIM)

CIM itself is a method of describing management data for systems, applications networks, devices, etc. CIM is based upon the concept of object-oriented modelling (OOM). Understanding CIM requires knowledge of OOM. OOM is based upon the concept of object-oriented programming. However, OOM is not a programming language, but a formal modelling language for the description of circumstances of the real world on an abstract level.

In OOM, real existing objects are represented by means of instances. An instance has certain properties, which are called attributes, and allows for execution of specific actions, which are called methods.

A class in OOM is the abstract description of an instance, i.e. it is the instance type. To illustrate: a Porsche Cabriolet in a car park is an object of the real world and is represented in OOM as an object instance. A Porsche Cabriolet is of the type car, thus it belongs to the class car. Thus, from an abstract point of view a Porsche Cabriolet – just like a BMW, Mercedes, etc. – is nothing more than a car. We hope that Porsche fans will forgive us for this comparison!

Classes can have subclasses. Subclasses inherit the attributes and methods of the parent class (Figure 10.8). Examples of subclasses of the class car are sports cars or station wagons. An inherited property in this case is that they have a chassis and four wheels. As we see, OOM can also be used to great effect for the description of non-computer-related circumstances.

In order to describe complex states of affairs between several classes, a further construct is required in OOM: the association (Figure 10.8). An association is a class that contains two or more references to other classes. In that way it represents a relationship between two or more objects. Such relationships exist between individual classes and can themselves possess properties. Let us consider the class 'person'. In the language of OOM 'man' and 'woman' are subclasses of the parent class 'person'. A relationship between

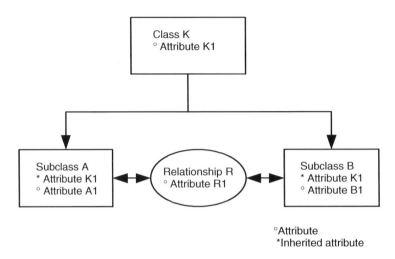

°Attribute
*Inherited attribute

Figure 10.8 Classes pass on their attributes to subclasses. Connections between classes are described as relationships, which again have their own attributes.

the class 'man' and the class 'woman' could be 'marriage'. A property of this relationship would be the date of the wedding, for example.

Modelling with CIM

Complex management environments can be described in abstract terms with the aid of the class and relationship constructs. An abstract description of a management environment using OOM is called a schema in CIM. CIM has three different types of schema: the Core Schema (also known as Core Model), the Common Schema (also known as Common Model) and further Extension Schemas (Figure 10.9).

The Core Schema defines the abstract classes and their relationships necessary for the description and analysis of complex management environments. The Core Schema specifies the basic vocabulary for management environments. For example, the Core Schema specifies that in a management environment elements to be managed exist, which themselves have logical and physical components. A strict differentiation is always made between logical and physical units or properties. Systems, applications or networks represent such elements to be managed and can be realised in CIM as extensions of this Core Schema. The Core Schema thus yields the conceptual template for all extensions.

The Common Schema builds upon the Core Schema to supply the abstract classes and relationships for all those components that the different management environments have in common, regardless of the underlying techniques or implementations. The abstract classes of the Common Schema all arise by inheritance from the classes of the Core Schema. The Common Schema defines the several submodels. The system model brings together all objects that belong in a management environment.

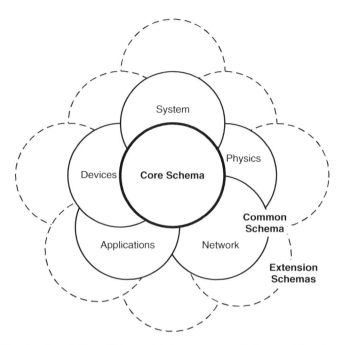

Figure 10.9   The basis of the CIM model is the Core Schema. The Common Schema supplies abstract classes and relationships on the basis of the Core Schema for all those components that the different management environments have in common. By means of Extension Schemas the abstract basic Core Schema and Common Schema can be concretised and expanded.

The device model represents the logical units of such a system that provide the system functions on a purely logical level in order to remain independent of the physical implementation. This makes sense because physical aspects of devices change but their importance on a logical level generally remains the same for the management of a system as a whole. For example, a Fibre Channel switch differs from an iSCSI switch only in terms of its physical properties. The logical property of acting as a connection node to connect other nodes in the network together is common to both components.

The application model defines the aspects that are required for the management of applications. The model is designed so that it can be used to describe both single location applications and also complex distributed software.

The network model describes the components of the network environment such as topology, connections and the various services and protocols that are required for the operation of and access to a network.

The physical model makes it possible to also describe the physical properties of a management environment. However, these are of low importance for the management, since the important aspects for the management are on a logical and not a physical level.

Extension Schemas permit the abstract basic Core Schema and Common Schema to be further concretised and expanded. The classes of the Extension Schema must be formed by inheritance from the classes of the Core and Common Schemas and the rules of the CIM model have to be applied. In this manner it is possible to use CIM for the description of the extremely different management data of the components of a storage network, which can be used for management by means of the following WBEM architecture.

WBEM architecture

Web Based Enterprise Management (WBEM) describes the architecture of a WBEM management system using the three pillars CIM, xmlCIM and CIM operations over HTTP. To this end, WBEM introduces additional elements to concretise the architecture (Figure 10.10). WBEM refers all objects to be managed as CIM managed objects. This can, for example, be a storage device or an application.

The CIM provider supplies the management data of a managed object. In the terminology of CIM this means that it provides the instances of the object that are defined in the CIM model for the managed device. The interface between CIM provider and CIM managed object is not described by WBEM.

Exactly, this interface being the starting point for the integration of other protocols such as SNMP. For instance, a CIM provider can manage a device via SNMP and represent it towards the management system as CIM managed object.

The CIM object manager (CIMOM) implements the CIM repository and provides interfaces for CIM provider and CIM clients (e.g. central management applications). The specification of the interfaces between CIM provider and CIMOM is also not part of WBEM, so manufacturer-specific mechanisms are used here too. The CIM repository contains templates for CIM models and object instances.

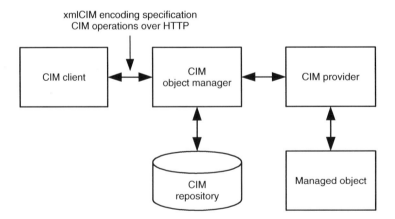

Figure 10.10   Data on a managed object is made available via a CIM provider. CIM clients can access this data by means of the CIM object manager.

The CIM client corresponds to a management system. The CIM client contacts the CIMOM in order to discover managed objects and to receive management data from the CIM provider. The communication between CIM client and CIMOM is based upon the techniques xmlCIM and CIM operations over HTTP described above. These should facilitate interoperability between CIM clients and CIMOMs of different manufacturers.

The CIM specification also provides a mechanism for sounding an alert in case of a state change of an object. The recognition of a state change such as creation, deletion, update or access of a class instance is called a Trigger. A Trigger will result in the creation of a short-living object called Indication which is used to communicate the state change information to a CIM client through the WBEM architecture. For that the CIM client needs to subscribe for indications with the CIMOM previously.

The use of CIM and WBEM for the management of storage networks

In the past, WBEM/CIM has proved itself useful in the management of homogeneous environments. For example, the management of Windows servers is based upon these techniques. Storage networks, on the other hand, contain components from very different manufacturers. Experience from the past has shown that WBEM and CIM alone are not sufficient to guarantee the interoperability between CIM clients, i.e. management systems, and CIMOMs in the field of storage networks. In the next section we will introduce the Storage Management Initiative Specification (SMI-S), a technique that aims to fill this gap.

10.4.3   Storage Management Initiative Specification (SMI-S)

At the beginning of this chapter we talked about the need for a central management system from where all the components of a storage network can be managed from different angles Section (10.1.4). The Storage Management Initiative Specification (SMI-S) should close the gaps of WBEM and CIM with its goal of defining an open and vendor-neutral interface (API) for the discovery, monitoring and configuration of storage networks. SMI-S should therefore define a standardised management interface for heterogeneous storage networks so that these networks can be managed in a uniform way. SMI-S was started in 2001 under the name of Bluefin by a group of manufacturers called Partner Development Process (PDP). Since August 2002 the further development of Bluefin has been in the hands of SNIA and will be continued there as the SMI-S within the framework of the Storage Management Initiative (SMI).

SMI-S is based upon the WBEM architecture and expands this in two directions. First, it refines the classes of the CIM Common Schemas to include classes for the management of storage networks. For example, it introduces classes for host, fabric, LUN, zoning, etc. Second, it extends the WBEM architecture by two new services: the Directory Manager and the Lock Manager (Figure 10.11).

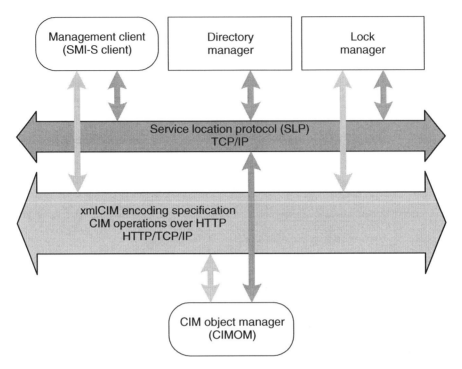

Figure 10.11 SMI-S expands the WBEM architecture: the Directory Manager and the Service Location Protocol (SLP) aim to simplify the location of resources (discovery) in the storage network. The Lock Manager helps to synchronise parallel accesses to protected methods of CIM objects.

The Directory Manager aims to simplify discovery in a storage network. SMI-S also defines how resources (physical and virtual) report to the Directory Manager by means of the Service Location Protocol (SLP, IETF standard since 1997), so that management systems can interrogate the resources in a storage network through a central service, i.e. the Directory Manager.

The Lock Manager aims to assist the concurrent access to resources from different management applications in a storage network. Access to CIM objects can be protected by locks, so that a transaction model can be implemented. To this end, SMI-S-compliant management systems must obtain the corresponding rights from the Lock Manager in the role of the lock management client, in order to be allowed to call up the protected methods.

An important aspect of SMI-S is the interoperability of the implementations of different manufacturers. The SNIA has set up the Interoperability Conformance Testing Program (ICTP) that defines standardised test standards for this purpose. This will enable

SMI-S-capable products to be tested for their interoperability with other SMI-S-capable devices and for their conformity to currently valid standards. ICTP comprises tests for SMI-S-capable CIM providers of storage components as well as for SMI-S-capable management systems. The web pages of SNIA (http://www.snia.org) show the current range of ICTP as well as the current test results.

Previously the cost of acquiring storage capacity was the main thing that pushed up costs. In the meantime the cost of the operation of storage has become higher than the actual acquisition costs. In our estimation, central management systems run on SMI-S are an important core technology for reducing the cost of operating storage networks. SMI-S was started in 2001 with basic management functions and since then has been gradually extended to cover more and more components and functions. As with any standard, SMI-S cannot take into account all device-specific peculiarities that exist. It will also take some time before new functions are integrated into SMI-S. Therefore, a central management system over SMI-S will also have to support proprietary interfaces in order to maximise the underlying storage components. Nevertheless, for the increasing number of basic functions there will be no way around using SMI-S.

10.4.4 CMIP and DMI

For the sake of completeness we will now list two further protocol standards for out-band management: the Common Management Information Protocol (CMIP) and the Desktop Management Interface (DMI). However, neither protocol has as yet made any inroads into storage networks and they have up until now been used exclusively for the monitoring of servers.

Common Management Information Protocol (CMIP)

At the end of the 1980s the CMIP was originally developed as a successor of SNMP and, together with the Common Management Information Services (CMIS), it forms part of the Open Systems Interconnect (OSI) Specification. Due to its complexity it is, however, very difficult to program and for this reason is not widespread today.

CMIP uses the same basic architecture as SNMP. The management information is also held in variables similar to the MIBs. However, in contrast to the MIBs in SNMP, variables in CMIP are comparatively complex data structures.

Like SNMP, CMIP provides corresponding operations for the reading and changing of variables and also incorporates messaging by means of traps. In addition, actions can be defined in CMIP that are triggered by the value change of a variable. CMIP has the advantage over SNMP that it has a proper authentication mechanism. The disadvantage of CMIP is that it is very resource-hungry during operation, both on the NMS side and also at the managed device.

Desktop Management Interface (DMI)

The DMI was also specified by the DMTF. It describes a mechanism by means of which management information can be sent to a management system over a network. The architecture of the DMI consists of a service layer, a database in the management information format (MIF), a management interface and a component interface (Figure 10.12). The service layer serves to exchange information between the managed servers and a management system. All properties of a managed server are stored in the MIF database. A DMI-capable management system can access a server and its components via the management interface. The component information is provided to the management interface by component interfaces. DMI thus provides an open standard for the management of servers, but is nowhere near as widespread as SNMP.

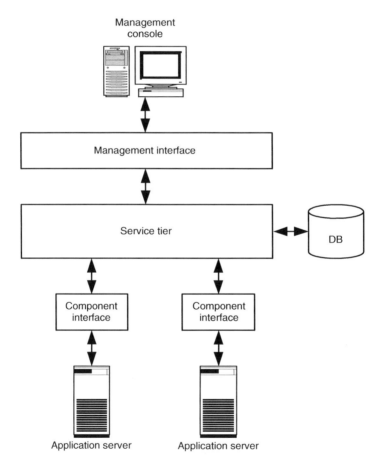

Figure 10.12 The DMI architecture allows a management system to access information of servers and their components.

10.5   OPERATIONAL ASPECTS OF THE MANAGEMENT OF STORAGE NETWORKS

In large heterogeneous environments the introduction of a management system appears indispensable for those wishing to take control of management costs and make full use of the storage network. For small environments, the implementation of a management system is recommended if the environment is expected to grow strongly in the medium term. Entry in a small environment offers the additional advantage that the management system grows with the environment and you have plenty of time to get used to the product in question. If the storage network reaches its critical size at a later date you will already be better prepared for the more difficult management. Because, by this time, the installed tools will already be well known, the optimal benefit can be drawn from them.

If you have the choice between standardised or proprietary mechanisms, then you should go for standardised mechanisms. Many device manufacturers have already built support for the standards in question into their products. Other manufacturers will follow this example. When purchasing new devices, the devices' support for standards is a critical selection criterion and should thus be checked in advance. When choosing a management system you should ensure corresponding support for the various standards. It should, however, also have interfaces for the support of proprietary mechanisms. The calling up of element managers from the management console is the minimum requirement here. Only thus can many older devices be integrated into the management system.

Which strategies should a management system use: in-band or out-band? This question cannot be answered in a straightforward manner since the success of a management system depends to a large degree upon the available interfaces and mechanisms of the devices used. In general, however, the following advantages and disadvantages of in-band and out-band management can be worked out.

The main advantage of the use of the in-band interface is that it is available as standard in the storage network. By the use of various protocol levels (transport and upper layer protocols) a great deal of detailed information about the storage network can be read. In environments where an additional out-band interface is not available or cannot be implemented, in-band monitoring may represent the only option for monitoring the status of devices.

The great disadvantage of in-band management is that a management agent connected to the storage network is required because the in-band management functions can only be used through such an agent. This can give rise to additional costs and sometimes increase the complexity of the storage network. On the developer side of management systems this naturally also increases the development and testing cost for suitable agent software.

Agent software can be used for additional services. These can be operating system-specific functions or more extensive functions such as a storage virtualisation integrated into the management system (Chapter 5). When using management agents in a Fibre Channel SAN, it should be noted that they are subject to the zoning problem.

This can be remedied by the measures described or – where already available – by more recent options such as fabric device management interfaces.

Out-band management has the advantage that it is not bound to the storage network infrastructure and technology in question. This dispenses with the necessity of support for the management system by in-band protocols such as Fibre Channel or iSCSI. Furthermore, abstract data models can be implemented with SNMP MIBs and CIM that are independent of the infrastructure. These must, however, be supplemented by infrastructure-specific data models. The fabric element MIB and the Fibre Channel management MIB are two examples in the field of SNMP.

A further advantage of the use of the out-band interface is that no dedicated management agent is required in order to gain access to management functions. A management system can communicate directly with the SNMP agents and CIM providers in question without having to make the detour via a management agent. In the event of problems with the management interface, this source of errors can be ruled out in advance. The corresponding costs and administrative effort associated with the management agent are therefore also not incurred.

A great disadvantage of out-band management is that up until now there has been no access to the operational services that are available in-band. Although this would be technically possible it has not yet been implemented. Finally, even the additional interface that is required can prevent the use of out-band management in cases where the implementation of a further interface is not possible or not desirable.

Therefore, a management system should, where possible, support both interfaces in order to get the best of both worlds. This means that it is also capable of integrating devices that only have one of the two interfaces. For devices that have access to both in-band and out-band interfaces an additional connection can be very helpful, particularly for error isolation. If, for example, the in-band connection to a device has failed, then an in-band management system would report both the failure of the line and also the failure of the device. A management system that operates both interfaces would still be able to reach the device out-band and thus trace the error to the failed connection.

10.6 SUMMARY

The previous chapters of the book covered elementary components and techniques of storage networks as well as their use. In this chapter we turned towards the management of storage networks.

At the beginning of this chapter we analysed the requirements to the management of storage networks from different point of views and derived from there the need for a central management system. Thereby it became apparent thus such a management system must provide five basic services: discovery, monitoring, configuration, analysis, and data control. The interfaces, which are used by such a management system, can be distinguished in in-band and out-band interfaces respectively proprietary and standardised interfaces. Then we discussed FC-GS for the in-band management of a Fibre Channel SAN and standardised out-band mechanisms such as SNMP, CIM/WBEM and SMI-S. It become clear that a management system for storage networks must support all type of interfaces, where SMI-S has am important role for the management of heterogeneous storage networks. At the end it can be concluded, that a storage-centric IT architecture with storage networks imposes higher requirements on management systems than do traditional server-centric IT architectures.

In the next chapter we continue the discussion of storage management: Removable media and large tape libraries are central components of large data centres. Considering the growing size of tape libraries and the increasing amount of tape cartridges, the management of removable media becomes more and more important. This is the topic of the next chapter.

11

Removable Media Management

Removable storage media, particularly magnetic tapes, are a central component of the storage architecture of large datacentres. Storage networks enable servers, and consequently, many different applications, to share media and libraries. Due to the growing use of these networks the management of removable media is becoming very important. In this chapter we will be looking at the characteristics of removable media management.

In the following section we first of all explain why, in spite of the ever-increasing capacity of hard disks and intelligent disk subsystems, removable media is indispensable (Section 11.1). Then we consider various types of removable media (Section 11.2) and libraries (Section 11.3), giving special consideration to media management. We then discuss the problems and requirements related to the management of removable media (Section 11.4). Finally, we introduce the IEEE 1244 Standard for Removable Media Management – an approach that describes both the architecture of a system for the management of removable media and also its communication with applications (Section 11.5).

11.1 THE SIGNIFICANCE OF REMOVABLE MEDIA

Articles with such titles as 'Tapes Have No Future' or 'Is Tape Dead?' keep appearing in the press. Some storage manufacturers proclaimed the end of tapes already years ago.

Storage Networks Explained: Basics and Application of Fibre Channel SAN, NAS, iSCSI, InfiniBand and FCoE, Second Edition
U. Troppens R. Erkens W. Müller-Friedt N. Haustein R. Wolafka © 2009 John Wiley & Sons, Ltd

The swan song about the end of tape is repeated in regular intervals, especially from the vendors of disk subsystems. After all, they said, hard disks (e.g. SATA disks) have now become so cheap that it is unnecessary to move data to other storage media. In our opinion, removable media is an important building block in the storage architectures of data centre.

In addition to their high capacity and low price, for many companies the fact that removable media can be stored separately from read and write devices and thus withdrawn from direct access is particularly relevant. Viruses, worms and other 'animals' are thus denied the possibility of propagating themselves uncontrollably, as they could on storage that is continuously available online. Furthermore, with removable media a very large quantity of data can be stored in a very small and possibly well-protected area at low storage costs. WORM (Write Once Read Multiple) properties, which are now available not only for optical media but also for magnetic tapes, additionally increase security. Furthermore, the requirement for storage capacity is increasing continuously. Progress in storage density and the capacity of cartridges can scarcely keep up with the ever-growing requirement, which means that the number of cartridges is also growing continuously.

For the film *The Lord of the Rings* alone, 160 computer animators generated and edited a data volume of one terabyte every day, which was stored to tape. At the end of the 3-year production period, the digital material for the final version of the film – 150 terabytes in size – was stored on tape.

The scientific-medical as well as medical and bio-informatics areas have been working with data volumes in petabytes for a long time already. This enormous demand for storage space cannot be met solely with the storage available online, such as hard disks. The cost of acquiring and operating solutions based solely on hard disks is so high that this type of storage cannot be economically justified by fast access times. There are not even any significant foreseeable changes in trends in the near future that would change this situation. Instead major efforts are still being directed towards the magnetic tape area, for example to increase storage density and thus to make the technology more competitive. The enormous cost difference between hard disks and magnetic tape will therefore remain almost constant, as has been the case in previous decades. In the end, as is also the case today, a mix of both types of media will continue to determine the storage architecture with the proportion of each media depending on individual requirements and needs. After all, in terms of robustness removable media has a clear advantage over hard disks: it is less sensitive towards surges and in the high-end tape media can have a life cycle of several decades if stored properly.

11.2   REMOVABLE MEDIA

Various types of removable media are currently in use. These are primarily magnetic tapes (Section 11.2.1), optical media such as CDs and DVDs and magneto-optical media (Section 11.2.2). In these sections we are primarily interested in how the special properties of the various media types should be taken into consideration in the management of removable media (Section 11.2.3).

11.2.1 Tapes

In contrast to disk which allow random access, tapes can only be accessed sequentially. The position of the head of a tape drive cannot, therefore, be chosen at will, but must be determined by the appropriate fast-forwarding and rewinding of the tape. This movement of the tape costs significantly more time than the movement of the head of a hard disk drive and an optimal speed can, therefore, only be achieved if as many associated data blocks as possible are read and written one after the other, i.e. sequentially (streaming).

Access to backup and archive data at will is often unnecessary. The speed at which the large quantities of data can be backed up and restored is likely to be a significantly more important factor than the random access to individual files. Backup and archiving applications available today utilise this special property of tapes by aggregating the data to be backed up into a stream of blocks and then writing these blocks onto tapes sequentially (Section 7.3.4).

11.2.2 CD, DVD and magneto-optical media

When writing to CDs, DVDs and magneto-optical media, a file system (e.g. ISO-9660) is generally applied. When writing to these media, the same limitations apply as for tapes, since only one application can write to the media at any one time. Normally, this application also writes a large portion – if not the whole – of the available storage capacity.

However, once these media have been written, applications can access them like hard disk drives. As a result, the applications have available to them the full support of the operating system for the read access to optical media, which is why in this case they behave like write-protected hard disks and can be shared accordingly. Magneto-optical media are generally readable and writeable on both sides.

Depending upon the drive, the cartridge may have to be turned over in order to access the second side. This property makes it necessary for management systems to be able to manage a second side of a cartridge and control the changing mechanism so that the cartridge can be turned over. Furthermore, the WORM properties must be suitably represented for these media.

11.2.3 Management features of removable media

In what follows we give an overview of the most important features and terms of removable media which must be modelled by a management system for removable media.

Cartridge   A cartridge is a physical medium upon which storage space is available. A cartridge can be moved and has one or more sides.

Scratch cartridge   A new cartridge without any content or an already used cartridge, the content of which is no longer of interest, and the entire storage capacity of which can be used for new purposes.

External cartridge label   A label that is applied to the outside of a cartridge and serves to identify the cartridge, for example, a mechanically readable barcode.

Internal cartridge label   A dataset in a certain format at a certain position on the tape media that serves to identify the cartridge.

Side   A physical part of a cartridge that provides storage space. A side contains one or more partitions. Tapes normally have only one side. DVDs and magneto-optical media are also available in double-sided variants.

Partition   Part of a side that provides storage space as a physical unit of the cartridge. Current (2009) removable media typically comprise only one partition for each side.

Volume   A volume is a logical data container. It serves to reserve storage space for applications on the media. A partition can hold as many volumes as desired. Please note that the term *volume* may have different meanings depending on the context it is used: in terms of the SNIA Shared Storage Model (Chapter 12) a tape volume is called a tape extent and may span multiple physical tape cartridges. In terms of backup software and mainframes a *volume* is often used synonymously with *cartridge*.

Drive   Removable media cartridges must be mounted into a drive for reading or writing.

Access handle   An identifier that an application can use to access the data of a volume. Under UNIX operating systems an access handle is equivalent to the name of a device-special file (for example: /dev/rmt0).

Mount request   The command to place a certain cartridge in a drive.

Audit trail   Audit trails consist of a series of data sets, which describe the processes that have been performed by a computer system. Audit trails are used primarily in security-critical fields in order to record and check access to data.

The management of removable media must model the logical and physical characteristics of cartridges. The currently (2009) most important removable media is magnetic tape, which is only usable on one side and cannot be partitioned. Nevertheless, with a view towards future developments of new storage technologies, management systems today should also be supporting new media that have multiple sides and partitions. For example, optical media often has two sides; holographic media will probably have even more sides. The data model for removable media management should therefore accept cartridges with an arbitrary number of sides. Each side can accept one or more partitions and each partition can accommodate any number of volumes (data containers).

11.3 LIBRARIES AND DRIVES

Tape libraries have slots in which the cartridges are stored when they are not being used. The libraries are equipped with one or more robots (media changers) that transport a cartridge from the slot to the drive as soon as a mount request is received for the cartridge.

Operating systems and applications that use removable storage media have to cope with a large selection of different library hardware (Section 11.3.1). Even if there is a considerable difference in the structure and performance of this hardware, certain general characteristics prevail: all libraries have drives that read and write cartridges (Section 11.3.2), a media changer (Section 11.3.3) as well as slots to receive the cartridges. The media changer takes the cartridges from the slots and transports them to the drives.

11.3.1 Libraries

In automated libraries the media changer is controlled over an interface. Depending on the equipment used, this interface is realised either as an in-band interface, such as SCSI, or an out-band interface (Section 11.3.3). The bandwidth of these automated libraries ranges from individual small auto loaders with one or two drives and a few slots all the way to large automated tape libraries with one or more media changers that can transport thousands of cartridges to tens or even hundreds of drives.

In addition to automated libraries, removable media management should also take into account manually operated libraries. In this case an operator takes over the functions of a media changer and places the cartridge into the appropriate drives. This enables the integration of even separate individual drives into the management system.

Depending on the level of abstraction, even a shelf or a safe filled with cartridges but without any drives can be considered a library. These libraries are also referred to as vaults or vaulting locations. Especially when both automated libraries and vaults are used, the level of abstraction chosen for the management of the media should allow the vaults to be handled as manual libraries without drives. This will enable the same procedures for auditing (querying all components, but mainly the cartridges in a library), exporting (removing a cartridge from a library) and importing (sending a cartridge to a library) to be applied to all libraries, irrespective of type.

11.3.2 Drives

Like hard drives, drives for removable media are currently equipped with a SCSI or Fibre Channel interface in the Open Systems environment and are connected to the storage network via these. In the mainframe field, ESCON and FICON are dominant.

As already mentioned, tape drives in particular can only work at full speed if they read and write many blocks one after the other (streaming). Although it is possible, and in the mainframe environment totally normal, to write individual files consisting of just one logical block to tape, or to read them from tape. Different drives are necessary for this than those used in the Open Systems backup operation. These enterprise drives have larger motors and can position the read-write heads significantly more quickly and precisely over a certain logical block.

11.3.3   Media changers

The task of media changers is to transport cartridges within a library. The start and destination of the transporting operation can be slots or drives. A library also uses an inventory directory in which all elements of the library with their attributes are noted. The media changer has access to this inventory directory.

Applications activate the media changers in automated libraries. The libraries are therefore equipped with appropriate interfaces that the applications use for sending commands and receiving acknowledgements. In addition to controlling transport operations, these interfaces are normally also used for requesting data in the inventory directories. The following information can therefore be requested over the media changer interface:

- Number and properties of the drives
- Number and properties of the slots
- Number and properties of the cartridges
- Number and properties of further media changers

In the following sections we will be taking a detailed look at certain types of interfaces. In the open systems environment a direct connection over SCSI or Fibre Channel is the most widely used in-band interface. In comparison, proprietary out-band interfaces are mostly used in the mainframe area.

In-band interface: SCSI media changer

For the operation of tape libraries the SCSI standard defines the SCSI media changer interface, which, depending on the hardware used, is either direct-attached using native SCSI cabling, or, if the library is part of a larger network, is connected over Ethernet (iSCSI and FCoE) or Fibre Channel cabling. Irrespective of the underlying transport protocol, the media changer is always controlled by utilising the SCSI media changer protocol (Section 3.3.8).

Three different methods have established themselves in this area (Figure 11.1). In the first case, the media changer is equipped with its own controller and is detectable as a separate device through its own SCSI target ID (independent media changer with its

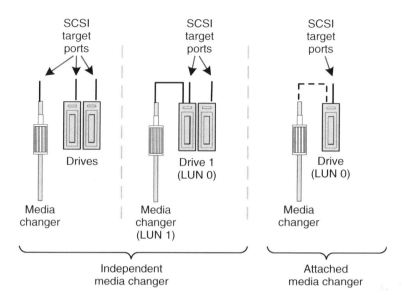

Figure 11.1 Independent media changers either possess their own SCSI target port or they can be reached as an additional device with a different LUN over the SCSI port of the drive. Attached media changers form a unit with the drive. In order to move the media changer, special SCSI commands such as READ ELEMENT STATUS ATTACHED and MOVE MEDIUM ATTACHED must be used.

own SCSI ID). In the two other cases, the media changer shares the controller with the drives. It is then either visible as an independent device with separate logical unit number (LUN) (independent media changer with its own LUN) or it is controlled through special commands over the LUN of the drive (attached media changer).

As is often the case in the IT world, there are two contrasting philosophies here, that reveal their specific advantages and disadvantages depending upon the application case. If the media changer shares the same controller with the tape drive, then the bandwidth available to the drive is reduced. However, as only a relatively small number of commands are transferred and carried out for media changers, the reduction of the bandwidth available for the drive is low. This is particularly true in the Open Systems environment, where tapes are predominantly used in streaming mode.

If, on the other hand, access is mainly file-based and if the files are located on several tapes, the ratio of media changer to drive commands increases correspondingly. In this case it can be worthwhile conducting the communication with the media changer over an additional controller. However, both this additional controller and the additional components required in the storage network make such a solution more expensive and involve additional management costs.

The addressing of the media changer over a second LUN of the drive controller has a further major advantage in addition to the low costs. Normally, several drives are fitted in a large library. Additional access paths make it possible to also control the media changer over the second LUN of a different drive controller. If a drive should fail, the media changer remains accessible via another drive. Furthermore, drives are often provided with a redundant power supply or the controllers possess an additional port, which can automatically be used if the first path fails.

Proprietary interfaces

In addition to SCSI interfaces, further interfaces have established themselves, particularly in the mainframe environment. These interfaces offer a higher level of abstraction than SCSI and often also a rudimentary management of the media. Typically, such interfaces are out-of-band, i.e. not accessible over the data path (SCSI or Fibre Channel connection), but instead over TCP/IP or RS-232 (Figure 11.2).

The commands that are exchanged over such interfaces are generally executed by a control unit that is fitted in the library. This control unit can usually accept and execute commands from several applications at the same time, without this leading to conflicts. Likewise, additional services such as the management of scratch pools can be made available to all applications.

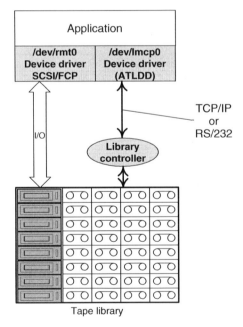

Figure 11.2 Tape library with a proprietary media changer interface. The media changer is controlled over a TCP/IP or an RS/232 interface, depending upon configuration.

11.4   PROBLEMS AND REQUIREMENTS IN RESPECT OF REMOVABLE MEDIA MANAGEMENT

The problems and requirements related to the integration of removable media management into storage networks can be divided into the areas 'storage media management' and 'sharing associated resources.' In large environments, removable media management must be able to catalogue hundreds of thousands of media. Not only are the media and its attributes stored but also the access to this media including data on errors and usage times.

In contrast to hard disk storage, removable media can be stored separately from the drives, which means that the system must be aware of the location of the medium at all times. As this location can be a manually managed depository or an automated library, special solutions are needed to reflect the respective requirements.

The second important field of application for removable media management in storage networks is the sharing of libraries, drives and media (Section 6.2.2). This tape library sharing among multiple applications connected to a storage network requires appropriate mechanisms for access control, access synchronisation and access prioritisation. These mechanisms control who may access which hardware and when so that potentially all applications can utilise all the resources available in the storage network.

To meet these requirements, storage networks increasingly need a management layer for removable media. This layer decouples applications from the hardware that is attached over the storage network (Figure 11.3). It controls and synchronises all access, remaining

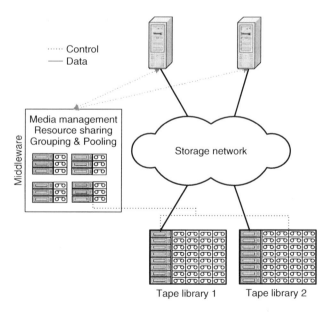

Figure 11.3   The middleware interface provides media management services, synchronises access to the hardware and permits the grouping of drives and cartridges.

as transparent as possible to the applications. As a central entity, this layer manages all resources and simplifies the sharing between heterogeneous applications, irrespective of which libraries are used.

Changes to tape hardware, such as the replacement of drives and the addition of further libraries, are encapsulated centrally so that these changes do not affect the applications themselves. Therefore, for example, the management functions for removable media that were previously embedded in the network backup system are now bundled together centrally. This kind of central management system for removable media thus represents a storage virtualisation layer within the storage network, thereby offering the advantages of storage virtualisation to removable media (Chapter 5).

We have already discussed the different options for implementing library and drive sharing for removable media into storage networks (Section 6.2.2). As we mentioned earlier, it is our opinion that the architecture that offers the best and most promising solution is the one that protects applications from the complex internal processes involved in management and sharing. The fact is that the management system is familiar with all components and how they interact together. This puts it in a position to exercise optimal control over the use of resources.

The individual problems and requirements in respect of removable media management can be listed as follows:

- Efficient use of all available resources by intelligent sharing.
- Access control: Applications and users must be authenticated. Applications and users may only be granted access to the media for which an appropriate authorisation exists.
- Access synchronisation: Access to libraries and drives must be synchronised.
- Access prioritisation: Priorities can be used should a number of access requests be executed to a resource or a drive at the same time.
- Grouping and pooling: It should be possible for media and drives to be combined dynamically into groups or pools for the purposes of simplifying management and sharing.
- Media tracking: There should be a guarantee at all times that each medium is retrievable even if it has been transferred to a different location.
- Life cycle management: Media runs through a life cycle. It is written, read, rewritten and after a certain period of time removed from circulation.
- Monitoring: Automated system monitoring.
- Reporting: Access to media must be logged. Audit trails should be possible.

We will take a detailed look at these problems and requirements in the course of this section.

11.4.1   Efficient use of the available resources

A great advantage of the use of well-designed storage networks is the fact that the available hardware resources are better utilised. In contrast to directly connected devices, the available storage space is available to many applications and can therefore be used significantly more effectively. In the field of removable media this is achieved by the better utilisation of the free storage capacity and the sharing of drives.

The disk storage pooling described in Section 6.2.1 is transferable to removable media one-to-one. In this case, free storage space should be taken to mean both unused removable media and also free slots for removable media, which must be kept in reserve due to the continuous growth in data. Ideally, this takes place in a cross-library manner. To this end, all free cartridges from all libraries are managed in a so-called scratch pool (Section 11.4.6), which is available to all applications, so that the remaining free storage capacity can be flexibly assigned.

What applies for the effective use of the free storage capacity also applies in the same way for the use of the drives. If drives are directly connected to servers they cannot be used by other servers, even if they currently have no cartridge loaded into them. By contrast, drives in storage networks can be assigned to the applications that need them at the time. Thus, it can be ensured that all drives that are installed are also actually used. In addition, a request queue can be used to increase the utilisation of the drives even further. We look at this technology in detail in Section 11.4.4.

11.4.2   Access control

Reliable control to prevent unauthorised access to media is indispensable. Users and applications must be authenticated. Successfully authenticated users can then be given suitable authorisation to access certain resources.

Authentication

Users, and also applications, that want to make use of removable media management services must be registered with the system. A sufficiently strict authentication mechanism should ensure that only users and applications that have been unambiguously identified can use the system.

Authorisation

Authorisation is necessary to prevent unauthorised users from being able to view, or even change, data belonging to other users. Authorisation can both apply for certain operations and also arrange access to certain objects (cartridges, partitions, volumes, drives, etc.). A successful authentication is a necessary prerequisite for authorisation.

By means of an appropriate authorisation, users or applications can be assigned the following rights regarding certain objects in the management system:

- Generation and deleting of objects (e.g. the allocation and deallocation of volumes);
- Read access to objects (e.g. read access to own volumes);
- Write access to objects (e.g. addition of cartridges to a scratch pool);
- Mount and unmount of cartridges, sides, partitions or volumes;
- Moving of cartridges within libraries;
- Import and export of cartridges;
- Activation and deactivation of libraries or drives.

The use of various authorisation levels allows access control to be modified according to the user's role. The following roles and activities are currently used in systems for the management of removable media and can be provided with different authorisations. This list serves as an example only. These roles and activities can also be assigned differently depending upon the specific requirements.

The *system administrator* is responsible for:

- Installation of the system
- Installation/deinstallation of libraries
- Management of users and applications
- Management of disk and cartridge groups

The *storage administrator* is responsible for:

- Management of disk and cartridge groups
- Controlling of the cartridge life cycle
- Planning of the future requirements for resources

The *library administrator* is responsible for:

- Management of drive and cartridge groups for individual libraries
- Planning of the future requirements for resources

The *library operator* is responsible for:

- Starting and stopping of the operation of individual libraries
- Monitoring the operation of libraries and its drives
- Starting and stopping of the operation of individual drives
- Import and export of cartridges into and out of libraries
- Execution of mount/unmount operations in manually operated libraries
- Moving cartridges within a library

The *users/applications* may:

- Allocate and de-allocate volumes to cartridges
- Mount and unmount volumes
- Read and write volumes
- List and display volumes that they have allocated
- List and display cartridges upon which volumes they have allocated have been put
- List and display scratch cartridges, which are included in cartridge groups to which there is an access right
- List and display drives that are included in drive groups to which there is an access right

The list makes clear that 'simple' users, for example, are not capable of managing cartridge groups or drive groups. If they were, it might be possible for a user to gain unauthorised access to foreign data. Likewise, a library operator should only have access to libraries for which he is responsible. In this manner, the smooth operation of the system as a whole can be ensured.

Authorisation for access to individual objects

It is currently still common to use authorisation procedures for entire cartridges only and not for their components. However, in order to be prepared for future developments, such as the terabyte and petabyte tape cartridge, a management system should, even now, have appropriately detailed access protection for subdivisions of cartridges such as sides, partitions and volumes.

All components of a cartridge are suitable for access control. The application purpose determines whether the user receives access to a side, a partition or a volume. An authorisation is always applicable to all elements of the authorised object. If, for example, the right to access a cartridge is granted, this right also applies to all sides, partitions and volumes of this cartridge.

It is not only cartridges that should be provided with access control. For example, it is a good idea to restrict applications' access to cartridges that are still available. To this end, the available cartridges are combined into one or more scratch pools. The applications are then granted the right to access only certain scratch pools. Common scratch pools increase the effective utilisation for free storage capacity. Dedicated scratch pools allow to reserve dedicated cartridges for certain applications or group of applications.

Drives are also suitable for access control. The grouping of drives is again an option here for simplifying management. The sharing of drives across all applications increases the utilisation of drives. Though, the assignment of certain drives to certain applications should still be possible, for instance, to reserve dedicated drives for important applications.

11.4.3 Access synchronisation

At a certain point in time, a drive can only be used by a single application. Therefore, it is required to synchronise the access from applications to drives. This type of synchronisation corresponds with dynamic tape library sharing and has already been described in Section 6.2.2. Tape library sharing enables applications to use all drives 'quasi' simultaneously. This sharing is similar to the sharing of a processor (CPU) where the operating systems allow a single processor to be made available to several processes one after the other for a limited duration.

11.4.4 Access prioritisation and mount request queuing

Access prioritisation and mount request queuing increase the effectiveness of access synchronisation. Despite tape library sharing, situations can arise in which more mount requests exist than drives are available. In this case, the management system should collect all remaining requests that cannot be processed immediately into a request queue (Figure 11.4). As soon as drives become available, the system can execute the next request and remove it from the queue. In the search for the next request for execution, a scheduler should evaluate the priorities of the request and adjust the sequence accordingly. Urgent tasks are granted a higher priority and are therefore dealt with more quickly. Depending on the implementation, this can also mean that an application is pre-empted from a drive and access to the cartridge must be interrupted.

Again depending on the implementation, request queues can be set up for each individual drive or for a group of drives. Request queues that are bound to a group of drives have the advantage that more drives are available to handle each request waiting in the queue. Consequently, these requests are normally processed more quickly than if they were waiting for one particular drive.

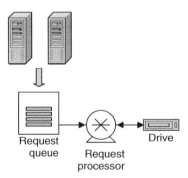

Request
queue   Request   Drive
                            processor

Figure 11.4   The mount requests are stored in a queue and processed one after the other by the request processor.

Mount request queuing increases the load on drives more so than tape library sharing. If the number of drives is not dimensioned adequately, then more time is needed to execute the mount requests in the queue. This is a typical time versus space optimisation problem. If a greater number of drives are available, then a larger number of mount requests can be completed in the same amount of time. However, if many mount requests are not urgent and can be left in a queue for a long period of time, fewer drives are needed and a cost saving is realised in terms of procurement and operation.

11.4.5 Grouping, pooling

Systems for the management of removable media must be able to deal with a great many cartridges and drives media and facilitate access to many applications. In order to plan and execute access control in a sensible manner and to guarantee its effective use, it is necessary to combine cartridges and drives into groups. Grouping also allows budgets for storage capacities or drives to be created, which are made available to the applications.

Scratch pools

A scratch pool contains unused cartridges that are available to authorised applications so that they can place volumes upon them. As soon as an application has placed a volume upon a cartridge from such a scratch pool, this cartridge is no longer available to all other applications and is thus removed from the scratch pool.

Scratch pools should be sized large enough, thus the management for removable media is capable at every point in time, to provide free storage capacity to applications. This is a basic requisite to meet backup windows. Thus scratch pools should be dynamically expandable, to expand the infrastructure for the always growing demand for storage capacity. To this end, an administrator must make additional storage space available to the system dynamically, whether by the connection of a new library or by the addition of previously unused cartridges. Ideally, this can be achieved without making changes to the applications that have already access to the scratch pool.

An adjustable minimum size (low water mark) makes the management of a scratch pool easier. If this threshold is reached, measures must be taken to increase the size of the pool, as otherwise there is the danger that the system will cease to be able to provide free storage space in the foreseeable future. The management system can help here by flexibly offering more options. Many actions are possible here, from the automatic enlargement of the scratch pools – as long as free media are available in the libraries – to the 'call home' function, in which an administrator is notified.

For large environments it is advantageous to define multiple scratch pools. For instance, it is a common requirement to separate usual cartridges and WORM cartridges. Or sometimes it is desired to reserve storage capacity for certain applications. Thus a storage administrator should be capable to provide a dedicated scratch pool for very important backup jobs where 'default' backup jobs receive their scratch cartridges from a default scratch pool. In that manner the backup application can choose which scratch pool to use.

The selection of scratch pools can be automated via priorities. If an application has access to multiple scratch pools, then the management system evaluates the priorities of the scratch pools to select one scratch pool for allocation a cartridge. In this way it is possible to reserve storage capacity for high priority applications without configuring all available scratch pools inside each application.

In addition, scratch pools can be designed for high availability, if they contain cartridges from two or more libraries (Figure 11.5). Thus in the event of the failure of individual libraries, scratch cartridges in the other libraries will remain usable. It is precisely in this case that the advantages of a storage network, together with intelligent management systems, fully come to bear in the optimal utilisation of resources that are distributed throughout the entire storage network.

Today (2009) complete cartridges are managed as the smallest unit of scratch pools. Applications receive access to the whole cartridge, because the different partitions of the same cartridge cannot be accessed concurrently. Considering the increasing capacity of tape media and the capability to partition such tapes, it could be desirable to collect just the free partitions – rather than the complete cartridges – into a scratch pool. Then it would be possible to manage the free storage capacity with a finer granularity and thus achieve an optimal utilisation of the total amount of available storage capacity.

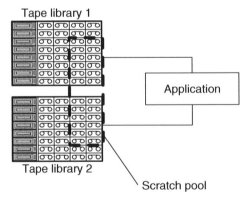

Figure 11.5 A scratch pool that contains the free cartridges from two libraries. A free cartridge is still available to the application even if one library has failed.

Drive pools

Similarly to cartridges, drives should be aggregated into drive pools and the drive pools should be tagged with a priority (Figure 11.6). Again the management system can evaluate authorisation and priorities of drive groups to select an eligible drive for an application.

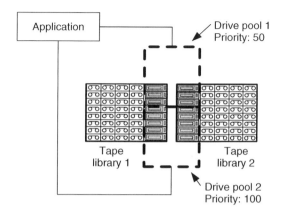

Figure 11.6 The application gets a drive from the drive pool with the highest priority.

This allows to preserve dedicated drives for high priority applications thus their mount requests can be satisfied at every point in time.

If several libraries are available, drive pools should include drives from several libraries in the same manner as scratch pools. This ensures that drives are still available even if one library has failed. At the very least, this helps when writing new data if free cartridges are still available. This is essential to meet backup windows or the free-up file systems from archive log files of data bases.

11.4.6 Media tracking and vaulting

In comparison to hard disks, removable media such as magnetic tape are a 'vagabond-type' of storage media: They can simply be removed from a library and transported to another location. The management system must know the storage location of each cartridge at all times. Under no circumstances may a cartridge go missing or be removed without authorisation from the system. Some business continuity strategies are based on the transfer of magnetic tapes to a secure location (Section 9.5.3). Removable media management must therefore continue to store all metadata for a cartridge, even if this happens to be located elsewhere at the moment. The metadata for a cartridge cannot be deleted then until the cartridge has been removed from an authorised application.

The term vaulting or vault management refers to the ability to manage cartridges that are not located in a library. Unlike cartridges that are in a library, cartridges that have been transferred to another location cannot be accessed immediately. Other locations such as well-protected premises and backup data centres are therefore also referred to as off-site locations. Removal media management should be able to manage cartridges in off-site locations and to monitor the transport of the cartridges between libraries and off-site locations (Figure 11.7). An automatic control mechanism must ensure that transferred cartridges actually arrive at the desired destination location.

The transfer of tapes to other repositories and the management of tapes in off-site repositories always require human support. In the process the location of the cartridges must continue to be updated in the management system. Removable media management therefore requires a management interface to support manual activities. For example, the management system can indicate to the operator which cartridges should be removed from a library and transported to an off-site repository. The operator accepts the respective requests and confirms them when they have been processed. The management system therefore always knows the location of all the cartridges, even those that have been transferred to a different storage location or are in transit to a new depository.

Depending on the configuration, many cartridges are regularly transported back and forth between primary storage and an off-site location. For example, a business continuity strategy could specify that backup media be transported daily to a remote bunker (Section 9.4.5). Policies help to automate these kinds of recurring manual activities. A movement policy for archive data could specify that tapes remain for a further 30 days in a tape library after they have been completely written, then be transferred to a safe and stored there for 10 years and finally destroyed.

Despite all the precautions taken, the actual storage location of cartridges does not always agree with the location noted in the system. Therefore, the inventory of libraries and off-site repositories should sometimes be counterchecked with the inventory in the removable media management (audit). Many removable media such as magnetic tapes are provided with a barcode. Tape libraries can use a laser to scan a barcode and automatically check their inventories with removable media management. In addition, many libraries use a SNMP trap to report when a door has been opened or closed again. An audit can be automatically initiated for such occurrences.

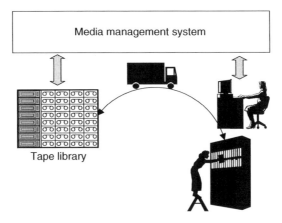

Figure 11.7   The integration of offline storage space requires manual support.

Central removable media management is the fulcrum for all access to removal media. No application should be able to remove itself from the control of a system nor be able to access or initiate media in an uncontrolled way. This is the only way to ensure that all processes are correctly logged and that all media can be found at all times. For some applications, such as the digital archiving of business data, these logs must be protected from manipulation to comply with regulatory requirements (Section 8.2.2).

11.4.7   Cartridge life cycle management

In contrast to hard disks where the medium and the write/read unit are closely wired together, the cartridges of removable media are completely detached from the drives. Therefore, in addition to the storage location of cartridges, removable media management must also manage their life cycles.

The life cycle of a cartridge starts with the declaration of cartridge in the management system (Initialisation). Then, a cartridge will be assigned to a scratch pool, which enables authorised applications to access the new cartridge. After a while, an application requests the cartridge (Allocation) and writes and reads data (Use). The application returns the cartridge when its data is no longer used (Deallocation). Now the cartridge can be assigned again to the scratch pool thus other applications can use its storage capacity, or the cartridge will be removed from the system when it reaches the end of its lifetime (Retirement).

The life cycle of a cartridge implies various states which the management system must track for each cartridge (Figure 11.8). At the beginning, a cartridge is 'undefined'. No further information is known about a cartridge. This is the initial state of a cartridge before it is taken into a management system.

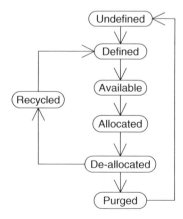

Figure 11.8   States in the life cycle of a cartridge.

As first step, a cartridge is announced to the management system thus its state changes to 'defined'. Information about the type, cartridge label, location and further metadata are given which describe the cartridge.

With the assignment to a scratch pool the cartridge becomes 'available'. Now the storage capacity of the cartridge is available for applications which are authorised for the respective scratch pool.

An application must reserve the storage capacity of a cartridge before its use. The allocation of storage capacity typically triggers the creation of a volume on a cartridge partition where an application should be able to freely choose the name of the volume. Normally the state of the first created volume determines the state of the whole cartridge. Once the first volume is created, the state of the respective cartridge changes to 'allocated'.

If the data on a volume is no longer of interest, then the application deallocates the volume to free its storage capacity. The whole cartridge remains allocated until it comprises at least one volume. To increase the storage utilisation an application should be capable to reallocate the same storage space. By no means should an allocated cartridge be assigned to another application or be removed from the system because this could lead to a data loss of the application which currently owns the cartridge. When the application deallocates the last volume of a cartridge, then the whole cartridge changes to 'deallocated'.

Once all volumes have been removed from a cartridge the entire storage space can be made available to other applications. Depending on the system configuration additional processing steps will be conducted, before a cartridge becomes 'recycled'. Some systems overwrite such a cartridge multiple times, thus the old data cannot be read by other applications which reuse the same cartridge later on. Other configurations just store the cartridge temporarily for some time before it is 'recycled' (retention). Quite often no processing is conducted at all, thus deallocated cartridges become immediately recycled.

Once a deallocated cartridge reaches the end of its lifetime, it will not be 'recycled' but completely 'purged' from the system. All metadata of the cartridge remain kept in the management system – for instance for formal audits to attest the regular compliance of past cartridge accesses. Not until the deletion of the metadata the cartridge turns again to the initial state 'undefined'.

In cartridge life cycle management, many tasks and state transition positively demand to be performed automatically. These include:

- The monitoring of retention periods;
- Transportation to the next storage location (movement or rotation);
- The copying of media when it reaches a certain age;
- The deletion of media at the end of the storage period;
- The recycling or automatic removal of the cartridge from the system.

The individual parameters for the automated tasks are specified by suitable policies for individual cartridges, or groups of cartridges to increase the grade of automation of cartridge life cycle management.

11.4.8 Monitoring

A management system for removable media has to support the administrators with the monitoring of drives and cartridges. Large set-ups comprise several thousand cartridges and by far more than one hundred drives. Automatic control of the system, or at least of parts of it, is therefore absolutely necessary for installations above a certain size. For removable media, in particular, it is important that monitoring is well constructed because in daily operation there is too little time to verify every backup. If errors creep in whilst the system is writing to tape this may not be recognised until the data needs to be restored – when it is too late. If there is no second copy, the worst conceivable incident for the datacenter has occurred: data loss!

Modern tape drives permit a very good monitoring of their state. This means that the number of read-write errors that cannot be rectified by the built-in firmware, and also the number of load operations, are stored in the drive. Ideally, this data will be read by the management system and stored so that it is available for further analysis.

A further step would be to have this data automatically analysed by the system. If certain error states are reached, actions can be triggered automatically so that at least no further error states are permitted. Under certain circumstances, errors can even be rectified automatically, for example by switching a drive off and back on again. In the worst case, it is only possible to mark the drive as defective so that it is not used further. In these tasks, too, a mechanism controlled by means of rules can help and significantly take the pressure off the administrator.

The data stored on the drives not only provides information on the drives themselves, but also on the loaded tapes. This data can be used to realise a tape quality management, which, for example, monitors the error rates when reading and writing and, if necessary, copies the data to a new tape if a certain threshold is exceeded and after that removes the defective cartridge from the system.

11.4.9 Reporting

A central management system for removable media precisely logs all actions regarding cartridges, drives, and libraries. The respective metadata is the basis for powerful reports such as:

- When was a cartridge incorporated into the system?
- Who allocated which volume to which cartridge when?
- Who accessed which volume when?
- Was this volume just read or also written?
- Which drive was used?
- Was this an authorised access or was access refused?

Laws and regulation stipulate formal audits to prove the regular compliance of certain applications such as digital archiving (Section 8.2.2). For this complete and not manipulable audit trails must be generated which lists all accesses from applications to cartridges. Individual entries of these audit trails should give information about who accessed a cartridge, for how long, and with what access rights.

Audit trails can also be used to derive usage statistics. Data about when the drives were used, and for how long they were used, is important in order to make qualitative statements about the actual utilisation of all drives. At any point in time, were sufficient drives available to carry out all mount requests? Are more drives available than the maximum amount needed at the same time over the last 12 months? The answers to such questions can be found in the report data. Like the utilisation of the drives, the available storage space is, of course, also of interest. Was enough free capacity available? Were there bottlenecks?

Usage statistics build the foundation for analysis of trends to forecast the future demand. How will the need for storage grow? How many drives will be required in 12 months? And how many slots and cartridges? A management system should be able to help in the search for answers to these questions to enable a timely planning and the implementation of respective expansions.

Just like the data on the use of resources, data regarding the errors that occurred during use is also of great importance for the successful use of removable media. Have the storage media of manufacturer X caused less read-write errors in the drives of manufacturer Y than the media of manufacturer Z? Appropriate evaluations help considerably in the optimisation of the overall performance of a system.

11.5 THE IEEE 1244 STANDARD FOR REMOVABLE MEDIA MANAGEMENT

As early as 1990, the IEEE Computer Society set up the 1244 project for the development of standards for storage systems. The Storage System Standards Working Group was also established with the objective of developing a reference model for mass storage systems (Mass Storage System Reference Model, MSSRM). This reference model has significantly influenced the design of some storage systems that are in use today. The model was then revised a few times and in 1994 released as the IEEE Reference Model for Open Storage Systems Interconnection (OSSI). Finally, in the year 2000, after further revisions, the 1244 Standard for Removable Media Management was released.

This standard consists of a series of documents that describe a platform-independent, distributed management system for removable media. It also defines both the architecture for a removable media management system and its interfaces towards the outside world. The architecture makes it possible for software manufacturers to implement very scalable, distributed software systems, which serve as generic middleware between application software and library and drive hardware. The services of the system can thus be consolidated in a central component and from there made available to all applications. The specification paid particular attention to platform-independence and the heterogeneous environment of current storage networks was thus taken into account amazingly early on.

Systems that build upon this standard can manage different types of media. In addition to the typical media for the computer field such as magnetic tape, CD, DVD or optical media, audio and video tapes, files and video disks can also be managed. In actual fact, there are no assumptions about the properties of a medium in IEEE 1244-compliant systems. Their characteristic features (number of sides, number of partitions, etc.) must be defined for each media type that the system is to support. There is a series of predefined types, each with their own properties. This open design makes it possible to specify new media types and their properties at any time and to add them to the current system.

In addition to neutrality with regard to media types, the standard permits the management of both automatic and manually operated libraries. An operator interface, which is also documented, and with which messages are sent to the appropriate administrators of a library, serves this purpose.

11.5.1   Media management system architecture

The IEEE 1244 standard describes a client/server architecture (Figure 11.9). Applications such as network backup systems take on the role of the client that makes use of the services of the removable media management system. The standard defines the required server components, the data model as well as the interfaces between these components and the applications.

In detail IEEE 1244 introduces the following components. The media manager serves as a central repository for all metadata and provides mechanisms for controlling and coordinating the use of media, libraries and drives. The library manager controls the library hardware on behalf of the media manager and transmits the properties and the content of the library to the media manager. The drive manager manages the drive hardware on behalf of the media manager and transmits the properties of the drives to the media manager.

The IEEE 1244 data model specifies all objects, with their respective attributes, that are necessary for modelling management systems for removable media (Table 11.1). IEEE 1244 provides for extending the objects through the addition of applications-specific attributes at runtime. This enables the object model to be adapted dynamically and flexibly to the respective task without components such as media manager or the protocols being changed.

In addition, the standard defines the interfaces for the communication with these components. Applications contact the media manager via the Media Management Protocol (MMP) to use the services and the resources of an IEEE 1244 compliant management system for removable media. The Library Management Protocol (LMP) defines the interfaces between the media manager and the library manager and the Drive Management Protocol (DMP) defines the interface between the media manager and the drive manager.

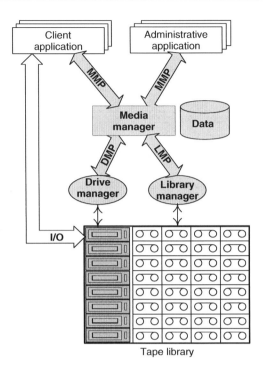

Figure 11.9   Architecture of the IEEE 1244 standard for removable media management.

These protocols use TCP/IP as the transport layer. As in popular Internet protocols such as HTTP, FTP or SMTP, commands are sent via TCP/IP in the form of text messages. These protocols can be implemented and used just as simply on different platforms.

The advantage of the IEEE 1244 architecture is that the media manager can implement components as a generic application, i.e. independently of the specific library and drive hardware used. The differences, in particular with the control, are encapsulated in the library manager or drive manager for the hardware in question. For a new tape library, therefore, only a new library manager component needs to be implemented that converts the specific interface of the library into LMP, so that this can be linked into an existing installation of an IEEE 1244 compliant management systems for removable media.

11.5.2   Media manager and MMP

From the view of the applications, the media manager works as a server that waits for requests from the applications. It distinguishes between privileged and non-privileged applications. Non-privileged applications, such as backup systems, can only work with

Table 11.1   The most important objects of the IEEE 1244 data model.

Object	Description
APPLICATION	Authorised client application. Access control is performed on the basis of applications. User management is not part of this standard, since it is assumed that it is not individual users, but applications that already manage their users, that will use the services of the media management system.
AI	Authorised instances of a client application. All instances of an application have unrestricted access to resources that are assigned to the application.
LIBRARY	Automatic or manually operated libraries.
LM	Library managers know the details of a library. The library manager protocol serves as a hardware-independent interface between media manager and library manager.
BAY	Part of a LIBRARY (contains DRIVES and SLOTS).
SLOT	Individual storage space for CARTRIDGEs within a BAY.
SLOTGROUP	Group of SLOTS to represent a magazine, for example, within a LIBRARY.
SLOTTYPE	Valid types for SLOTs, for example 'LTO', 'DLT or 3480', 'QIC' or 'CDROM'.
DRIVE	Drives, which can mount CARTRIDGEs for writing or reading.
DRIVEGROUP	Groups of drives.
DRIVEGROUPAPPLICATION	This object makes it possible for applications to access drives in a DRIVEGROUP. This connection can be assigned a priority so that several DRIVEGROUPs with different priorities are available. The media manager selects a suitable drive according to priority.
DM	Drive manager. Drive managers know the details of a drive and make this available to the media manager. The drive manager protocol serves as a hardware-independent interface between media manager and drive manager.
CARTRIDGE	Removable media.
CARTRIDGEGROUP	Group of CARTRIDGEs.
CARTRIDGEGROUPAPPLICATION	This object makes it possible for applications to access CARTRIDGEs in a CARTRIDGEGROUP. This connection can be assigned a priority so that several CARTRIDGEGROUPs with different priorities are available. The media manager selects a suitable CARTRIDGE according to priority in order to allocate a VOLUME, if no further entries are made.

(continued overleaf)

Table 11.1 (*continued*).

Object	Description
CARTRIDGETYPE	Valid types for CARTRIDGEs are for example 'LTO-1', 'LTO-2' or '3592'.
SIDE	CARTRIDGEs can have more than one SIDE, for example magneto-optical disks.
PARTITION	SIDEs can be divided into several separately allocable partitions.
VOLUME	Logical unit with a name to which an application connects an allocated partition.
MOUNTLOGICAL	VOLUMEs currently being accessed.
DRIVECARTRIDGEACCESS	Object for the tracking of all accesses to cartridges and drives. Contains time, duration and errors of the accesses. Can be used to compile error statistics.
SESSION	Describes an existing connection of an application to the media management system.
TASK	Commands, that are currently waiting for the resources to become available.
MESSAGE	Operator or error message of the system.
REQUEST	Request to an operator.

objects that have been granted an appropriate authorisation. Privileged applications, which are usually administrative applications, are allowed to execute all actions and manipulate all objects. They are mainly used to define non-privileged applications into the system and to establish the appropriate access controls.

The Media Management Protocol (MMP) is used by the applications to make use of the media management services of an IEEE 1244-compatible system. MMP is a text-based protocol, which exchanges messages over TCP/IP. The syntax and semantics of the individual protocol messages are specified in the MMP specification IEEE 1244.3. MMP permits applications to allocate and mount volumes, read and write metadata and to manage and share libraries and drives platform-independently. Due to the additional abstraction levels, the application is decoupled from the direct control of the hardware. Thus, applications can be developed independently of the capability of the connected hardware and can be made available to a large number of different types of removable media.

MMP is an asynchronous protocol. Applications send commands as MMP messages to the media manager. The media manager first acknowledges the acceptance of a command, passes it on for processing and then sends a second message with the result of the command. An application can send further commands as soon as it receives a message acknowledging acceptance of the previous command. Applications must therefore provide the commands with task identifiers so that they can relate the responses from the media manager accordingly.

Example 1   An application wants to mount the volume with the name backup-1999-12-31. To this end it, sends the following command to the media manager:

```
mount task["1"]
        volname ["backup-1999-12-31"]
        report [MOUNTLOGICAL."MountLogicalHandle"];
```

The media manager has recognised the command and accepted it for processing and therefore sends the following response:

```
response task["1"] accepted;
```

Now the media manager will transport the cartridge containing the volume into a drive to which the application has access. Once the cartridge has been successfully inserted, the media manager sends a response to the application that could look like this:

```
response task["1"] success text ["/dev/rmt0"];
```

This is how the application learns which path (device handle) to use for reading and describing the drive with the requested volume.

The media manager stores all commands in a task queue until all resources required for execution are available. Once all the resources are available, the media manager removes the command from the task queue and executes it. If several commands are present that require the same resources, the media manager selects the next command to be carried out on the basis of priorities or on a first come, first served basis. All other commands remain in the task queue until the resources in question become free again. In this manner, libraries, drives and also cartridges can be shared. Commands that are in the task queue can be removed again using the Cancel command.

Example 2   An application wants to mount volume SU1602. Therefore, it sends the following command to the media manager:

```
mount task["1"]
        volname[volname "SU1602"]
        report[MOUNTLOGICAL."MountLogicalHandle"];
```

The media manager has recognised the command and accepted it for processing. It therefore sends the following response:

```
response task["1"] accepted;
```

Now the application can send another command. For example, a list of names of all defective drives can be requested.

```
show task["2"] report[DRIVE."DriveName"]
        match[streq(DRIVE."DriveStateHard" "broken")];
```

The media manager has recognised the command and accepted it for processing. Therefore, it sends the following response:

```
response task["2"] accepted;
```

As all the drives are currently in use, the show command is executed before the mount command. The response from the media manager could appear as follows:

```
response task["2"] success text["LTO-DRIVE-3"
                                 "LTO-DRIVE-7"
                                 "LTO-DRIVE-8"];
```

After some time all resources needed for the mount command become available. The media manager instructs the responsible library manager to send the cartridge on which the volume resides to an appropriate drive. Then it instructs the responsible drive manager to load the cartridge and to generate a task identifier for the drive (access handle). The application then receives this MountLogical Handle (task identifier) from the media manager and uses it to access the volume:

```
response task["1"] success text["/dev/tape/LTO123"];
```

One command after another is then processed in this way until the task queue is empty.

In Example 2 the show command is used together with the match clause to display all the defective drives. 'Show' and 'match' form a powerful query language that can be used to generate complex reports. Graphical user interfaces can offer predefined reports that are translated into the corresponding MMP commands.

Example 3 In the spring of 2003 an administrator suspected that drive LTO123 was indicating a high error rate. Now it wants to display all the cartridges mounted in the mentioned drive since 2003 where an error occurred during the mount. The list should be sorted in descending order according to the number of write errors. The corresponding show command would look like the following:

```
show task["1"]
  report[DRIVECARTRIDGEACCESS."CartridgePCL"]
  match[
    and(
      streq(DRIVECARTRIDGEACCESS."DriveName""LTO123")
      strgt(DRIVECARTRIDGEACCESS."TimeUnmount"
                                "2003 01 01 00 00 00 000")
    or(
      numgt(DRIVECARTRIDGEACCESS."HardReadErrorCount" "0")
      numgt(DRIVECARTRIDGEACCESS."SoftReadErrorCount" "0")
      numgt(DRIVECARTRIDGEACCESS."HardWriteErrorCount" "0")
      numgt(DRIVECARTRIDGEACCESS."SoftWriteErrorCount" "0")
```

```
        )
      )
    ]
    order[numhilo(DRIVECARTRIDGEACCESS."HardWriteErrorCount")];
```

A characteristic of IEEE 1244 is the extension of the data model during runtime. Applications can add optional attributes to any object in order to file vendor-specific and applications-specific characteristics of objects in the media manager. For example, a backup system could create a list of all files on a cartridge as an attribute or a video rental shop could assign the number of the last customer who rented a particular video cassette (cartridge). MMP provides the appropriate attribute command.

Example 4   An application wants to set the applications-specific attribute 'color' to the value 'red' for the drive with the name LTO123. It sends the following command to the media manager:

```
    attribute task["3"] set[DRIVE."Color" "red"];
            match[streq(DRIVE."DriveName" "LTO123")];
```

The media manager has also recognised this command, accepted it for processing and then sends the following response:

```
    response task["3"] accepted;
```

The media manager applies this attribute dynamically if it has never been applied to this drive before. Otherwise only the value of the attribute is changed. Lastly, the media manager sends the following message back to the application in order to signal that processing has been successful:

```
    response task["3"] success;
```

11.5.3   Library manager and drive manager

IEEE 1244 specifies the interface between media manager and library manager over the Library Management Protocol (LMP). The library manager receives the LMP commands of the media manager and converts them into the specific commands of the respective library model. From the view of the media manager, a uniform abstract interface that hides the characteristics of the respective hardware therefore exists for all libraries. This enables new hardware to be integrated into the management system over a suitable library manager without the need for changes to the overall system.

In a similar way, the Drive Management Protocol (DMP) specifies the interface between media manager and drive manager, thereby enabling different drive hardware to be activated in a uniform way. However, the drive managers must take into account the specific characteristics of the different client platforms on which the applications using the media management system are being run. If an application is being run on a Unix-compatible platform, the drive manager must provide the name of a device-special file, such as /dev/rmt0, for access to the drive. Under windows this kind of drive manager must supply a Windows-specific file name, such as \\.\TAPE0\.

Like MMP, the LMP and DMP protocols are based on text messages that are exchanged over TCP/IP. Both protocols are used to control hardware, transport tapes (LMP) and insert tapes into a drive (DMP). The syntax and semantics of the individual LMP messages are described in specification IEEE 1244.5. The DMP messages are specified in IEEE 1244.4.

11.6 SUMMARY

Strong growth in data quantities and increasing cartridge quantities call for improved solutions for the efficient management of removable media in the storage network. The requirements of such a system in the storage network are largely comparable with the requirements that are also made of disk subsystems. However, due to the special properties of removable media, such as the separate storage of medium and drive, there are additional problems to master.

We have described the following requirements for the management of removable media in detail:

- Centralised administration and control of all resources such as media, drives and libraries, make efficient management significantly easier.

- Intelligent sharing of libraries, drives and scratch pools makes the efficient use of the available resources easier and can be used to implement highly available solutions (cross-library drive and scratch pools).

- Authentication and authorisation of users/applications facilitate adherence to data protection guidelines and the allocation of resources.

- Mount request management, including request queuing, increases the robustness of the system as a whole and permits prioritised processing and allocation of mount requests to drives.

- Uninterrupted tracking of the media ensures that at every point in time the location of a medium is known and thus protected against data loss.

- Automated life cycle management of the removable media greatly frees up the administrators from tasks that occur repeatedly.

- Full integration of online and offline locations permits security concepts in which the removable media must be stored in a well-protected location.

- Monitoring provides the automatic recording and processing of errors in the system. Rule-based mechanisms can perform suitable actions for error rectification or can signal manual intervention.

Unfortunately, there are currently (2009) still no cross-manufacturer and cross-application management layers and interfaces for removable media. As a result, each manufacturer puts their own proprietary system, which is often part of a backup solution, on the market. Some of these systems are not capable of using several instances of the connected hardware cooperatively and thus do not fully exploit the potential of resource sharing in storage networks. Other solutions are not very scalable, which is why they can only sensibly be used up to a certain number of drives or cartridges. With different applications from several manufacturers, a cooperative use of all removable media resources is currently (2009) almost impossible.

The IEEE 1244 Standard for Media Management Systems provides the basis for effectively managing removable media in the storage network and intelligently sharing resources and thus describes the architecture for an asymmetrical virtualisation of removable media. Due to the use of text-based protocols over TCP/IP, a lot of platforms can easily make use of the services of a system based upon this standard, in a similar way to the widespread Internet protocol (SMTP, HTTP). The standard addresses the most important problems in dealing with removable media in the storage network and provides tools for the solution. Of particular importance is the fact that not only interfaces are defined for applications (MMP) and hardware (LMP and DMP), but an architecture is also described that fits in optimally with these interfaces.

12

The SNIA Shared Storage Model

The fact that there is a lack of any unified terminology for the description of storage architectures has already become apparent at several points in previous chapters. There are thus numerous components in a storage network which, although they do the same thing, are called by different names. Conversely, there are many systems with the same name, but fundamentally different functions.

A notable example is the term 'data mover' relating to server-free backup (Section 7.8.1) in storage networks. When this term is used it is always necessary to check whether the component in question is one that functions in the sense of the 3rd-party SCSI Copy Command for, for example, a software component of backup software on a special server, which implements the server-free backup without 3rd-party SCSI Copy Command.

This example shows that the type of product being offered by a manufacturer and the functions that the customer can ultimately expect from this product are often unclear. This makes it difficult for customers to compare the products of individual manufacturers and find out the differences between the alternatives on offer. There is no unified model for this with clearly defined descriptive terminology.

For this reason, in 2001 the Technical Council of the Storage Networking Industry Association (SNIA) introduced the so-called Shared Storage Model in order to unify the terminology and descriptive models used by the storage network industry. Ultimately, the SNIA wants to use the SNIA Shared Storage Model to establish a reference model, which will have the same importance for storage architectures as the seven-tier Open System Interconnection (OSI) model has for computer networks.

Storage Networks Explained: Basics and Application of Fibre Channel SAN, NAS, iSCSI, InfiniBand and FCoE, Second Edition
U. Troppens R. Erkens W. Müller-Friedt N. Haustein R. Wolafka © 2009 John Wiley & Sons, Ltd

In this chapter, we would first like to introduce the disk-based Shared Storage Model (Section 12.1) and then show, based upon examples (Section 12.2), how the model can be used for the description of typical disk storage architectures. In Section 12.3 we introduce the extension of the SNIA model to the description of tape functions. We then discuss examples of tape-based backup architectures (Section 12.4). Whilst describing the SNIA Shared Storage Model we often refer to text positions in this book where the subject in question is discussed in detail, which means that this chapter also serves as a summary of the entire book.

12.1 THE MODEL

In this book we have spoken in detail about the advantages of the storage-centric architecture in relation to the server-centric architecture. The SNIA sees its main task as being to communicate this paradigm shift and to provide a forum for manufacturers and developers so that they can work together to meet the challenges and solve the problems in this field. In the long run, an additional reason for the development of the Shared Storage Model by SNIA was the creation of a common basis for communication between the manufacturers who use the SNIA as a platform for the exchange of ideas with other manufacturers. Storage-centric IT architectures are called shared storage environments by the SNIA. We will use both terms in the following.

First of all, we will describe the functional approach of the SNIA model (Section 12.1.1) and the SNIA conventions for graphical representation (Section 12.1.2). We will then consider the model (Section 12.1.3), its components (Section 12.1.4) and the layers 'file/record layer' and 'block layer' in detail (Section 12.1.5 to Section 12.1.8). Then we will introduce the definitions and representation of concepts from the SNIA model, such as access paths (Section 12.1.9), caching (Section 12.1.10), access control (Section 12.1.11), clustering (Section 12.1.12), data (Section 12.1.13) and resource and data sharing (Section 12.1.14). Finally, we will take a look at the service subsystem (Section 12.1.15).

12.1.1 The functional approach

The SNIA Shared Storage Model first of all describes functions that have to be provided in a storage-centric IT architecture. This includes, for example, the block layer or the file/record layer. The SNIA model describes both the tasks of the individual functions and also their interaction. Furthermore, it introduces components such as server ('host computer') and storage networks ('interconnection network').

Due to the separation of functions and components, the SNIA Shared Storage Model is suitable for the description of various architectures, specific products and concrete installations. The fundamental structures, such as the functions and services of a shared storage environment, are highlighted. In this manner, functional responsibilities can be assigned to individual components and the relationships between control and data flows in the storage network worked out. At the same time, the preconditions for interoperability between individual components and the type of interoperability can be identified. In addition to providing a clear terminology for the elementary concepts, the model should be simple to use and, at the same time, extensive enough to cover a large number of possible storage network configurations.

The model itself describes, on the basis of examples, possible practicable storage architectures and their advantages and disadvantages. We will discuss these in Section 12.2 without evaluating them or showing any preference for specific architectures. Within the model definition, however, only a few selected examples will be discussed in order to highlight how the model can be applied for the description of storage-centric environments and further used.

12.1.2 Graphical representations

The SNIA Shared Storage Model further defines how storage architectures can be graphically illustrated. Physical components are always represented as three-dimensional objects, whilst functional units should be drawn in two-dimensional form. The model itself also defines various colours for the representation of individual component classes. In the black and white format of the book, we have imitated these using shades of grey. A coloured version of the illustrations to this chapter can be found on our home page http://www.storage-explained.com. Thick lines in the model represent the data transfer, whereas thin lines represent the metadata flow between the components.

12.1.3 An elementary overview

The SNIA Shared Storage Model first of all defines four elementary parts of a shared storage environment (Figure 12.1):

1. *File/record layer*
   The file/record layer is made up of database and file system.
2. *Block layer*
   The block layer encompasses the storage devices and the block aggregation. The SNIA Shared Storage Model uses the term 'aggregation' instead of the often ambiguously used term 'storage virtualisation'. In Chapter 5, however, we used the term 'storage virtualisation' to mean the same thing as 'aggregation' in the SNIA model, in order to avoid ambiguity.

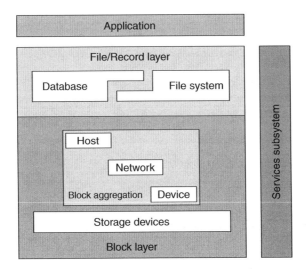

Figure 12.1 The main components of the SNIA Shared Storage Model are the file/record layer, the block layer and the services subsystem. Applications are viewed as users of the model.

3. *Services subsystem*
   The functions for the management of the other components are defined in the services subsystem.

4. *Applications*
   Applications are not discussed further by the model. They will be viewed as users of the model in the widest sense.

12.1.4 The components

The SNIA Shared Storage Model defines the following components:

- *Interconnection network*
  The interconnection network represents the storage network, i.e. the infrastructure, that connects the individual elements of a shared storage environment with one another. The interconnection network can be used exclusively for storage access, but it can also be used for other communication services. Our definition of a storage network (Section 1.2) is thus narrower than the definition of the interconnection network in the SNIA model.
    The network must always provide a high-performance and easily scaleable connection for the shared storage environment. In this context, the structure of the interconnec-

tion network – for example redundant data paths between two components to increase fault-tolerance – remains just as open as the network techniques used. It is therefore a prerequisite of the model that the components of the shared storage environment are connected over a network without any definite communication protocols or transmission techniques being specified.

In actual architectures or installations, Fibre Channel, Fast Ethernet, Gigabit Ethernet, InfiniBand and many other transmission techniques are used (Chapter 3). Communication protocols such as SCSI, Fibre Channel FCP, TCP/IP, RDMA, CIFS or NFS are based upon these.

- *Host computer*
  Host computer is the term used for computer systems that draw at least some of their storage from the shared storage environment. According to SNIA, these systems were often omitted from classical descriptive approaches and not viewed as part of the environment. The SNIA shared storage model, however, views these systems as part of the entire shared storage environment because storage-related functions can be implemented on them.

  Host computers are connected to the storage network via host bus adapters or network cards, which are operated by means of their own drivers and software. Drivers and software are thus taken into account in the SNIA Shared Storage Model. Host computers can be operated fully independently of one another or they can work on the resources of the storage network in a compound, for example, a cluster (Section 6.4.1).

- *Physical storage resource*
  All further elements that are connected to the storage network and are not host computers are known by the term 'physical storage resource'. This includes simple hard disk drives, disk arrays, disk subsystems and controllers plus tape drives and tape libraries. Physical storage resources are protected against failures by means of redundant data paths (Section 6.3.1), replication functions such as snapshots and mirroring (Section 2.7) and RAID (Section 2.5).

- *Storage device*
  A storage device is a special physical storage resource that stores data.

- *Logical storage resource*
  The term 'logical storage resource' is used to mean services or abstract compositions of physical storage resources, storage management functions or a combination of these. Typical examples are volumes, files and data movers.

- *Storage management functions*
  The term 'storage management function' is used to mean the class of services that monitor and check (Chapter 10) the shared storage environment or implement logical storage resources. These functions are typically implemented by software on physical storage resources or host computers.

12.1.5 The layers

The SNIA Shared Storage Model defines four layers (Figure 12.2):

  I. Storage devices
 II. Block aggregation layer
III. File/record layer
    IIIa. Database
    IIIb. File system
IV. Applications

Applications are viewed as users of the model and are thus not described in the model. They are, however, implemented as a layer in order to illustrate the point in the model to which they are linked. In the following we'll consider the file/record layer (Section 12.1.6), the block layer (Section 12.1.7) and the combination of both (Section 12.1.8) in detail.

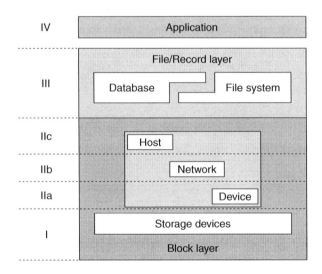

Figure 12.2   The SNIA Shared Storage Model defines four layers.

12.1.6 The file/record layer

The file/record layer maps database records and files on the block-oriented volume of the storage devices. Files are made up of several bytes and are therefore viewed as byte vectors in the SNIA model. Typically, file systems or database management systems take over these functions. They operate directories of the files or records, check the access, allocate storage space and cache the data (Chapter 4). The file/record layer thus works on volumes that are provided to it from the block layer below. Volumes themselves consist of several arranged blocks, so-called block vectors. Database systems map one or more records, so-called tuple of records, onto volumes via tables and table spaces:

$$\text{Tuple of records} \longrightarrow \text{tables} \longrightarrow \text{table spaces} \longrightarrow \text{volumes}$$

In the same way, file systems map bytes onto volumes by means of files:

$$\text{Bytes} \longrightarrow \text{files} \longrightarrow \text{volumes}$$

Most database systems can also work with files, i.e. byte vectors. In this case, block vectors are grouped into byte vectors by means of a file system – an additional abstraction level. This additional layer can cost performance, but it simplifies the administration. Hence, in real installations both approaches can be found (Section 4.11).

The functions of the file/record layers can be implemented at various points (Figure 12.3, Section 5.6):

- *Exclusively on the host*
  In this case, the file/record layer is implemented entirely on the host. Databases and the host-based file systems work in this way.

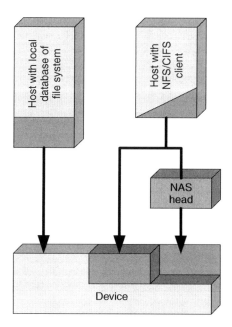

Figure 12.3 The functions of the file/record layer can be implemented exclusively on the host or distributed over a client and a server component.

- *Both in the client and also on a server component*
  The file/record layer can also be implemented in a distributed manner. In this case the functions are distributed over a client and a server component. The client component is realised on a host computer, whereas the server component can be realised on the following devices:

  - *NAS/file server*
    A NAS/file server is a specialised host computer usually with a locally connected, dedicated storage device (Section 4.2.2).

  - *NAS head*
    A host computer that offers the file serving services, but which has access to external storage connected via a storage network. NAS heads correspond with the devices called NAS gateways in our book (Section 4.2.2).

In this case, client and server components work over network file systems such as NFS or CIFS (Section 4.2).

12.1.7 The block layer

The block layer differentiates between block aggregation and the block-based storage devices. The block aggregation in the SNIA model corresponds to our definition of the virtualisation on block level (Section 5.5). SNIA thus uses the term 'block aggregation' to mean the aggregation of physical blocks or block vectors into logical blocks or block vectors.

To this end, the block layer maps the physical blocks of the disk storage devices onto logical blocks and makes these available to the higher layers in the form of volumes (block vectors). This either occurs via a direct (1:1) mapping, or the physical blocks are first aggregated into logical blocks, which are then passed on to the upper layers in the form of volumes (Figure 12.4). In the case of SCSI, the storage devices of the storage device layer exist in the form of one or more so-called logical units (LUs).

Further tasks of the block layer are the labelling of the LUs using so-called logical unit numbers (LUNs), caching and access control. Block aggregation can be used for various purposes, for example:

- *Volume/space management*
  The typical task of a volume manager is to aggregate several small block vectors to form one large block vector. On SCSI level this means aggregating several LUs to form a large volume, which is passed on to the upper layers such as the file/record layer (Section 4.1.4).

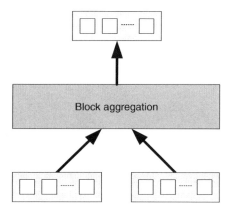

Figure 12.4 The block aggregation layer aggregates physical blocks or block vectors into logical blocks or block vectors.

- *Striping*
  In striping, physical blocks of different storage devices are aggregated to one volume. This increases the I/O throughput of the read and write operations, since the load is distributed over several physical storage devices (Section 2.5.1).

- *Redundancy*
  In order to protect against failures of physical storage media, RAID (Section 2.5) and remote mirroring (Section 2.7.2) are used. Snapshots (instant copies) can also be used for the redundant storage of data (Section 2.7.1).

The block aggregation functions of the block layer can be realised at different points of the shared storage environment (Section 5.6):

- *On the host*
  Block aggregation on the host is encountered in the form of a logical volume manager software, in device drivers and in host bus adapters.

- *On a component of the storage network*
  The functions of the block layer can also be realised in connection devices of the storage network or in specialised servers in the network.

- *In the storage device*
  Most commonly, the block layer functions are implemented in the storage devices themselves, for example, in the form of RAID or volume manager functionality.

In general, various block aggregation functions can be combined at different points of the shared storage environment. In practical use, RAID may, for example, be used in the disk subsystem with additional mirroring from one disk subsystem to another via the volume manager on the host computer (Section 4.1.4). In this set-up, RAID protects against the failure of physical disks of the disk subsystem, whilst the mirroring by means of the volume manager on the host protects against the complete failure of a disk subsystem. Furthermore, the performance of read operations is increased in this set-up, since the volume manager can read from both sides of the mirror (Section 2.5.2).

12.1.8 Combination of the block and file/record layers

Figure 12.5 shows how block and file/record layer can be combined and represented in the SNIA shared storage model:

- *Direct attachment*
  The left-hand column in the figure shows storage connected directly to the server, as is normally the case in a server-centric IT architecture (Section 1.1).

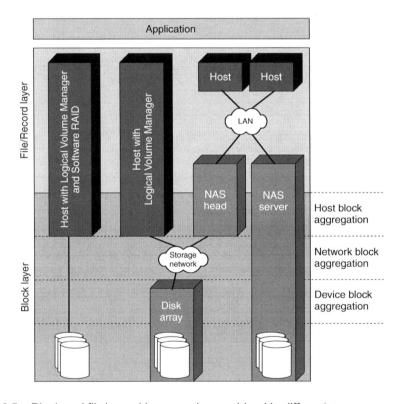

Figure 12.5 Block and file/record layer can be combined in different ways.

- *Storage network attachment*
  In the second column we see how a disk array is normally connected via a storage network in a storage-centric IT architecture, so that it can be accessed by several host computers (Section 1.2).

- *NAS head (NAS gateway)*
  The third column illustrates how a NAS head is integrated into a storage network between storage network attached storage and a host computer connected via LAN.

- *NAS server*
  The right-hand column shows the function of a NAS server with its own dedicated storage in the SNIA Shared Storage Model.

12.1.9 Access paths

Read and write operations of a component on a storage device are called access paths in the SNIA Shared Storage Model. An access path is descriptively defined as the list of components that are run through by read and write operations to the storage devices and responses to them. If we exclude cyclical access paths, then a total of eight possible access paths from applications to the storage devices can be identified in the SNIA Shared Storage Model (Figure 12.6):

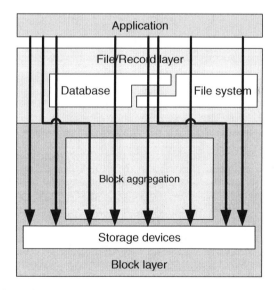

Figure 12.6 In the SNIA Shared Storage Model, applications can access the storage devices via eight possible access paths.

1. Direct access to a storage device.
2. Direct access to a storage device via a block aggregation function.
3. Indirect access via a database system.
4. Indirect access via a database system based upon a block aggregation function.
5. Indirect access via a database system based upon a file system.
6. Indirect access via a database system based upon a file system, which is itself based upon a block aggregation function.
7. Indirect access via a file system.
8. Indirect access via a file system based upon a block aggregation function.

12.1.10   Caching

Caching is the method of shortening the access path of an application – i.e. the number of the components to be passed through – to frequently used data on a storage device. To this end, the data accesses to the slower storage devices are buffered in a faster cache storage. Most components of a shared storage environment can have a cache. The cache can be implemented within the file/record layer, within the block layer or in both.

In practice, several caches working simultaneously on different levels and components are generally used. For example, a read cache in the file system may be combined with a write cache on a disk array and a read cache with pre-fetching on a hard disk (Figure 12.7). In addition, a so-called cache-server (Section 5.7.2), which temporarily stores data for other components on a dedicated basis in order to reduce the need for network capacity or to accelerate access to slower storage, can also be integrated into the storage network.

However, the interaction between several cache storages on several components means that consideration must be given to the consistency of data. The more components that use cache storage, the more dependencies arise between the functions of individual components. A classic example is the use of a snapshot function on a component in the block layer, whilst another component stores the data in question to cache in the file/record layer. In this case, the content of the cache within the file/record layer, which we will assume to be consistent, and the content of a volume on a disk array that is a component of the block layer can be different. The content of the volume on the array is thus inconsistent. Now, if a snapshot is taken of the volume within the disk array, a virtual copy is obtained of an inconsistent state of the data. The copy is thus unusable. Therefore, before the snapshot is made within the block layer, the cache in the file/record layer on the physical volume must be destaged, so that it can receive a consistent copy later.

12.1.11   Access control

Access control is the name for the technique that arranges the access to data of the shared storage environment. The term access control should thus be clearly differentiated from

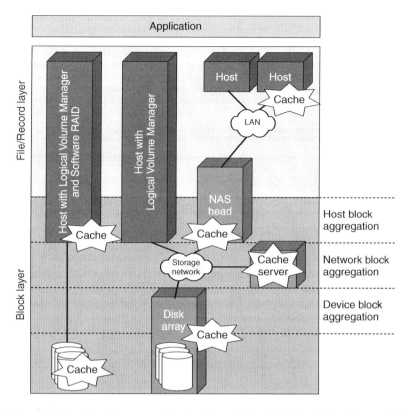

Figure 12.7 Caching functions can be implemented at different levels and at different components of a shared storage environment.

the term access path, since the mere existence of an access path does not include the right to access. Access control has the following main objectives:

- *Authentication*
  Authentication establishes the identity of the source of an access.

- *Authorisation*
  Authorisation grants or refuses actions to resources.

- *Data protection*
  Data protection guarantees that data may only be viewed by authorised persons.

All access control mechanisms ultimately use a form of secure channel between the data on the storage device and the source of an access. In its simplest form, this can be a check to establish whether a certain host is permitted to have access to a specific storage device.

Access control can, however, also be achieved by complex cryptographic procedures, which are secure against the most common external attacks. When establishing a control mechanism it is always necessary to trade off the necessary protection and efficiency against complexity and performance sacrifices.

In server-centric IT architectures, storage devices are protected by the guidelines on the host computers and by simple physical measures. In a storage network, the storage devices, the network and the network components themselves must be protected against unauthorised access, since in theory they can be accessed from all host computers. Access control becomes increasingly important in a shared storage environment as the number of components used, the diversity of heterogeneous hosts and the distance between the individual devices rise.

Access controls can be established at the following points of a shared storage environment:

- *On the host*
  In shared storage environments, access controls comparable with those in server-centric environments can be established at host level. The disadvantage of this approach is, however, that the access rights have to be set on all host computers. Mechanisms that reduce the amount of work by the use of central instances for the allocation and distribution of rights must be suitably protected against unauthorised access. Database systems and file systems can be protected in this manner. Suitable mechanisms for the block layer are currently being planned. The use of encryption technology for the host's network protocol stack is in conflict with performance requirements. Suitable offload engines, which process the protocol stack on the host bus adapter themselves, are available for some protocols.

- *In the storage network*
  Security within the storage network is achieved in Fibre Channel SANs by zoning and virtual storage networks (Virtual SAN [VSAN], Section 3.4.2) and in Ethernet-based storage networks by so-called virtual LANs (VLAN). This is always understood to be the subdivision of a network into virtual subnetworks, which permit communication between a number of host ports and certain storage device ports. These guidelines can, however, also be defined on finer structures than ports.

- *On the storage device*
  The normal access control procedure on SAN storage devices is the so-called LUN masking, in which the LUNs that are visible to a host are restricted. Thus, the computer sees only those LUNs that have been assigned to it by the storage device (Section 2.7.3).

12.1.12 Clustering

A cluster is defined in the SNIA Shared Storage Model as a combination of resources with the objective of increasing scalability, availability and management within the shared

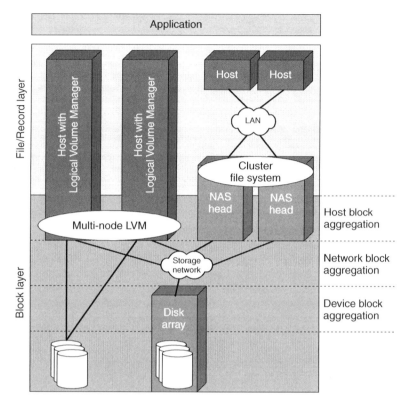

Figure 12.8 Nodes of a cluster share resources via distributed volume managers or cluster file systems.

storage environment (Section 6.4.1). The individual nodes of the cluster can share their resources via distributed volume managers (multi-node LVM) and cluster file systems (Figure 12.8, Section 4.3).

12.1.13 Storage, data and information

The SNIA Shared Storage Model differentiates strictly between storage, data and information. Storage is space – so-called containers – provided by storage units, on which the data is stored. The bytes stored in containers on the storage units are called data. Information is the meaning the semantics – of the data. The SNIA Shared Storage Model names the following examples in which data – container relationships arise (Table 12.1).

Table 12.1 Data–container relationships.

Relationship	Role	Remark
User	Data	Inputs via keyboard
Application	Container	Input buffer
User	Data	Input buffer file
File system	Container	Byte vector
File system	Data	A file
Volume manager	Container	Blocks of a volume
Volume manager	Data	Mirrored stripe set
Disk array	Container	Blocks of a logical unit

12.1.14 Resource and data sharing

In a shared storage environment, in which the storage devices are connected to the host via a storage network, every host can access every storage device and the data stored upon it (Section 1.2). This sharing is called resource sharing or data sharing in the SNIA model, depending upon the level at which the sharing takes place (Figure 12.9).

If exclusively the storage systems – and not their data content – are shared, then we talk of resource sharing. This is found in the physical resources, such as disk subsystems and tape libraries, but also within the network.

Data sharing denotes the sharing of data between different hosts. Data sharing is significantly more difficult to implement, since the shared data must always be kept consistent, particularly when distributed caching is used.

Heterogeneous environments also require additional conversion steps in order to convert the data into a format that the host can understand. Protocols such as NFS or CIFS are used in the more frequently used data sharing within the file/record layers (Section 4.2).

For data sharing in the block layer, server clusters with shared disk file systems or parallel databases are used (Section 4.3, Section 6.2.3).

12.1.15 The service subsystem

Up to now we have concerned ourselves with the concepts within the layers of the SNIA Shared Storage Model. Let us now consider the service subsystem (Figure 12.10). Within the service subsystem we find the management tasks which occur in a shared storage environment and which we have, for the most part, already discussed in Chapter 10.

In this connection, the SNIA Technical Council mention:

• Discovery and monitoring
• Resource management

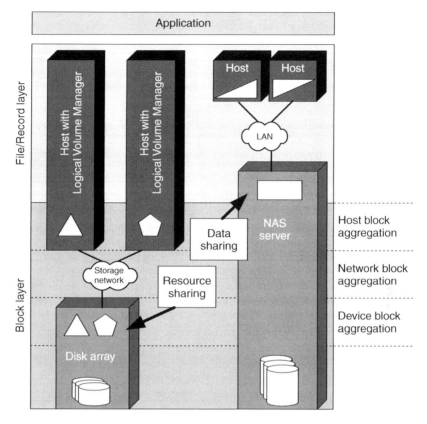

Figure 12.9 In resource sharing, hosts share the physical resources – in this case a disk array – which make a volume available to each host. In data sharing, hosts access the same data – in this case the NAS server and its data.

- Configuration
- Security
- Billing (charge-back)
- Redundancy management, for example, by network backup
- High availability
- Capacity planning

The individual subjects are not yet dealt with in more detail in the SNIA Shared Storage Model, since the required definitions, specifications and interfaces are still being developed. At this point we expressly refer once again to the check list in the Appendix B, which reflects a cross-section of the questions that crop up here.

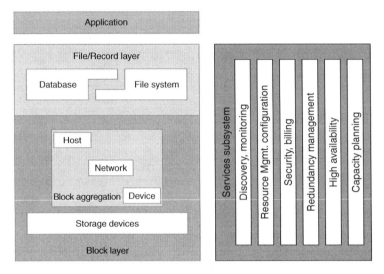

Figure 12.10   In the services subsystem, the SNIA defines the management tasks in a shared storage environment.

12.2   EXAMPLES OF DISK-BASED STORAGE ARCHITECTURES

In this section we will present a few examples of typical storage architectures and their properties, advantages and disadvantages, as they are represented by the SNIA in the Shared Storage Model. First of all, we will discuss block-based architectures, such as the direct connection of storage to the host (Section 12.2.1), connection via a storage network (Section 12.2.2), symmetric and asymmetric storage virtualisation in the network (Section 12.2.3 and Section 12.2.4) and a multi-site architecture such as is used for data replication between several locations (Section 12.2.5). We then move on to the file/record layer and consider the graphical representation of a file server (Section 12.2.6), a NAS head (Section 12.2.7), the use of metadata controllers for asymmetric file level virtualisation (Section 12.2.8) and an object-based storage device (OSD), in which the position data of the files and their access rights is moved to a separate device, a solution that combines file sharing with increased performance due to direct file access and central metadata management of the files (Section 12.2.9).

12.2.1   Direct attached block storage

Figure 12.11 shows the direct connection from storage to the host in a server-centric architecture. The following properties are characteristic of this structure:

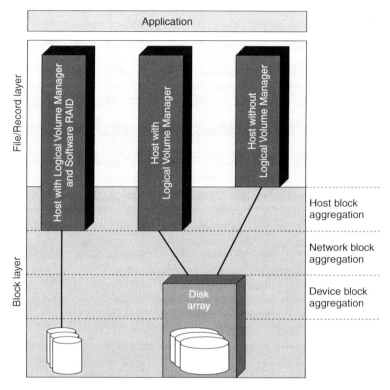

Figure 12.11   In direct attachment, hosts are connected to storage devices directly without connection devices such as switches or hubs. Data sharing and resource sharing is not possible without additional software.

- No connection devices, such as switches or hubs, are needed.
- The host generally communicates with the storage device via a protocol on block level.
- Block aggregation functions are possible both in the disk subsystem and on the host.

12.2.2   Storage network attached block storage

The connection from storage to host via a storage network can be represented in the Shared Storage Model as shown in Figure 12.12. In this case:

- Several hosts share several storage devices.
- Block-oriented protocols are generally used.
- Block aggregation can be used in the host, in the network and in the storage device.

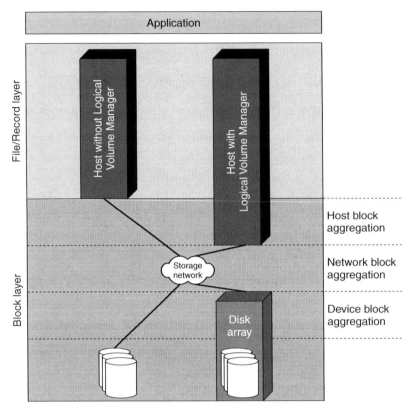

Figure 12.12   In storage connected via a storage network, several hosts share the storage devices, which are accessed via block-oriented protocols.

12.2.3   Block storage aggregation in a storage device: SAN appliance

Block aggregation can also be implemented in a specialised device or server of the storage network in the data path between hosts and storage devices, as in the symmetric storage virtualisation (Figure 12.13, Section 5.7.1). In this approach:

- Several hosts and storage devices are connected via a storage network.
- A device or a dedicated server – a so-called SAN appliance – is placed in the data path between hosts and storage devices to perform block aggregation, and data and metadata traffic flows through this.

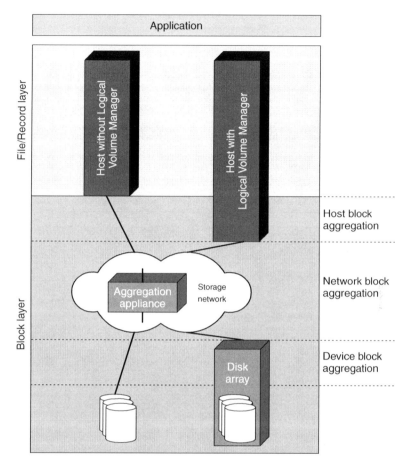

Figure 12.13   In block aggregation on a specialised device or server in the storage network, a SAN appliance maps between logical and physical blocks in the data path in the same way as symmetric virtualisation.

12.2.4   Network attached block storage with metadata server: asymmetric block services

The asymmetric block services architecture is identical to the asymmetric storage virtualisation approach (Figure 12.14, Section 5.7.2):

- Several hosts and storage devices are connected over a storage network.
- Host and storage devices communicate with each other over a protocol on block level.
- The data flows directly between hosts and storage devices.

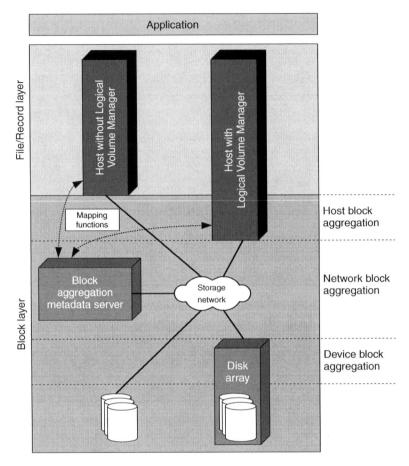

Figure 12.14 In an asymmetric block services architecture a metadata server outside the data path performs the mapping of logical to physical blocks, whilst the data flows directly between hosts and storage devices.

- A metadata server outside the data path holds the information regarding the position of the data on the storage devices and maps between logical and physical blocks.

12.2.5 Multi-site block storage

Figure 12.15 shows how data replication between two locations can be implemented by means of Wide Area Network (WAN) techniques. The data can be replicated on different layers of the model using different protocols:

- Between volume managers on the host;

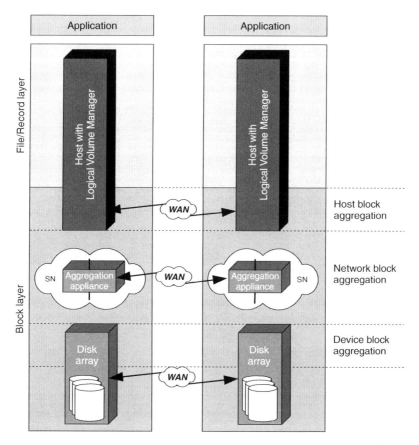

Figure 12.15   Data replication between two locations by means of WAN technology can take place at host level between volume managers, at network level between specialised devices, or at storage device level between disk arrays.

- Between specialised devices in the storage network; or
- Between storage systems, for example disk subsystems.

If the two locations use different network types or protocols, additional converters can be installed for translation.

12.2.6   File server

A file server (Section 4.2) can be represented as shown in Figure 12.16. The following points are characteristic of a file server:

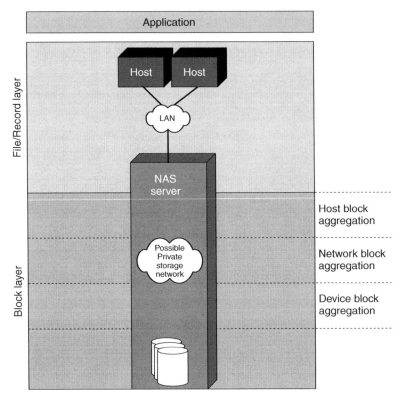

Figure 12.16   A file server makes storage available to the hosts via a LAN by means of file sharing protocols.

- The combination of server and normally local, dedicated storage;
- File sharing protocols for the host access;
- Normally the use of a network, for example, a LAN, that is not specialised to the storage traffic;
- Optionally, a private storage network can also be used for the control of the dedicated storage.

12.2.7   File server controller: NAS heads

In contrast to file servers, NAS heads (Figure 12.17, Section 4.2.2) have the following properties:

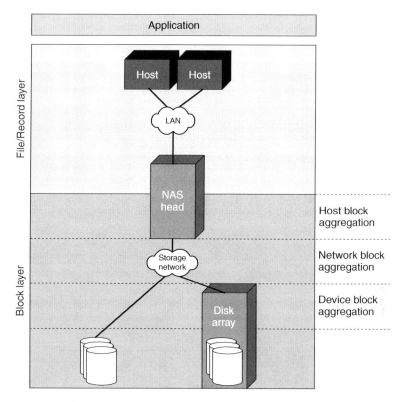

Figure 12.17 A NAS head separates the storage devices from the hosts and thereby achieves better stability and a more efficient use of resources.

- They separate storage devices from the controller on the file/record layer, via which the hosts access.
- Hosts and NAS heads communicate over a file-oriented protocol.
- The hosts use a network for this that is generally not designed for pure storage traffic, for example a LAN.
- When communicating downwards to the storage devices, the NAS head uses a block-oriented protocol.

NAS heads have the advantage over file servers that they can share the storage systems with other hosts that access them directly. This makes it possible for both file and block services to be offered by the same physical resources at the same time. In this manner, IT architectures can be designed more flexibly, which in turn has a positive effect upon scalability.

12.2.8 Asymmetric file services: NAS/file server metadata manager

A file server metadata manager (Figure 12.18) works in the same way as asymmetric storage virtualisation on file level (Section 5.7.2):

- Hosts and storage devices are connected via a storage network.
- A metadata manager positioned outside the data path stores all file position data, i.e. metadata, and makes this available to the hosts upon request.

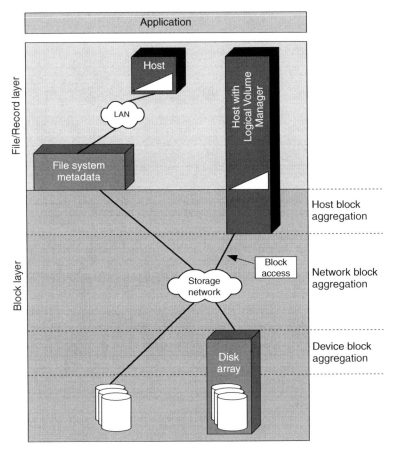

Figure 12.18 A file server metadata manager holds all position data of the files on the storage devices and makes this available to the hosts upon request. Then the hosts can exchange their useful data with the storage devices directly over the storage network. In addition, a metadata manager can offer classical file sharing services in a LAN.

- Hosts and metadata manager communicate over an expanded file-oriented protocol.
- The actual user data then flows directly between hosts and storage devices by means of a block-oriented protocol.

This approach offers the advantages of fast, direct communication between host and storage devices, whilst at the same time offering the advantages of data sharing on file level. In addition, in this solution the classic file sharing services can be offered in a LAN over the metadata manager.

12.2.9   Object-based storage device (OSD)

The SNIA Shared Storage Model defines the so-called object-based storage device (OSD). The idea behind this architecture is to move the position data of the files and the access rights to a separate OSD. OSD offers the same advantages as a file sharing solution, combined with increased performance due to direct access to the storage by the hosts, and central metadata management of the files. The OSD approach functions as follows (Figure 12.19):

- An OSD device exports a large number of byte vectors instead of the LUNs used in block-oriented storage devices. Generally, a byte vector corresponds to a single file.
- A separate OSD metadata manager authenticates the hosts and manages and checks the access rights to the byte vectors. It also provides appropriate interfaces for the hosts.
- After authentication and clearance for access by the OSD metadata manager, the hosts access the OSD device directly via a file-oriented protocol. This generally takes place via a LAN, i.e. a network that is not specialised for storage traffic.

12.3   EXTENSION OF THE SNIA SHARED STORAGE MODEL TO TAPE FUNCTIONS

The SNIA Shared Storage Model described previously concentrates upon the modelling of disk-based storage architectures. In a supplement to the original model, the SNIA Technical Council defines the necessary extensions for the description of tape functions and backup architectures.

The SNIA restricts itself to the description of tape functions in the Open Systems environment, since the use of tapes in the mainframe environment is very difficult to model and differs fundamentally from the Open Systems environment. In the Open Systems field, tapes are used almost exclusively for backup purposes, whereas in the field of mainframes tapes are used much more diversely. Therefore, the extension of the SNIA model concerns itself solely with the use of tape in backup architectures.

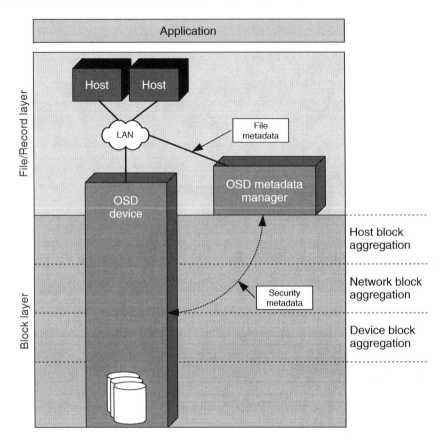

Figure 12.19 Object-based storage devices offer file sharing and facilitate direct I/O between hosts and storage. A metadata manager authenticates the hosts and controls access.

Only the general use of tapes in shared storage environments is described in the model. The SNIA does not go into more depth regarding the backup applications themselves. We have already discussed network backup in Chapter 7. More detailed information on tapes can be found in Section 11.2.1.

First of all, we want to look at the logical and physical structure of tapes from the point of view of the SNIA Shared Storage Model (Section 12.3.1). Then we will consider the differences between disk and tape storage (Section 12.3.2) and how the model is extended for the description of the tape functions (Section 12.3.3).

12.3.1   Logical and physical structure of tapes

Information is stored on tapes in so-called tape images, which are made up of the following logical components (Figure 12.20):

- *Tape extent*
  A tape extent is a sequence of blocks upon the tape. A tape extent is comparable with a volume in disk storage. The IEEE Standard 1244 (Section 9.5) also uses the term volume but it only allows volumes to reside exactly on one tape and not span multiple tapes.
- *Tape extent separator*
  The tape extent separator is a mark for the division of individual tape extents.
- *Tape header*
  The tape header is an optional component that marks the start of a tape.
- *Tape trailer*
  The tape trailer is similar to the tape header and marks the end of a tape. This, too, is an optional component.

In the same way as logical volumes of a volume manager extend over several physical disks, tape images can also be distributed over several physical tapes. Thus, there may be precisely one logical tape image on a physical tape, several logical tape images on a physical tape, or a logical tape image can be distributed over several physical tapes. So-called tape image separators are used for the subdivision of the tape images (Figure 12.21).

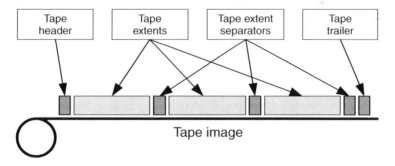

Figure 12.20   Logically, a tape image is made up of tape extents and tape extent separators. A tape header and trailer may optionally mark the start and end of a tape image respectively.

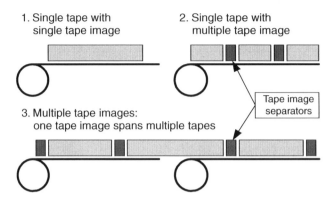

Figure 12.21   Physically, a tape image can take up on precisely one tape (1), several tape images can share a tape (2), or a tape image can extend over several tapes (3). Tape image separators separate the individual tape images.

12.3.2   Differences between disk and tape

At first glance, disks and tapes are both made up of blocks, which are put together to form long sequences. In the case of disks these are called volumes, whilst in tapes they are called extents. The difference lies in the way in which they are accessed, with disks being designed for random access, whereas tapes can only be accessed sequentially. Consequently, disks and tapes are also used for different purposes. In the Open Systems environment, tapes are used primarily for backup or archiving purposes. This is completely in contrast to their use in the mainframe environment, where file structures – so-called tape files – are found that are comparable to a file on a disk. There is no definition of a tape file in the Open systems environment, since several files are generally bundled to form a package, and processed in this form, during backup and archiving. This concept is, therefore, not required here.

12.3.3   Extension of the model

The SNIA Shared Storage Model must take into account the differences in structure and application between disk and tape and also the different purposes for which they are used. To this end, the file/record layer is expanded horizontally. The block layer, which produces the random access to the storage devices in the disk model, is exchanged for a sequential access block layer for the sequential access to tapes. The model is further supplemented by the following components (Figure 12.22):

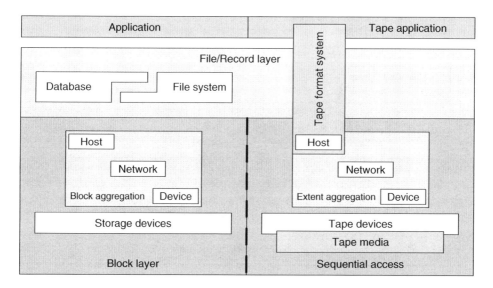

Figure 12.22   The extension of the SNIA model to tape functions expands the file/record layer in the horizontal direction, exchanges the block layer for a sequential access block layer and adds the required components of a tape architecture.

- *Tape media and tape devices*
  Tape media are the storage media upon which tape images are stored. A tape devices is a special physical storage resource, which can process removable tape media. This differentiation between media and devices is particularly important in the context of removable media management (Chapter 11). The applicable standard, IEEE 1244, denotes tape media as cartridge and tape device as drive.

- *Tape applications*
  The SNIA model concentrates upon the use of tapes for backup and archiving. Special tape applications, for example, backup software, are used for backup. This software can deal with the special properties of tapes.

- *Tape format system*
  In the tape format system, files or records are compressed into tape extents and tape images. Specifically in the Open Systems environment, the host generally takes over this task. However, access to physical tape devices does not always have to go through the tape format system. It can also run directly via the extent aggregation layer described below or directly on the device.

- *Extent aggregation layer*
  The extent aggregation layer works in the same way as the block aggregation layer (Section 12.1.7), but with extents instead of blocks. However, in contrast to the random

access of the block aggregation layer, access to the physical devices takes place sequentially. Like the access paths, the data flows between the individual components are shown as arrows.

12.4 EXAMPLES OF TAPE-BASED BACKUP TECHNIQUES AND ARCHITECTURES

First of all, we want to examine four examples that illustrate backup techniques. At the forefront are the access paths and the interaction of the individual components with the UNIX tool *tar* in the file backup (Section 12.4.1), file system volume backup using *dump* (Section 12.4.2), the volume backup using *dd* (Section 12.4.3) and the use of virtual tapes (Section 12.4.4).

We then concentrate upon the data flow between the individual components of a backup architecture with the disk, first of all discussing the two classical approaches to back up to tape: tape connected directly to the host (Section 12.4.5) and the data flow in a backup over LAN (Section 12.4.6). We then consider typical approaches for tape sharing in a shared storage environment, such as tape library sharing (12.4.7) and tape library partitioning (Section 12.4.8).

Next we see how tape virtualisation by means of a virtual tape controller (Section 12.4.9) and supplemented by a disk cache (Section 12.4.10) changes the data flow. In addition to a virtual tape controller, a data mover can also be positioned in the storage network to permit the realisation of server-free backup. As in LAN-free backup, in addition to the LAN and the backup server this also frees up the host performing the backup (Section 12.4.11).

We will then look at two variants of the Network Data Management Protocol (NDMP) local backup with local (Section 12.4.12) and external (Section 12.4.13) storage. Finally, we will consider an architecture in which the NDMP is used with a data mover for the realisation of server-free backup (Section 12.4.14).

12.4.1 File backup

The example shows how a file backup using the UNIX tool *tar* functions (Figure 12.23):

1. *Tar* reads files from the file system.
2. *Tar* compresses the files in the integral tape format system.
3. It finally writes them to tape.

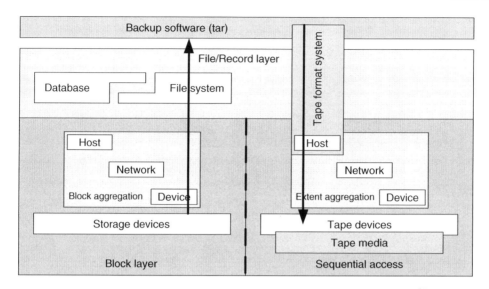

Figure 12.23 *Tar* carries out a file backup by reading data from the file system, then compressing it in the integral tape format system and writing it to tape. In the restore case, the access paths are reversed.

In the restore case the access paths are turned around:

1. *Tar* reads the file packages from tape.
2. *Tar* extracts them by means of the integral tape format system.
3. It writes them into the file system.

12.4.2   File system volume backup

Using the file system backup tool *dump* it is possible to use the file system to back up a logical volume – and thus the files contained within it – bypassing the file system (Figure 12.24). The meta-information of the file system is also backed up, so that it is possible to restore individual files later. *Dump*, like *tar*, has an integral tape format system for the compression and extraction of the files during backup or restore.

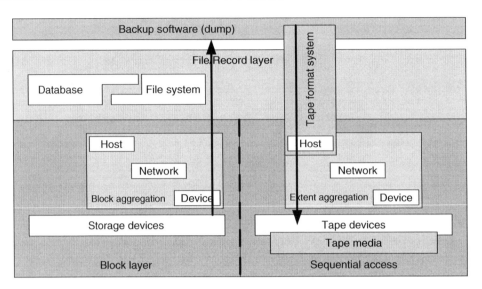

Figure 12.24   With *dump*, files can be backed up directly from a logical volume, bypassing the file system. As is the case for *tar*, an integral tape format system looks after the compression and extraction during restore or backup.

12.4.3 Volume backup

The program *dd* represents the simplest way of creating a copy of a logical volume and writing it directly to tape (Figure 12.25). *dd* writes the information it has read to tape 1:1 without previously sending it through a tape format system. The restore can be represented in a similar way by reversing the access paths.

12.4.4 File backup to virtual tape

The concept of virtual tapes can also be described using the SNIA model. Figure 12.26 uses the example of the *tar* command to show how a disk-based storage system is used to emulate a virtual tape. The sequential tape access of *tar* is diverted via the tape format system in the extent aggregation layer to the block aggregation layer of a disk storage system, where random access can take place.

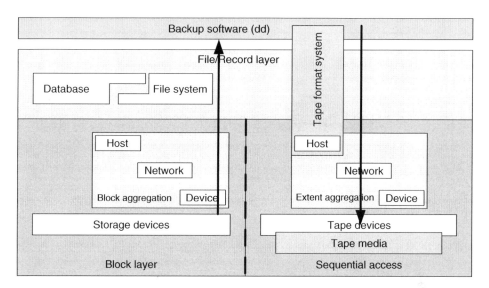

Figure 12.25   The *dd* program creates a copy of a logical volume on tape without the use of a tape format system.

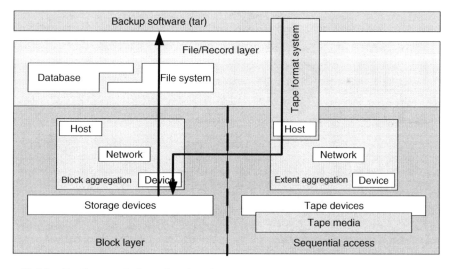

Figure 12.26   By the emulation of a virtual tape, the sequential access of the *tar* command in the extent aggregation layer is diverted into the block aggregation layer of a disk-based storage system, which permits random access.

12.4.5   Direct attached tape

The simplest backup architecture is the direct connection of the tape to the host, in which the data flows from the disk to the tape library via the host (Figure 12.27).

12.4.6   LAN attached tape

LAN attached tape is the classic case of a network backup (Section 7.2), in which a LAN separates the host to be backed up from the backup server, which is connected to the tape library. The backup data is moved from the host, via the LAN, to the backup server, which then writes to the tape (Figure 12.28).

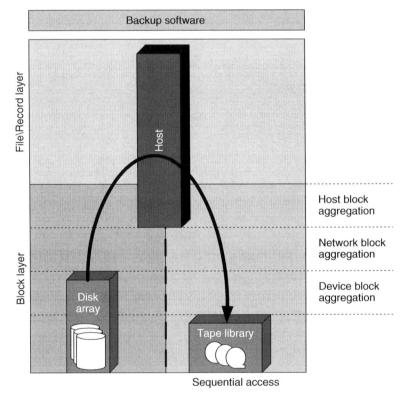

Figure 12.27   In direct attached tape the data flows from the disk to the tape library via the host, as shown by the arrow.

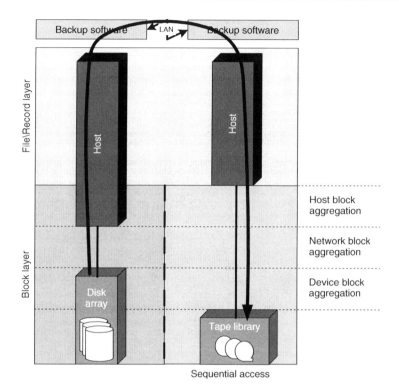

Figure 12.28 In classical network backup, the data must be moved from the host to be backed up, via the LAN, to the backup server, which then writes to the tape.

12.4.7 Shared tape drive

In tape library sharing, two hosts use the same tape drives of a library. In this approach, the hosts dynamically negotiate who will use which drives and tape media. To achieve this, one server acts as library master, all others as library clients (Section 7.8.6). The library master co-ordinates access to the tapes and the tape drives (Figure 12.29). In this manner, a LAN-free backup can be implemented, thus freeing up the LAN from backup traffic (Section 7.8.2).

12.4.8 Partitioned tape library

In library partitioning a library can be broken down into several virtual tape libraries (Section 6.2.2). Each host is assigned its own virtual library to which it works. In this

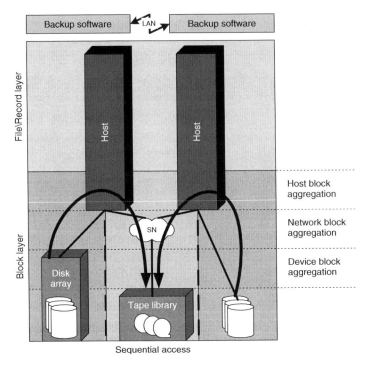

Figure 12.29 A shared tape drive facilitates the implementation of LAN-free backup, which frees the LAN from backup traffic.

manner, several backup servers can work to the library's different tape drives simultaneously. The library co-ordinates the parallel accesses to the media changer (Figure 12.30) independently.

12.4.9 Virtual tape controller

Additional backup functionality now comes into play in the storage network! A virtual tape controller in the storage network permits the virtualisation of tape devices, media and media changer. Thus, different interfaces can be implemented and different tape devices emulated. However, the backup data still runs directly from the hosts to the drives (Figure 12.31).

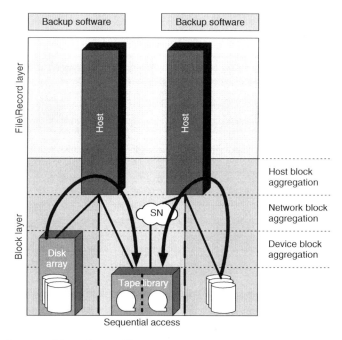

Figure 12.30 In a partitioned tape library, several hosts work to virtual tape libraries that consist of different physical tape drives, but which share a common robot.

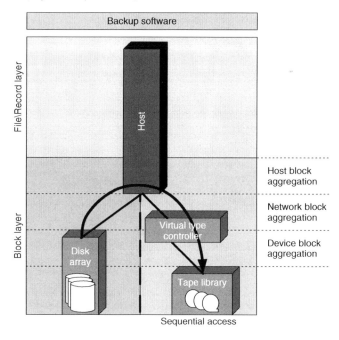

Figure 12.31 A virtual tape controller virtualises tape devices, media and media changer.

12.4.10 Virtual tape controller with disk cache

The approach using a virtual tape controller can be expanded to include an additional disk cache (Figure 12.32). This yields the following three-stage process for a backup:

1. First of all, the host reads the data to be backed up from disk.
2. This data is first written to a disk belonging to the virtual tape controller, the so-called disk cache.
3. Finally, the data is moved from the disk cache to tape.

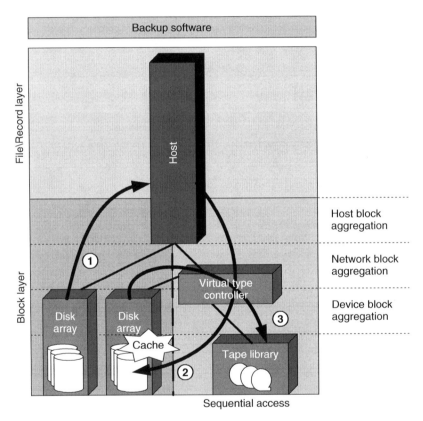

Figure 12.32   If the virtual tape controller is extended to include a disk cache, the backup software can benefit from the higher disk performance.

In this manner, a backup can benefit from the higher performance of the disk storage. This is especially useful when backed up data must be restored: Most restore requests deal with data which was backed up within the last 1 or 2 days.

12.4.11   Data mover for tape

With an additional data mover in the storage network that moves the data from disk to tape, server-free backup can be implemented. This frees up both the LAN and also the participating hosts from backup traffic (Section 7.8.1). The backup servers only have to control and check the operations of the data mover (Figure 12.33).

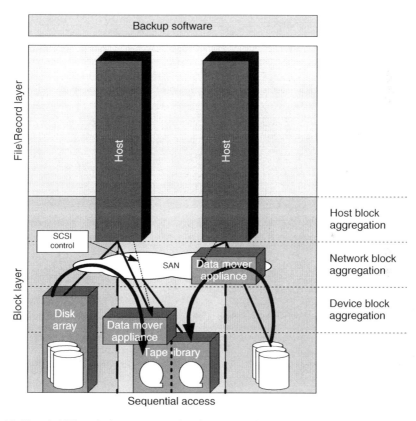

Figure 12.33   Additional data movers in the network implement the server-free backup, which frees both the LAN and the hosts from backup traffic at the same time.

12.4.12   File server with tape drive

Figure 12.34 shows the implementation of the NDMP local backup (Section 7.9.4). In this approach, the file server itself transports the data from disk to tape, which in this case is even locally connected. External backup software checks this process and receives the meta information of the backed up data via a LAN connection by means of the NDMP protocol.

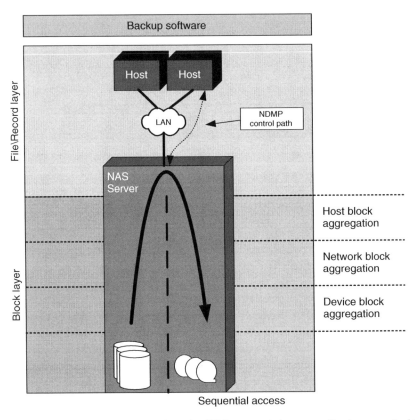

Figure 12.34   In the NDMP local backup the NAS server takes over the transport of the data from disk to tape, which in this case is even locally connected.

12.4.13   File server with external tape

If the NAS server in Section 12.4.12 is exchanged for a NAS head with external disk and tape storage, then the backup software additionally checks the functions of the tape library on the host. Again, additional meta-information on the backed up information flows from the NAS head to the backup server (Figure 12.35).

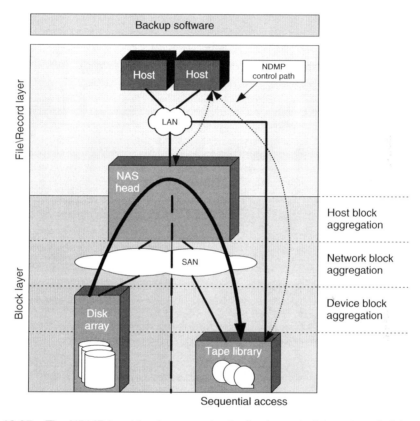

Figure 12.35   The NDMP local backup can also be implemented for external disk and tape storage on a NAS head.

12.4.14 File server with data mover

An additional data mover in the storage network (Figure 12.36), which takes over the transport of the backup data from the NAS head with external storage, also implements server-free backup (Section 7.8.1) on file server level. LAN and backup software are already freed from data transport by the use of NDMP (Section 7.9.4).

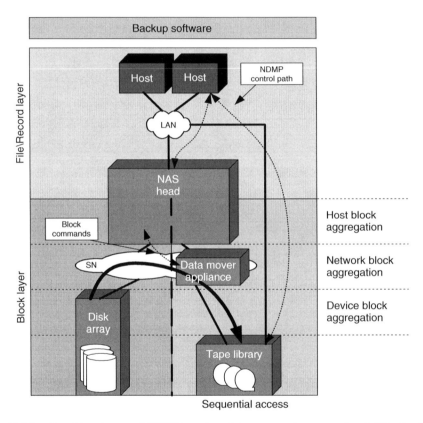

Figure 12.36 Combined use of NDMP and a data mover frees up the LAN, the backup server due to NDMP, and frees up the NAS head from the transport of the backup data by the implementation of server-free backup at file server level.

12.5 SUMMARY

The SNIA Shared Storage Model permits architectures to be described and compared with one another in a value-neutral manner and discussed using a consistent vocabulary. This makes it easier for manufacturers to present the differences between their products and competing products to the customer on the basis of a common vocabulary. The customer interested in the actual functionality finds it easier to compare and choose between different product alternatives. He benefits from the function-centred approach of the SNIA Shared Storage Model, which puts the entire functionalities of the Shared Storage environment in the foreground and only highlights the components on the basis of which these are implemented as a secondary consideration.

13

Final Note

Dear Readers, we have finished a joint journey of around 500 written pages. We hope that you have enjoyed reading this book and learning from it as much as we have had in writing it. Until now we are the only ones who have been doing the talking. So what about you? Was there anything in particular that you liked? Is there anything that we could improve upon in the next edition? Have you found any errors that we should be correcting?

This book is already impressive in size. Nevertheless, some topics have to be expanded or added in order to make it complete. Which topics would you particularly like to see added in a new edition?

We look forward to receiving feedback from you on this book. Please write to us at authors@storageexplained.com or leave a comment on our webpage www.storage-explained.com.

Storage Networks Explained: Basics and Application of Fibre Channel SAN, NAS, iSCSI, InfiniBand and FCoE, Second Edition
U. Troppens R. Erkens W. Müller-Friedt N. Haustein R. Wolafka © 2009 John Wiley & Sons, Ltd

Glossary

3rd-Party SCSI Copy Command   The 3rd-Party SCSI Copy Command is the specification for the use of the SCSI XCOPY command in order to copy blocks from one storage device to another within a storage network. This command is, for example, the basis for server-free backup.

64b/66b-encoding   Variant of 8b/10b-encoding used for 10-Gigabit networks with various cable types.

8b/10b encoding   An encoding procedure that converts an 8-bit data byte sequence into a 10-bit transmission word sequence that is optimised for serial transmission. The 8b/10b encoding is used, for example, for Fibre Channel, Gigabit Ethernet and InfiniBand.

Access control   The granting or refusal of a request for access to services or resources based upon the identity of the requester.

Access path   Descriptively defined as the list of components that are run through by read and write operations to the storage devices and responses to them.

ACL   An access control list (ACL) allows the control of access rights in a computer system. The control of the access is usually based on a user name or on groups and grants certain rights, such as permission to read or change a file.

Active   An active configuration means a component that is not designed with built-in redundancy.

Storage Networks Explained: Basics and Application of Fibre Channel SAN, NAS, iSCSI, InfiniBand and FCoE, Second Edition
U. Troppens R. Erkens W. Müller-Friedt N. Haustein R. Wolafka   © 2009 John Wiley & Sons, Ltd

Active/active An active/active configuration describes a component designed with built-in redundancy, in which both subcomponents are used in normal operation. We differentiate between active/active configurations with and without load sharing.

Active/passive An active/passive configuration describes a component designed with built-in redundancy, in which the second component is not used in normal operation (stand-by).

Agent In the fields of storage networks and system management the client software of a client-server application is very often referred to as the agent. For example, we talk of the backup agent for an application-specific backup client in a network backup system or the SNMP agent for the management of storage devices.

Aggregation The combining of multiple similar and related objects or operations into a single one. Two or more disks can be aggregated into a single virtual disk or in a RAID array.

AIIM The Association for Information and Image Management (AIIM) is an international organisation that is concerned with the subject of enterprise content management (ECM). Its main goal is to merge the interests of users and ECM vendors and to establish standards.

AL_PA The Arbitrated Loop Physical Address (AL_PA) is the address of a device (host bus adapter or switch) in a Fibre Channel Arbitrated Loop.

API An application programming interface (API) is an interface that is provided by a software system so that other programs can use its functions and services. In contrast to an application binary interface (ABI), an API offers the use of the interface at the source code level.

Appliance A device for the execution of a very specific task. Appliances differ from normal computers due to the fact that their software has generally been modified for this very specific purpose.

Application server-free backup Application server-free backup refers to the backup of application data with the aid of an instant copy generated in the disk subsystem and a second server, so that the load for the backup is offloaded from the application server to the second server.

Arbitrated Loop One of the three Fibre Channel topologies. The other two are point-to-point and fabric.

Archive bit The archive bit is a bit in the metadata of a file, which can be used to accelerate the realisation of the incremental-forever strategy.

Archiving Archiving is a intelligent process where data which usually does not change any more is moved into an archiving system. Data access, search, processing and protection must be guaranteed over a long life cycle.

Asymmetric storage virtualisation   Asymmetric storage virtualisation is the form of storage virtualisation within a storage network in which the data flow is separated from the control flow. The data flow runs directly between the servers and storage devices whereas the control flow, i.e. the control of the virtualisation by a configuration entity, travels outside the data path.

Autoloader   Small automatic tape library with few slots and usually just one drive.

Backup   The goal of backup is to generate one or more copies of data that can be used in the event of a failure to restore the original data.

Backup window   Time window that is particularly favourable for the backup of the data of an application. For some applications the backup window specifies the maximum period of time that is available for the backup of data.

Bare Metal Restore   Alternative expression for 'image restore'.

BC   See Business continuity.

Binary encoding   Physical encoding procedure for the transmission of data.

Block aggregation   The bringing together of physical blocks or block vectors to form logical blocks or block vectors (block-based storage virtualisation). Two or more physical disks can thus be aggregated to form one virtual disk.

Block layer   Component of the SNIA Shared Storage Model that includes block-based storage devices and block aggregation.

Block level   This expression refers to the physical or virtual blocks of hard disks and tapes. For example, we talk of backup or storage virtualisation on block level.

Block level incremental backup   Block level incremental backup describes the capability of a network backup system to incrementally back up only those subsections (blocks) of files or of entire files systems that have changed since the previous backup.

Block orientation   Storage devices and I/O protocols that are organised in blocks are called block-oriented, for example hard disks, SCSI, iSCSI and Fibre Channel FCP. File orientation represents an alternative to this.

Bluefin   See SMI-S.

Bus   Physical I/O medium with several lines for parallel signal transmission.

Business continuity   Business continuity (BC) describes technical and organisational strategies for ensuring continuous access to business-critical data and applications even in crisis situations.

Cache   Fast storage, in which data accesses to slower storages are buffered.

Cache server Describes a component in a network that temporarily stores data for other components in order to reduce the consumption of network capacity or to provide damping for accesses to slower storage.

Cartridge Physical medium on which storage capacity is available. The storage capacity can be distributed over several sides.

CAS Content Addressable Storage is a storage system that uses a cryptographic hash or CRC that is calculated for an entity of data as an address to reference this entity of data. When data is stored in a CAS, the system calculates the hash and returns it to the application. The application uses the hash to address or reference the data for subsequent access. CAS systems enable efficient data integrity checks in which a cryptographic hash is calculated again for the data and it is compared to the content address. Based on the cryptographic hash, CAS systems also provide data deduplication functions (see Deduplication).

CDP Continuous data protection (CDP) is a backup technique that backs up data immediately at the point when the data changes, capturing every version of data that is generated.

CEE Some vendors use the term Converged Enhanced Ethernet (CEE) to refer to Data Center Bridging (DCB) and FCoE. However, some vendors include additional protocols or features to their definition of CEE than other vendors, thus customers must check carefully as to what are included in a vendor's version of CEE and what are not.

CIFS Common Internet File System (CIFS), the network file system from Microsoft for Windows operating systems.

CIM The Common Information Model (CIM) is an object-oriented description of systems, applications, networks and devices. CIM is a significant component of the Web Based Enterprise Management (WBEM), a standard developed by the Distributed Management Task Force (DMTF) for the management of storage networks, which is currently viewed as the successor to the Simple Network Management Protocol (SNMP).

Class 1, Class 2, Class 3 Different service classes for transmission in a Fibre Channel network.

CLI A command line interface (CLI) is a mechanism for monitoring and controlling hardware and software components by typing commands and receiving text-based results. See also GUI.

Cluster A compound of the same type of resources. The term 'cluster' is often used without being defined more precisely. Sometimes the term 'cluster' also denotes a single node of such a compound. Therefore when talking about clusters you should always ask precisely what is meant by the term.

CMIP The Common Management Information Protocol (CMIP) was designed at the end of the 1980s as the successor to the Simple Network Management Protocol (SNMP). In practice, however, CMIP is hardly ever used.

CN Congestion Notification (CN, IEEE 802.1Qau) is part of the DCB protocol family. It propagates congestion situations from the switches to the end devices to throttle the sending rate of end devices.

CNA An Ethernet card (Network Interface Controller, NIC) which supports conventional TCP as well as the new DCB and FCoE protocols thus being suitable for I/O consolidation is also called Converged Network Adapter (CAN).

Co-location Co-location describes the capability of a network backup system to write (co-locate) several incremental backups of one or more servers onto just a few tapes, so that the number of tape mounts is reduced if the data has to be restored.

Cold backup Cold backup describes the backup of a database that has been shut down for the duration of the backup.

Common Scratch pool Group of cartridges, the storage capacity of which is (or has once again become) completely available and to which all applications have access so that they can reserve a cartridge from it for the purpose in question.

Community name The Simple Network Management Protocol (SNMP) has no secure authentication mechanisms. Instead, so-called community names are used. Two components (for example an SNMP agent and an SNMP-capable management system) can only communicate with each other if they are configured with the same community name.

Consistency group A consistency group combines a number of instant copy or remote mirroring relationships into one unit. Consistency groups increase the consistency of the data that is distributed over several virtual hard disks and are copied using instant copy or remote mirroring.

Copy-on-demand Copy-on-demand is an implementation variant of instant copies and snapshots in which source data is not copied until it has been changed on the source.

CRC A cyclic redundancy code (CRC) represents a checksum that is calculated for an entity of data and enables the data to be checked for integrity. However, the checksum is relatively weak, which means that different data is more likely to produce the same checksum. Thus deliberate manipulation of the data can succeed in producing the same checksum again, which contradicts the data integrity. CRC is normally used to detect corruptions during data transfer. Compared to CRC, cryptographic hashes, such as Message Digest 5 (see MD5) and Secure Hash Algorithms, are more robust in regard to manipulation and are used to check the integrity of data that is stored for longer periods of time.

Credit The credit model is a procedure for the realisation of flow control. Fibre Channel differentiates between buffer-to-buffer credit for link flow control and end-to-end credit for the flow control between two end devices.

Cryptographic Hash A cryptographic hash is a unique checksum that is calculated for an entity of data using cryptographic hash functions such as Message Digest 5 (MD5), Secure Hash Algorithm (SHA) or Whirlpool. A given hash value is most likely unique for a given set of data and can be used to prove the integrity of the data. However, a low probability exists that two distinct sets of data can result in the same hash value, which is called hash collision. Cryptographic hashes are also used in data deduplication techniques (see Deduplication).

Cut-through routing   Cut-through routing is the capability of a switch, a director or a router to forward incoming data packets before they have been fully received.

CWDM   Coarse Wavelength Division Multiplexing (CWDM) uses similar procedures to DWDM. The two techniques differ mainly in the division of the frequency ranges and the number of payload streams that they can transmit over a single fibre-optic cable. See also DWDM.

D2D   Disk-to-disk (D2D) refers to the class of backup methods in which data is copied from one hard disk or disk subsystem to a second one.

D2D2T   Disk-to-disk-to-tape (D2D2T) refers to the class of backup methods in which data is copied from one hard disk or disk subsystem to a second one and from there additionally to tape storage.

DAFS   The Direct Access File System (DAFS) is a network file system that is based upon the Virtual Interface Architecture (VIA) and Remote Direct Memory Access (RDMA). DAFS aims to achieve lightweight and very fast file access within a data centre.

DAS   Direct Attached Storage (DAS) is storage that is directly connected to a server without a storage network, for example over SCSI or SSA.

Data backup   See Backup.

Data copying   A variant of data sharing, in which common data is copied for each applications.

Data integrity   Data integrity means that the stored data is always the same as its original state. Methods such as cyclic redundancy code (CRC) and cryptographic hashes are used to determine whether violations of the data integrity exist. Error-correcting methods are used to restore the data integrity if possible.

Data migration   Data migration describes the transfer of data from a source location to a target location. The meaning of 'location' depends on the context and can be anything from storage media, such as magnetic tapes and hard disks, and file systems to applications such as a Document Management System (DMS). During the data migration the copy of the data at the source location is usually invalidated or deleted, whereas the copy of the data at the target is considered active.

Data scrubbing   As a background process, high-end RAID controllers and disk subsystems regularly read all the blocks of a RAID array, thereby detecting defective blocks before they are actively accessed by applications. Data scrubbing significantly reduces the probability of data loss.

Data sharing The use of common data by several applications.

Data shredding Data shredding is a method of physically destroying data on a storage medium without destroying the medium itself. In the process, randomly generated data patterns are usually written over the data that is to be destroyed. This type of data shredding is only possible with rewritable storage media.

DCB The IEEE refers Data Center Bridging (DCB) to a set of protocols which enhances Ethernet to make it a suitable transmission technology for storage traffic.

DCBX Data Center Bridging Exchange (DCBX) refers to an enhancement of the Link Layer Discovery Protocol (LLDP, IEEE 802.1AB-2005) to manage components which support the DCB protocol family. In the second half of 2009 it is still open whether DCBX will become a seperate standard or whether it will be integrated into Enhanced Transmission Selection (ETS).

DCE Some vendors use the term Data Center Ethernet (DCE) to refer to Data Center Bridging (DCB) and FCoE. However, some vendors include additional protocols or feature to their definition of DCE than other vendors thus customers must check carefully what is included in a vendor's version of DCE and what not.

Deduplication Data deduplication is a process in which identical data is identified and only one instance of identical data is stored. Other instances of identical data are referenced to the stored instance. This reduces the amount of stored data. The identification process might utilise cryptographic hashes such as MD5 or SHA to identify two identical sets of data. The deduplication process must maintain a list of identity characteristics (such as cryptographic hashes) for the data that has been stored. The process must also maintain a list of references for identical sets of data.

Degraded RAID array A RAID array where a physical hard disk has failed is also referred to as a degraded array. With most RAID procedures all data remains intact in spite of disk loss. However, the performance in a degraded array can sink dramatically because some of the data blocks have to be reconstructed.

DICOM DICOM stands for Digital Imaging and Communications in Medicine. DICOM is a worldwide open standard for the exchange of digital images in medicine. DICOM standardises the format for the storage of image data as well as the communications protocol for the exchange of the images. Almost all suppliers of medical image-making systems support this standard. DICOM is used as a communication protocol by Picture Archiving and Communication Systems for communication with the modalities (see PACS).

Digital signature A digital signature is a private key-encrypted cryptographic hash that is calculated for a given data object. The digital signature for a data object serves two purposes: (1) it is used to prove the integrity of the data object via the cryptographic hash; and (2) it is used to identify the originator of the data object via the private encryption key. A digital signature for a given data object is created by an originator that calculates the cryptographic hash for the object and this hash is encrypted using the originator's private key. For validation of the integrity of the data object the encrypted hash must be decrypted using the originator's private key and the hash is then compared against a newly calculated hash for the data object. The encryption and decryption of the hash can also be based on asymmetric encryption, whereby the originator encrypts the hash using its private key and an associated public key is used to decrypt the hash. Secure electronic signatures can be generated through the use of digital signatures (also see electronic signatures).

Director A director is a switch with a higher fault-tolerance than that of a simple switch as a result of redundant components.

Disaster recovery Disaster recovery (DR) describes those measures that are needed in order to restore an IT operation in a controlled and predictable manner after an outage.

Discovery Discovery is the automatic detection of all resources (hardware, network topologies, applications) used in a storage network or more generally in a computer network.

Disk subsystem A disk subsystem is a collection of hard disks installed in a common enclosure. We differentiate between JBODs, RAID systems and intelligent disk subsystems. The storage capacity of a disk subsystem is between several Terabytes and 1 Petabyte (2009).

DMAPI The Data Management Application Programming Interface is an interface provided by some file systems to enable applications to intercept file system operations and perform additional file system independent tasks. One example for such an application is hierarchical storage management (HSM). Not all file systems provide the DMAPI interface.

DMI The Desktop Management Interface (DMI) is a protocol for the management of servers specified by the Distributed Management Task Force (DMTF). DMI is seldom used in comparison to the Simple Network Management Protocol (SNMP).

DMS A document management system (DMS) is used to manage archive data. It permits the search, linking and processing of archive data. In the context of this book the term DMS is used as a synonym for the archive management application sometimes also called Enterprise Content Management System (see ECM).

DMTF The Distributed Management Task Force (DMTF) is an association of manufacturers with the objective of driving forward the standardisation of the management of IT systems.

DR see Disaster recovery.

Dual SAN Dual SAN denotes the installation of two storage networks that are completely separate from each other. Dual SANs have the advantage that even in the event of a serious fault in a storage network (configuration error or defective switch, which floods the storage network with corrupt frames) the connection over the other storage network is maintained.

DWDM Dense Wavelength Division Multiplexing (DMWM) increases the capacity of a fibre-optic cable by assigning several incoming optical signals (= payload streams) to certain optical frequency ranges. Metaphorically speaking, each payload stream is transmitted in a different colour. Since the signals are only optically transformed, there are no limitations with regard to data rates or data formats of the payload streams. As a result, very different payload streams such as Fibre Channel, ESCON, Gigabit Ethernet and Sonet/SDH can be transmitted simultaneously over a single fibre-optic cable.

ECC Error-Correcting Encoding (ECC) specifies coding that enables the correction of errors in a set of data. In addition to the actual data, redundant data is generated and stored together with the data. If an error occurs, the original data can be restored using the redundant data. However, this is only possible to a certain extent, which means that only a limited number of errors can be corrected.

ECM The terminology Enterprise Content Management (ECM) is not used consistently in the literature or by suppliers. Sometimes ECM is used with the same meaning as Document Management System (DMS). On the other hand, the Association for Information and Image Management (AIIM), for example, interprets ECM as technologies for the collection, management, storage, retention and provision of information and documents to support the business processes of a company. With this definition a DMS is part of the ECM solution that manages documents. In this book we concern ourselves with the archiving side of DMS and ECM so that both terms overlap and we therefore make no further distinction between the two.

Electronic signature An electronic signature is interpreted as data that enables the identification of the signer or creator of the data and can be used to check the integrity of the signed electronic data. Digital signatures in conjunction with digital certificates are often used to generate and evaluate electronic signatures. Technically, an electronic signature therefore serves the same purpose as a person's own handwriting on a document.

Electronic vaulting See Vaulting.

Element manager The element manager is a device-specific management interface that is classified as an out-band interface. It is often realised in the form of a GUI or web interface.

Emulated loop Facilitates communication between private loop devices of a Fibre Channel arbitrated loop and devices in a Fibre Channel fabric.

Enhanced shared-nothing cluster Server clusters of up to several ten servers. Enhanced shared-nothing clusters can react to load peaks with a delay.

ENode FCoE end devices (server, storage) are also called FCoE Node or just ENode.

ERP An enterprise resource planning (ERP) system supports a company by planning the maximum use of in-house resources such as capital, equipment and personnel for its business processes.

Error handler Component of a network backup system. The error handler helps to prioritise and filter error messages and to generate reports.

ESCON The Enterprise System Connection (ESCON) is a serial I/O technology for mainframes.

ETS Enhanced Transmission Selection (ETS, IEEE 802.1Qaz) is part of the DCB protocol family. It allows to share the bandwidth of an Ethernet port among priority groups whilst it assures a configurable minimum bandwidth for each priority group.

Exchange An exchange is a logical communication connection between two Fibre Channel devices.

External storage Storage (hard disks, tape drives), which is located outside the computer enclosure.

Fabric The most flexible and scalable of the three Fibre Channel topologies.

Fabric login (FLOGI) Fabric login denotes the registration of an N-Port into a fabric topology. It establishes a session between the N-Port and the corresponding F-Port of a Fibre Channel switch.

FC Abbreviation for Fibre Channel.

FCIA The Fibre Channel Industry Association (FCIA) is an association of manufacturers from the field of Fibre Channel technology.

FCIP Tunnelling protocol that transports the Fibre Channel traffic between two Fibre Channel devices via TCP/IP.

FCN A Fibre Channel Name (FCN) is a 64-bit identifier for a Fibre Channel component, which in contrast to a WWN is not unique worldwide. It has become common practice to refer to WWNs and FCNs simply as WWNs.

FCoE Fibre Channel over Ethernet (FCoE) is a protocol mapping for transferring Fibre Channel frames via Ethernet. Ethernet needs certain enhancements to be suitable for FCoE which are referred as Data Center Bridging (DCB).

FCP The Fibre Channel Protocol (FCP) is the protocol mapping that maps the SCSI protocol onto the Fibre Channel transmission technology.

Fiber Alternative name for fiber-optic cable.

Fibre Channel A technology that can realise both storage networks and data networks. Fibre Channel is currently the predominant technology for the realisation of storage networks. We differentiate between three network topologies: arbitrated loop, fabric and point-to-point.

Fibre Channel SAN A Fibre Channel network that is used as a storage network. Or the other way around: A storage network that is realised with Fibre Channel.

FICON Fibre Connection (FICON) is the mapping of the ESCON protocol on Fibre Channel.

File level The files of a file system are the object of the processing. For example, we talk of backup on file level or storage virtualisation on file level.

File orientation Storage devices and I/O protocols are called file-oriented if they are organised in files or file fragments, for example NAS servers, NFS, CIFS and HTTP. An alternative to this is block orientation.

File/record layer Component of the SNIA Shared Storage Model that maps the database records and files on the block-oriented volumes of the storage device.

FIP The FCoE Initialisation Protocol (FIP) complements FCoE for the discovery and initialisation of FCoE capable devices.

Flow control Mechanism for the regulation of the data stream between a sender and a receiver. The flow control ensures that the transmitter only sends data at a speed that the receiver can process it.

Forward recovery Forward recovery, sometimes also called 'roll forward', denotes the restoring of a database using a backup copy plus archive log files generated after the backup copy and the active log files that are still present.

FPMA Fabric Provided MAC Address (FPMA) is a mechanism where the FCoE switch generates a MAC address for the FCoE Line End Point (FCoE LEP) of an FCoE end device (ENode).

Frame The data packets that are transmitted in a Fibre Channel network are called frames.

GDPdU The GDPdU (Grundsätze zum Datenzugriff und der Prüfbarkeit von digitalen Unterlagen) is a set of guidelines for German auditors and tax offices that incorporates standards from the German tax and trade laws. GDPdU essentially specifies how digital data must be archived in order to be compliant with these laws. German companies and companies trading in Germany must follow these standards. Similar to the GDPdU, the U.S. Securities and Exchange Commission (SEC) has released regulations specifying legal requirements for data archiving in the U.S.

GUI A graphical user interface (GUI) is a window-based interface for monitoring and controlling hardware and software components. See also CLI.

HA see High availability.

Hard zoning In hard zoning only the end devices that lie in at least one common zone can communicate with each other. Hard zoning is often confused with port zoning.

HBA A host bus adapter (HBA) is another term for an adapter card that is fitted in a server. Examples of host bus adapters are SCSI controllers, Fibre Channel cards and iSCSI cards.

HCA Host channel adapter (HCA) denotes the connection point of a server to an Infini-Band network.

Hierarchical storage management (HSM) Hierarchical storage management (HSM) denotes the automatic movement of data that has not been used for a long time from fast storage to slower but cheaper storage, for instance from disk to tape. Thereby the movement of the data is transparent to users and applications. HSM is commonly a subfunction of network backup systems.

High availability High availability (HA) refers to the capability of a system to maintain IT operations despite the failure of individual components or subsystems.

HIS A Hospital Information System (HIS) is used to manage data, such as family history data of patients, test results, medical procedures performed and billing information, in hospitals. HIS systems are not normally used to manage picture information because this is usually handled by Picture Archiving and Communication Systems (see PACS).

HL7 Health level 7 (HL7) is an international standard for data exchange between data capturing and processing systems in the healthcare industry. The '7' in the name relates to layer 7 of the ISO/OSI reference model for communication (ISO7498-1) and expresses that the communication is specified at the applications layer. HL7 offers interoperability between hospital information systems (HIS), medical office management systems, laboratory information systems, billing systems for medical services and systems that function as electronic patient files.

Host I/O bus The host I/O bus represents the link between system bus and I/O bus. The most important representative of the host I/O bus is the PCI bus.

Hot backup Hot backup denotes the backup of a database during operation.

Hot spare disks In a RAID configuration (RAID array, intelligent disk subsystem) a spare disk is called a hot spare disk.

HSM See Hierarchical storage management.

Hub A component that is not visible to end devices, which simplifies the physical cabling of a network. In Fibre Channel networks the ring (physical) of the arbitrated loop (logical) is simplified to a star shape (physical).

I/O bus Physical communication connection between servers and storage devices, for example SCSI, Fibre Channel or iSCSI. Originally, parallel buses were used for this such as SCSI or IDE. For historical reasons, serial I/O techniques such as SSA, Fibre Channel or iSCSI are also often called I/O buses.

I/O path The path from CPU and main memory to the storage devices via system bus, host I/O bus and I/O bus.

iECM Interoperable Enterprise Content Management (iECM) is a standard developed by Association for Information and Image Management (AIIM) for the coupling and communication between multiple Document Management System (DMS) and Enterprise Content Management (ECM) systems.

IETF The Internet Engineering Task Force (IETF) is a committee that standardises the protocols for the Internet. These include TCP/IP-based protocols such as FTP, HTTP, NFS, iSCSI, FCIP, iFCP and iSNS.

iFCP Internet FCP (iFCP), a standard with the objective of replacing the network layer in a Fibre Channel SAN with a TCP/IP network.

ILM Information Life cycle Management (ILM) comprises processes, tools and methods that have the aim of ascertaining the value of information and data and adapting the cost to store this information and data to the established value. ILM accounts for the fact that the value of data changes during its lifetime.

Image restore Image restore (also known as Bare Metal Restore) denotes the restoration of a server or a hard disk partition (Windows) or a volume (Unix) from a previously generated copy of a hard disk partition or volume.

In-band management We talk of in-band management if the management of a resource takes place over the same interface over which the actual data is transmitted. Examples of this are the SCSI Enclosure Services (SES) and the corresponding services of the Fibre Channel FCP protocol.

In-band virtualisation Alternative name for 'symmetric virtualisation'.

Incremental-forever strategy The incremental-forever strategy relates to the capability of a network backup system to calculate the last state of the file system from continuous incremental backups of a file system by means of database operations. A complete backup of the file system is only necessary the first time. After this, only incremental backups are performed. The metadata database in the backup server helps to immediately recreate the last state of the file system when restoring the file system.

InfiniBand New transmission technology that aims to replace the parallel PCI-bus with a serial network. InfiniBand may be used for interprocess communication, client-server communication and server-storage communication.

Instant copy Instant copy is the capability of a storage system to virtually copy large data sets within a few seconds.

Internal storage Storage (hard disks, tape drives) located inside the enclosure of the computer.

IPFC IP over Fibre Channel (IPFC), the protocol mapping that makes it possible to use a Fibre Channel network for IP data traffic.

IP storage General term for storage networks that use TCP/IP as a transmission technique. IP storage includes the protocols iSCSI, FCIP and iFCP.

iSCSI Internet SCSI (iSCSI) is the protocol mapping of SCSI on TCP/IP.

iSCSI SAN A storage network that is realised with iSCSI.

iSER iSCSI Extension for RDMA (iSER) is an application protocol for RDMA over TCP. iSER enables to transmit the SCSI data traffic via the quick and CPU-friendly RDMA over TCP instead of via TCP.

iSNS The Internet Storage Name Service (iSNS) defines a name service that is used by different IP storage standards such as iSCSI and iFCP.

ISL The inter-switch link (ISL) is a connection cable between two Fibre Channel switches.

Java Content Respository The Java Content Repository (JCR) is an interface specification in the programming language Java that specifies the interface and protocol between an archiving application and a Document Management System (DMS). JCR is based on the Java Specification Requests JSR-170 and JSR-283.

JBOD Just a Bunch of Disks (JBOD) is the term for a disk subsystem without a controller.

Jitter As a result of physical influences, incoming signal steps at the receiver are not the same length. This bucking within the signal sequence is called jitter.

Job scheduler Component of a network backup system. It controls which data is backed up when.

Journaling Journaling of a file system describes a method in which the file system – in a similar way to a database – first writes changes to a log file and only then enters them in the actual data area. Journaling significantly reduces the time for a file system check after a system crash.

JSR-170, JSR-283 Java Specification Request 170 (JSR-170) specifies Version 1.0 of the Java Content Repository (JCR) and JSR-283 the extended Version 2.0.

K28.5 symbol Special transmission symbol of the 8b/10b encoding, which does not represent a data byte. The K28.5 symbol includes a special bit sequence that does not occur in a bit sequence generated with 8b/10b encoding even across symbol boundaries. The K28.5 symbols scattered in a data stream allows to synchronise transmitter and receiver.

Label A label is both the sticker on the cartridge, which often has a barcode upon it, and a storage area on the tape that holds metadata.

LAN Local Area Network (LAN), a data network with low geographic extension (maximum several tens of kilometres).

LAN-free backup Backup method of a network backup system in which the backup client copies the data directly to the backup medium via the storage network bypassing the backup server and the LAN.

Latency Latency describes the time duration that passes before the input signal becomes visible in an expected output reaction.

Library partitioning Tape library partitioning statically divides a physical tape library into several logical (=virtual) tape libraries, which are perceived as independent libraries by the connected servers.

Library sharing In tape library sharing several applications dynamically share the tapes and the drives of a tape library.

Link Physical connection cable in a Fibre Channel network.

LIP The loop initialisation primitive sequence (LIP) describes the procedure for the initialisation of a Fibre Channel arbitrated loop. During the LIP procedure the data traffic on the arbitrated loop is interrupted.

Loop Abbreviation for Fibre Channel arbitrated loop.

LUN The SCSI protocol and its derivates such as Fibre Channel FCP and iSCSI address subcomponents of a device (SCSI target) by means of the Logical Unit Number (LUN). It has become common practice to also call these subcomponents LUN. Examples of LUNs are physical or virtual hard disks exported from a disk subsystem and the tape drives and the media changer of a tape library.

LUN masking LUN masking limits the visibility of disks exported by a disk subsystem. Each computer sees only the disks that are assigned to it. LUN masking thus works as a filter between the disks exported from the disk subsystem and the accessing computers.

LUN zoning Alternative term for LUN masking. Often used in the context of more modern switches that offer zoning on the basis of LUNs and thus facilitate LUN masking in the storage network.

Magneto-optical disk (MOD) A magneto-optical disk (MOD) is a rotating storage medium that is magnetically written and optically read out. An MOD essentially consists of two layers: a reflection layer and a magnetised layer. When the disk is written, the magnetised layer is heated up through an optical laser and a magnetic field provides for the direction of the magnetisation. The direction of the magnetisation has an influence on the polarisation of the light, based on the magneto-optical Kerr effect. When the disk is read, a laser with a low temperature is used and, depending on the direction of the magnetisation, reads out light with varying degrees of polarisation that are interpreted as data bits and bytes.

MAN   Metropolitan Area Network (MAN), a data network with average geographic extension (maximum several hundred kilometres).

Managed hub   Fibre Channel hub with additional management functions.

Management console   Central point, from which all aspects of a storage network, or all aspects of an IT system in general, can be monitored and managed.

Manchester encoding   Encoding procedure that generates at least one signal change for every bit transmitted.

MD5   Message Digest Algorithm 5 (MD5) is a widely used cryptographic hash function (see Cryptographic hash) that generates a 128-bit hash value that is used as a checksum. MD5 checksums are used, for example, to test data integrity and authenticity and also for data deduplication (see Deduplication). Secure Hash Algorithm (SHA) is a modern version of cryptographic hash functions and is more robust against collisions because the hash values are longer (up to 2,048 bits in 2009).

Media changer   Mechanical transport device that can transport media between slots and drives of a tape library.

Media manager   The term "media manager" is used in multiple ways. In network backup systems it refers to the component which managers the hard disks and the tapes upon which the backed up objects (files, file systems, images) are stored. The IEEE 1244 standard refers the server component of a management system for removable media as media manager.

Metadata   General data that contains information about actual data is referred to as metadata or meta information. Metadata can therefore be interpreted as additional data that is associated with the actual data. Metadata must also be stored either at the same storage location as the actual data or in a different location such as a database. For example, metadata associated with a file that is stored in a file system includes the name of the file, the size of the file, access rights or other file properties. Usually metadata for files is also stored in the file system. Another example is the full-text index for a file, which also represents metadata and is usually stored in a database to enable efficient searches. Metadata can also contain control information for data – for example, to interpret the data or to manage the life cycle.

Metadata controller (MDC)   The metadata controller (MDC) is a management and synchronisation entity in a distributed application. For example, we talk of the metadata controller of a shared disk file system or of the metadata controller of the storage virtualisation.

Metadata database   The metadata database is the brain of a network backup system. It includes approximately the following entries for every object backed up: name, computer of origin, date of last change, data of last backup, name of the backup medium, etc.

mFCP Metro FCP (mFCP) is an iFCP variant, which in contrast to iFCP is not based upon TCP but on UDP.

MIB The term management information base (MIB) stems from SNMP jargon. An MIB is a hierarchically constructed collection of variables, which describes the management options of a resource (server, storage device, network component, application).

MIB file File that contains an MIB description.

Microfilm Microfilms are dramatically reduced photographic copies of information that were originally captured on paper. Microfilms are based on a transparent medium and the information on the medium can be made visible through the use of light. The principle involved is similar to that of transparencies and slides.

Mirroring Mirroring of data on two or more hard disks (RAID 1).

Modality A modality refers to any type of device in the medical field that produces images, such as X-rays, ultrasound and computer tomography. Modalities capture the images and usually transfer them to a Picture Archiving and Communication System (see PACS) that stores and manages the data.

Monitoring Monitoring denotes the monitoring of all resources used in the storage network (hardware, network topology, applications).

MTBF Mean Time Between Failure (MTBF) indicates the average period of time between two sequential errors in a particular component or system.

MTTF Mean Time to Failure (MTTF) indicates the average period of time between the recovery of a component or of a system and the occurrence of a new failure.

MTTR Mean Time to Repair (MTTR) indicates the average period of time before a component or a system is restored after a failure.

Multipathing Multipathing is the existence of several I/O paths between server and storage system. The objectives are to increase fault-tolerance by means of redundant I/O paths, to increase the I/O throughput by means of the simultaneous use of several I/O paths, or both at the same time.

Name server In general, the term name server is used to describe an information service in distributed systems. In the case of Fibre Channel the name server (here Simple Name Server) manages information about all N-Ports connected in a fabric such as their WWPN, WWNN, Node_ID and supported service classes and application protocols.

NAS Network Attached Storage (NAS) refers to the product category of preconfigured file servers. NAS servers consist of one or more internal servers, preconfigured disk capacity and usually a stripped-down or special operating system.

NDMP The Network Data Management Protocol (NDMP) defines the interface between the client and the server of a network backup system. The objective of the NDMP is to improve and standardise the integration of NAS servers in a network backup system.

Network Management System (NMS) In SNMP jargon a Network Management System is an application that monitors and manages components by means of the SNMP protocol.

Network File System Network file systems are the natural extension of local file systems: end users and applications can access directories and files over a network file system that physically lie on a different computer – the file server. Examples of network file systems are the Common Internet File System (CIFS), the Network File System (NFS) and the Direct Access File System (DAFS).

Network backup system Network backup systems can back up heterogeneous IT environments incorporating several thousand computers largely automatically.

NFS Network File System (NFS) is the network file system originally developed by SUN Microsystems, which is currently supplied as standard with all Unix systems.

NIC Network Interface Controller (NIC), Network Interface Card (NIC); both terms for network cards.

NRO The Network Recovery Objective (NRO) is an elementary parameter for business continuity. If indicates the maximum allowed time after an outage to restart the network (LAN and WAN).

Off-site location An off-site location is a remote location at which a second copy of data that has been backed up by means of a network backup system is stored. The second copy of the data in the off-site location serves to protect against major catastrophes.

OOM Object-oriented modelling (OOM) is an object-oriented specification language, which is used for the description of the Common Information Model (CIM).

Open Systems Open Systems signifies the world of the non-mainframe server. Unix, Windows, OS/400, Novell and MacOS belong to the Open System world. Incidentally, for us 'Unix' also covers the Linux operating system, which is sometimes listed separately in such itemisations.

Ordered set 8b/10b encoded group of four transmission words that begins with the K28.5 symbol.

Out-band management Out-of-band management (out-band management for short) signifies the management of a resource by means of a second interface, which exists in addition to the data path. An example of out-band management would be the management of a Fibre Channel switch by means of an Ethernet connection and SNMP.

Out-band virtualisation Alternative term for 'asymmetric virtualisation'.

PACS A Picture Archiving and Communications System (PACS) manages and stores digital images that are captured at modalities (see Modality) of imaging medicine.

Parity Parity is a binary cross-check sum or check sum. RAID 4 and RAID 5, for example calculate and store additional parity blocks, with which the data stored upon a hard disk can be reconstructed after its failure.

Partition Part of a side, which provides storage capacity as a physical unit of the cartridge.

PCI Peripheral Component Interconnect (PCI) is currently the predominant technology for host I/O buses.

PFC Priority Flow Control (PFC, IEEE 802.1Qau) is part of the DCB protocol family. It refines the concepts of the Ethernet PAUSE frame to priority groups thus each priority group can be suspended individually without disabling the whole port.

Point-in-time restore Point-in-time restore signifies the capability of a network backup system to recreate any desired earlier state of a file system.

Point-to-point The simplest of the three Fibre Channel topologies, which solely connects two end devices (server, storage) together.

Port A port denotes the physical interface of a device (servers, storage devices, switches, hubs, etc.) to a storage network.

Port login (PLOGI) Port login denotes the establishing of a connection (session) between two Fibre Channel end devices. Port login exchanges service parameters such as service class and end-to-end credit. It is an absolute prerequisite for further data exchange.

Port zoning Zoning variant, in which the zones are defined by means of port addresses. Port zoning is often confused with hard zoning.

Prefetch hit rate The prefetch hit rate describes the success rate of a cache in shifting data from a slower storage device before a different component demands precisely this data from the cache.

Private loop A Fibre Channel arbitrated loop that is not connected to a fabric.

Private loop devices A private loop device is a device connected to a Fibre Channel arbitrated loop that does not master the fabric protocol. It is not capable of communicating with end devices in the fabric via a Fibre Channel switch connected to the loop.

Protocol converter A protocol converter connects two incompatible interfaces and translates between them.

Protocol mapping The Fibre Channel standard denotes the mapping of an application protocol such as SCSI or IP on the Fibre Channel transport layer (FC-2, FC-3) as protocol mapping.

Process login (PRLI) Process login describes the establishing of a connection (session) between two processes on the FC-4 layer of Fibre Channel.

Public loop A Fibre Channel arbitrated loop, which is connected to a fabric via a switch.

Public loop devices   A public loop device denotes a device connected to a Fibre Channel arbitrated loop, which in addition to the loop protocol also masters the fabric protocol. It can communicate with end devices in the fabric via a Fibre Channel switch connected to the loop.

Quickloop   Implementation variant of the emulated loop by the company Brocade.

RAID   Originally RAID was the abbreviation for 'Redundant Array of Inexpensive Disks'. Today RAID stands for 'Redundant Array of Independent Disks'. RAID has two primary objectives: to increase the performance of hard disks by striping and to increase the fault-tolerance of hard disks by redundancy.

RDMA   Remote Direct Memory Access (RDMA) enables processes to read and write the memory areas (RAM) of processes that are running on another computer. RDMA is based on VI. RDMA is aimed at the lightweight and very fast interprocess communication within a data centre.

RDMA over TCP   Standardised RDMA variant that uses TCP as the transmission medium.

Real time data sharing   Variant of data sharing in which several applications work on the same data set concurrently.

Regulatory compliance   When archiving is subject to legal conditions and requirements, reference is made to regulatory-compliant archiving. In practice, the terminology 'regulatory compliance' is used with different meanings. We interpret 'regulatory compliance' as the obligation to maintain legal requirements governing the protection of data from deletion and manipulation.

Remote mirroring   Remote mirroring signifies the capability of a block-based storage system (e.g. a disk subsystem) to copy data sets to a second storage system without the involvement of a server.

Replication   Replication denotes automatic copying and synchronisation mechanisms on file level.

RIS   A radiological information system (RIS) is a data processing system that manages information from the radiology area. This information is of an administrative and medical nature and encompasses such things as patients' medical history, the scheduling of radiology equipment, test results and the data used for invoicing. In terms of its function, an RIS is closely related to hospital information systems.

RNIC   RDMA enabled NIC (network interface controller), a network card that supports RDMA over TCP and, in addition to RDMA, most likely also realise the functions of a TCP/IP offload engine (TOE).

Roll forward   See Forward recovery.

Rolling disaster   With a rolling disaster, parts of an IT infrastructure fail gradually but not the entire data centre at the same time. This produces an inconsistency in the available data and IT services because some applications may still be running while others have already ceased operation.

RPO   Recovery Point Objective (RPO) is an elementary parameter for business continuity. It indicates the maximum amount of data loss tolerated in a crisis.

RSCN   The Registered State Change Notification (RSCN) is an in-band mechanism in Fibre Channel networks, by means of which registered end devices are automatically informed of status changes of network components and other end devices.

RTO   Recovery Time Objective (RTO) is an elementary parameter for business continuity. It indicates the maximum length of time that is tolerated for the restart of IT operations after an outage.

SAFS   SAN Attached File System (SAFS), an alternative term for shared disk file system.

SAN   SAN is an abbreviation for two different terms. Firstly, SAN is the abbreviation for 'Storage Area Network'. Very often 'storage area networks' or 'SANs' are equated with Fibre Channel technology. The advantages of storage area networks can, however, also be achieved with alternative technologies such as for example iSCSI. In this book we therefore do not use the term SAN or 'Storage Area Network' alone. For general statements on storage area networks we use the term 'storage network'. Otherwise, we always state the transmission technology with which a storage area network is realised, for example Fibre Channel SAN or iSCSI SAN.

Secondly, SAN is an abbreviation for 'System Area Network'. A system area network is a network with a high bandwidth and low latency, which serves as a connection between computers in a distributed computer system. In this book we have never used the abbreviation SAN to mean this. However, it should be noted that the VIA standard uses the abbreviation SAN in this second sense.

SAN router   Alternative name for a Fibre Channel-to-SCSI bridge.

SAS   Serial Attached SCSI is an I/O technique that links individual hard disks and tape drives that sequentially transmit the conventional parallel SCSI protocol and, therefore, achieve higher transmission rates than SCSI.

SATA   Serial ATA (SATA) is an economical I/O technology for disk attachment that transmits the conventional parallel ATA protocol serially and thus permits higher transmission rates than IDE/ATA.

Scratch pool   Group of cartridges, the storage capacity of which is (or has once again become) completely available.

Scratch tape   A new tape without content or a tape the content of which is no longer of interest and the whole of the storage capacity of which can be used for new purposes.

SCSI The Small Computer System Interface (SCSI) is an important technology for I/O buses. The parallel SCSI cables are increasingly being replaced by serial I/O techniques such as Fibre Channel, TCP/IP/Ethernet, SATA, SAS and InfiniBand. The SCSI protocol, however, lives on in the new serial techniques, for example as Fibre Channel FCP or as iSCSI.

SDK A software development kit (SDK) is a collection of programs and documentation for a specific software. SDKs are designed to simplify the development of applications that are run on the software concerned. Sometimes the use of an SDK is essential to the development of such applications.

SDP The Socket Direct Protocol (SDP) maps the socket API of TCP/IP on RDMA, so that protocols based upon TCP/IP such as NFS and CIFS do not need to be modified. Users of SDP benefit both from the simplicity of the protocol and also from the low latency and low CPU load obtained with RDMA.

SEC The U.S. Securities and Exchange Commission (SEC) is an independent agency of the United States government that enforces the federal securities laws and regulates the securities industry/stock market. As such, the SEC Regulations 17 CFR 240.17a-3 and 17 CFR 240.17a-4 stipulate the records retention requirements for the securities broker-dealer industry.

SES The SCSI Enclosure Services (SES) are an in-band management interface for SCSI devices.

Sequence A sequence is a large data unit in the FC-2 layer of Fibre Channel that is transmitted from transmitter to receiver in the form of one or more frames.

Server-centric IT architecture In a server-centric IT architecture, storage devices are only connected to individual servers. Storage only ever exists in relation to the servers to which it is connected. Other servers cannot directly access the data; they must always go through the server to which the storage is connected.

Server consolidation Server consolidation is the replacement of many small servers by a more powerful large server.

Server-free backup Backup method of a network backup system, in which the data is copied from the source disk to the backup medium via the storage network without a server being connected in between. Server-free backup makes use of the 3rd-Party SCSI Copy Command.

Services subsystem Component of the SNIA Shared Storage Model in which the management tasks of a shared storage environment are brought together.

SHA A Secure Hash Algorithm (SHA) refers to a group of cryptographic hash functions (see Cryptographic hash).

Shared disk file system Shared disk file systems are a further development of local file systems in which several computers can directly access the hard disks of the file system at the same time via the storage network. Shared disk file systems must synchronise the write accesses to shared disks in addition to the functions of local file systems.

Shared-everything cluster The shared-everything cluster is the cluster configuration that permits the greatest flexibility and the best load balancing. In shared-everything clusters, several instances of an application run on different computers, with all instances providing the same services towards the outside. A corresponding load balancing software ensures that all instances are loaded to the same degree.

Shared-nothing cluster Shared-nothing clusters are a configuration of two servers in which in the event of the failure of one computer the remaining computer takes over the tasks of the failed computer in addition to its own.

Shared-null configuration The shared-null configuration is a server or an application that is not designed with built-in redundancy. If the server fails the application is no longer available.

Shared storage environment SNIA term for storage-centered IT architectures.

Side Part of a cartridge that provides storage capacity. A side contains one or more partitions. Tapes normally possess only one side. DVDs and magneto-optical media are also available in double-sided variants. Holographic storage may provide even more than two sides.

Single point of failure Single point of failure signifies a subcomponent of a system, the failure of which leads to the failure of the entire system. Fault-tolerant systems such as server clusters or high-end disk subsystems must not have any single points of failure.

Skew Skew means the divergence of signals that belong together in a parallel bus.

SLA Service Level Agreements (SLAs) describe in detail and in a measurable form what an IT service customer requires of an IT service provider. These IT service agreements are usually part of an official contract between the IT provider and the customer.

Slot Storage location for cartridges that are not being accessed.

SMI-S The Storage Management Initiative Specification (SMI-S) is a further development of WBEM and CIM by SNIA, which is specially tailored to the management of storage networks. Amongst other things, the standardised refinement of the CIM classes aims to guarantee the interoperability of management systems for storage networks.

Snapshot A snapshot means an instant copy within a file system or a volume manager.

SNIA Storage Networking Industry Association (SNIA), an association of manufacturers in the field of storage and storage networks.

SNMP The Simple Network Management Protocol (SNMP) is a standard that was originally developed for the management of IP networks. SNMP is now a widespread standard for the management of IT systems that is also used for the management of storage networks.

Soft zoning   Soft zoning describes a zoning variant that restricts itself to the information of the name server. If an end device asks the name server for further end devices in the Fibre Channel network then it is only informed of the end devices with which it lies in at least one common zone. However, if an end device knows the address of a different device, with which it does not lie in a common zone, then it can nevertheless communicate with the other device. Soft zoning is often confused with WWN zoning.

SoIP   Storage over IP (SoIP), the name of a product from the former corporation Nishan Technologies. According to the manufacturer these products are compatible with various IP storage standards.

SPB   Shortest Path Bridging (SPB, IEEE 801.1aq) is an approach to define routing mechanism for unicast and for multicast frames which supports redundant links and parallel VLAN configurations.

SPMA   Server Provided MAC Address (SPMA) is a mechanism where the management software of a server generates a MAC address for the FCoE Line End Point (FCoE LEP) of an FCoE end device (ENode).

SRM   Storage Resource Management (SRM) is the category of software products that unifies storage virtualisation and storage management.

SRP   The SCSI RDMA Protocol (SRP) allows the transfer of SCSI via RDMA bypassing TCP or any other intermediate protocol.

SSA   Serial Storage Architecture, an alternative I/O technology to SCSI.

SSH   Secure Shell (SSH) is a network protocol that, in contrast to TELNET, enables a user to register with a remote computer over an encrypted network connection and to execute programs there.

SSP   A Storage Service Provider (SSP) is a business model in which a service provider (the SSP) operates a storage network, which is used by many customers. Originally it was hoped that this would result in cost benefits. In practice this business model has failed. However, it is very likely that this approach will experience a renaissance in a modified form with the increasing use of the web architecture and so-called cloud computing.

Storage-centric IT architecture   In contrast to server-centric IT architecture, in storage-centric IT architecture, storage exists completely independently of any computers. A storage network installed between the servers and the storage devices allows several servers to directly access the same storage device without a different server necessarily being involved.

Storage consolidation   Storage consolidation means the replacement of a large number of small storage systems by one more powerful large storage system.

Storage gateway   Alternative term for a Fibre Channel-to-SCSI bridge.

Storage Hierarchie   Storage Hierarchie (Tiered Storage) denotes a storage architecture which comprises multiple storage types such as disk and tape or disks with varying performance characteristics. Storage virtualisation and applications such as network backup and archiving systems can leverage a storage hierarchie and move the data between the tiers to store the data cost effective.

Storage networks   The idea behind storage networks is to replace the SCSI cable between servers and storage devices by a network, which is installed alongside the existing LAN as an additional network and is primarily used for the data exchange between computers and storage devices.

Storage virtualisation   Storage virtualisation (often just called virtualisation) is generally used to mean the separation of storage into the physical implementation in the form of storage devices and the logical representation of the storage for the use by operating systems, applications and users. A differentiation is made between three levels of storage virtualisation: (1) virtualisation within a storage system, for example in a RAID disk subsystem or an intelligent disk subsystem, (2) virtualisation in the form of an own virtualisation entity in the storage network and (3) virtualisation on the server by host bus adapter, volume manager, file systems and databases. A further differentiation is made with regard to the granularity of the virtualisation (virtualisation on block level and virtualisation on file level) and, for the virtualisation in the storage network, we also differentiate between symmetric and asymmetric virtualisation.

Streaming   The reading or writing of large quantities of data to a tape, whereby the data is written in one go without stopping, rewinding and restarting the tape.

Striping   Distribution of data over two or more hard disks (RAID 0).

Support matrix   In heterogeneous storage networks, numerous components from extremely different manufacturers come together. In the support matrix, manufacturers of hardware and software components state which components from other manufacturers their components will work with.

Switch   The switch is the control centre in networks such as Ethernet and the Fibre Channel fabric. It realises the routing of frames and services such as name server and zoning.

Switched hub   A special kind of Managed hub, which in addition allow for the direct communication between two end devices, so that several end devices can communicate with each other in pairs within a Fibre Channel arbitrated loop at the same time.

Symmetric storage virtualisation   Symmetric storage virtualisation is the form of storage virtualisation within a storage network in which the data flow between servers and storage devices plus the control flow – i.e. the control of the virtualisation by a virtualisation instance – take place in the data path.

System bus   The I/O bus in a computer that connects, amongst other things, the CPUs to the main memory (RAM).

Tape library partitioning   Tape library partitioning (library partitioning for short) divides a physical tape library statically into several logical (= virtual) tape libraries, which are perceived as independent libraries by the connected servers.

Tape library sharing   In tape library sharing (library sharing) several applications dynamically share the tapes and drives of a tape library.

Tape mount   The inserting of a tape in a tape drive.

Tape reclamation   In a network backup system over time more and more data is left on a tape that is no longer required. With current technology it is difficult to write new data to these gaps on tapes that have become free. In tape reclamation the data that is still valid from several such tapes with gaps is copied onto one new tape so that these tapes can be rewritten.

Target   The SCSI protocol calls the device connected to an SCSI bus a target. Examples of targets are servers, disk subsystems and tape libraries.

Target ID   Target_ID is the name for the address of a device (target), which is connected to an SCSI bus.

TCA   InfiniBand calls the connection point of a server to an InfiniBand network a Target Channel Adapter (TCA). The complexity of a TCA is low in comparison to an Host Channel Adapter (HCA).

TCP/IP offload engine (TOE)   A network card that realises the TCP/IP protocol stack completely in firmware on the network card. TOEs significantly reduce the CPU load for TCP/IP data traffic.

TELNET   TELNET is a network protocol that, in contrast to SSH, enables a user to register with a remote computer over a non-encrypted network connection and to execute programs there. All data – including the passwords for setting up the connection – are transmitted in clear text.

Three-tier architecture   Further development of the client-server architecture, in which the data, applications and the user interface are separated into different layers.

Tiered Storage   see Storage hierarchy.

Translated Loop   Implementation variant of the emulated loop from CNT/Inrange.

Trap   A trap is a mechanism with which a resource managed by SNMP (or to be more precise its SNMP agent) informs a management system for storage networks or a general management system of state changes.

Trap recipient   The trap recipient is the recipient of SNMP messages (traps). To set up a trap recipient the IP address of the computer that is to receive the trap is entered on the SNMP agent.

TRILL The Transparent Interconnection of Lots of Links (TRILL) is an approach for enabling Ethernet to operate multiple parallel links in parallel which is needed to support the multipathing concept of storage networks.

Twin-tailed SCSI cabling Cabling method in which the storage devices are connected to two servers via a SCSI bus for the benefit of fault-tolerance.

UDO Ultra Density Optical (UDO) was developed by the British supplier Plasmon and is considered the successor to MODs (see Magneto-optical disk) for professional company use. Technically, UDO is based on phase-change technology and the current UDO Version 2 is marketed as having 60 GBytes and was presented at CeBit 2007. Since September 2004 UDO has been based on the cross-vendor standard ISO/IEC 17345.

ULP Upper level protocol (ULP). Application protocol of a Fibre Channel network. Examples of ULPs are SCSI and IP.

Unmanaged hub Fibre Channel hub without management functions.

Vaulting Vaulting is the transfer of data to a secure site. With conventional vaulting, the data is first backed up on tape and the tapes are then moved to a remote location. With more recent techniques the data is copied directly over WAN connections to a different data centre (electronic vaulting).

VI The Virtual Interface (VI) denotes a communication connection in the Virtual Interface Architecture (VIA).

VIA The Virtual Interface Architecture (VIA) is a system-level I/O technology, which facilitates the lightweight and fast data exchange between two processes that run on different servers or storage devices within a data centre.

VI NIC VI-capable network card. Today VI-capable network cards exist for Fibre Channel, Ethernet and InfiniBand.

Virtualisation See Storage virtualisation.

Voice over IP (VoIP) VoIP is the transmission of telephone calls via IP data networks.

Volume A volume is a logical data container. It serves to reserve storage capacity on storage devices for applications.

Volume level Backup mode in which an entire volume (e.g. disk, partition of a disk or logical volume) is backed up as a single object.

Volume manager Virtualisation layer in the server between disk and file system that can bring together several physical hard disks to form one or more logical hard disks.

VPN A virtual private network (VPN) enables secure transmission over an insecure network. The users of a VPN can exchange data the same way as in an internal LAN. The connection over the insecure network is usually encrypted. The authentication of

VPN users can be guaranteed through the use of passwords, public keys and digital certificates.

VSAN A virtual SAN (VSAN) makes it possible to operate several virtual Fibre Channel fabrics that are logically separate from one another over one physical Fibre Channel network. In addition, separate fabric services such as name server and zoning are realised for every virtual storage network.

WAN Wide Area Network (WAN), a data network with large geographical extension (several thousand kilometres).

WBEM The Web Based Enterprise Management (WBEM) is a standard developed by the Distributed Management Task Force (DMTF) for IT infrastructure management, which is currently viewed as the successor to the Simple Network Management Protocol (SNMP). WBEM uses web techniques. A significant part of WBEM is the Common Information Model (CIM).

Web architecture Further development of the three-tier architecture to a five-tier architecture for the flexible support of Internet and e-business applications. The representation layer is broken down into the web server and the web browser and the data layer is broken down into the organisation of the data (databases, file servers) and storage capacity for data (disk subsystems and tape libraries).

WebGUI A web graphical user interface (WebGUI) is a GUI implemented using web techniques.

WORM Write once read many (WORM) means that data can only be written once and then only read. It is not possible to change or overwrite the data that complies with regulatory requirements (see Regulatory compliance). Traditionally, storage media such as CDs and DVDs – where data bits are burned – was used as WORM Media. In the newer implementations rewritable media (such as hard disk and tapes) is also being used, whereby additional control software provides WORM protection.

Write order consistency With asynchronous remote mirroring, write order consistency ensures that data that is distributed over multiple virtual hard disks is updated on the destination disks in the same sequence as on the primary disks despite the asynchronous remote mirroring.

WWN A World Wide Name (WWN) is a 64-bit identifier for a Fibre Channel component, which in contrast to FCN is unique worldwide. In practice it has become common practice to call WWNs and FCNs simply WWNs.

WWN zoning Zoning variant in which the zones are defined by WWNs. WWN zoning is often confused with soft zoning.

WWNN The World Wide Node Name (WWNN) is the WWN for a device (server, storage device, switch, director) in a Fibre Channel network.

WWPN The World Wide Port Name (WWPN) is the WWN for a connection port of a device (server, storage device, switch, director) in a Fibre Channel network.

XAM The Extensible Access Method (XAM) is an interface standard that specifies the communication between a document management system (DMS) and archive storage. XAM was developed under the umbrella of the Storage Networking and Industry Association (SNIA) with the participation of multiple storage vendors. SNIA approved the first version of the XAM architecture and specification in July 2008.

XCOPY SCSI command that realises the 3rd-Party SCSI Copy Command.

Zoning Subdivision of a network into virtual subnetworks, which can overlap.

Annotated Bibliography

When we began writing this book in April 2001 there were hardly any books about storage networks. Since then a couple of books have appeared on this subject. In the following we introduce a selection of the sources (books, white papers and websites) that have been helpful to us when writing this book, in addition to our daily work. That means, the following list represents our subjective list of readings – there may be a lot of other useful resources available as well.

GENERAL SOURCES

Marc Farley *Building Storage Networks* (*2nd Edition*), McGraw-Hill 2001. In our opinion the first comprehensive book on storage networks. When we started to work with storage networks in mid-2000 Farley's book quickly became the 'storage bible' for us. We still use this book as a reference today. The book gives a particularly good overview of the fundamental technologies for storage networks.

Tom Clark *Designing Storage Area Networks: A Practical Reference for Implementing Fibre Channel and IP SANs* (*2nd Edition*), Addison-Wesley 2003. This book gives a good overview about techniques for storage networking and their application.

InfoStor (*http://is.pennet.com*), a manufacturer-neutral technical journal on storage and storage networks. For us the companion website is the first port of call for new developments such as IP storage, RDMA, SMI-S or InfiniBand. At InfoStor you can also

Storage Networks Explained: Basics and Application of Fibre Channel SAN, NAS, iSCSI, InfiniBand and FCoE, Second Edition
U. Troppens R. Erkens W. Müller-Friedt N. Haustein R. Wolafka © 2009 John Wiley & Sons, Ltd

order a free weekly e-mail newsletter with up-to-date information on storage and storage networks.

Storage (*http://searchstorage.techtarget.com*), the website SearchStorage of the Storage magazine is also an important place to go. The site provides actual news, solid background information and a free email newsletter.

Storage Networking Industry Association *(SNIA, http://www.snia.org):* The SNIA is an association of manufacturers, system integrators and service providers in the field of storage networks. The website includes a directory of all SNIA members, which at the same time gives a good overview of all important players in the field of 'storage networks'. Furthermore, the SNIA website provides a couple of other useful information including white papers, a dictionary, presentations, a link collection and a regular newsletter. In addition to that, Europeans can subscribe to the freely distributed *SNS Europe* magazine at *http://www.snia-europe.com*.

IBM Redbooks (*http://www.redbooks.ibm.com):* IBM makes technical expertise and material on its products freely available via IBM Redbooks. Many IBM Redbooks deal with the integration, implementation and operation of realistic customer scenarios. They should thus be viewed as a supplement to the pure handbooks. Many Redbooks also deal with product-independent subjects such as RAID or the fundamentals of storage networks. IBM Redbooks can be downloaded free of charge from the website.

Wikipedia (*http://www.wikipedia.org):* The free online encyclopedia is a useful source to look up many basic definitions.

Storage Explained (*http://www.storage-explained.com):* The homepage of this book. On this page we will publish corrections and supplements to this book and maintain the bibliography. In addition to this, we have provided the figures from this book and presentations to download.

INTELLIGENT DISK SUBSYSTEMS

We know of no comprehensive representation of disk subsystems. We have said everything of importance on this subject in this book, so that the next step would be to look at specific products. On the subject of **RAID** we have drawn upon Marc Farley's *Building Storage Networks*, Jon William Togo's *The Holy Grail of Data Storage Management* and various IBM Redbooks. By the way, all three sources also provide information about tape and tape libraries. In a blog hosted by Adaptec we found a spreadsheet from Tom Treadway (*http://treadway.us/SA_Images/MTTDL.xls*). We used that spreadsheet to calculate the failure probability of RAID 5 which we presented in Table 2.18.

I/O TECHNIQUES

SCSI: With regard to SCSI, two sources were important to us. Firstly, we must again mention Marc Farley's *Building Storage Networks* and the book by Robert Kembel below. In addition to that you may refer to *"The SCSI Bench Reference"* by Jeffrey Stai.

Robert Kembel *Fibre Channel: A Comprehensive Introduction*, Northwest Learning Associations, 2000. This book is the first book of a whole series on Fibre Channel. It explains the Fibre Channel standard in bits and bytes and also includes an interesting section on 'SCSI-3 Architectural Model (SAM)'.

IP Storage – iSCSI and related subjects: With regard to IP storage we have drawn upon Marc Farley's *Building Storage Networks*, various articles from InfoStor (*http://is. pennet.com*) and the relevant standards of the Internet Engineering Task Force (IETF) on *http://www.ietf.org*. More and more iSCSI products are coming onto the market, so that more and more information on this subject can be found on the websites of relevant manufacturers. A book has now appeared – Tom Clark's *IP SANs: An Introduction to iSCSI, iFCP, and FCIP Protocols for Storage Area Networks* – that leaves no questions on this subject unanswered.

Fibre Channel over Ethernet (FCoE) and Data Center Bridging (DCB): Silvano Gai's *Data Center Networks and Fibre Channel over Ethernet (FCoE)* provides an excellent overview about these new technologies. In addition to that we referred to the INCITS T11 committee (*http://www.t11.org/fcoe*), the IEEE DCB Task Group (*http://www.ieee802.org/1/pages/dcbridges.html*), the FCIA (*http://www.fibrechannel. org/FCoE.html*), the IETF (*http://www.ietf.org*), and the SNIA (*http://www.snia.org*).

InfiniBand, Virtual Interface Architecture and RDMA: For InfiniBand, Marc Farley's *Building Storage Networks* and InfoStor (*http://is.pennet.com*) should again be mentioned. We should also mention the homepage of the InfiniBand Trade Association (*http://www.infinibandta.org*), the homepage of the Virtual Interface Architecture (*http://www.viarch.org*, in 2007 this page was no longer accessible), the homepage of the RDMA Consortium (*http://www.rdmaconsortium.org*) and various white papers from the homepages of relevant manufacturers.

LAN and WAN techniques: We confess that our coverage of LAN techniques like TCP/IP and Ethernet and of WAN techniques like Dark Fiber, DWDM and SONET/SDH must be improved. As general introduction in computer networks and LAN techniques we recommend Andrew S. Tanenbaum's *Computer Networks*. With regard to WAN techniques we recommend the Lightreading's Beginner's Guides at *http://www.lightreading. com/section.asp?section_id=29*. Lightreading is also a very good starting point for upcoming WAN techniques like the Resilient Packet Ring (RPR) and the Generic Framing Procedure (GFP).

FILE SYSTEMS

For **basic information** on modern file systems we recommend that you take a look at the handbooks and white papers of relevant products. Particularly worth a mention are the Veritas File System from Symantec Veritas (*http://www.veritas.com*) and the Journaled File System from IBM (*http://www.ibm.com*, *http://www.redbooks.ibm.com*).

A good **comparison of NFS and CIFS** can be found in Marc Farley's *Building Storage Networks*. He gives a very good description of the difficulties of integrating the two protocols in a NAS server. Further information on NAS servers can also be found on the websites of the relevant manufacturers.

For **GPFS** we primarily used the two IBM Redbooks *Sizing and Tuning GPFS* by Marcello Barrios *et al.* and *GPFS on AIX Clusters: High Performance File System Administration Simplified* by Abbas Farazdel *et al.*

With regard to **DAFS** we referred to Marc Farley's *Building Storage Networks*, the DAFS homepage (*http://www.dafscollaborative.com*, in 2007 this page was no longer accessible) as well as the standard itself at the Internet Engineering Task Force (*http://www.ietf.org*). At print of the second edition the standard was no longer available at the IETF. Also helpful were articles by Boris Bialek on *http://www.db2magazin.com* and by Marc Farley on *http://storagemagazine.techtarget.com*.

STORAGE VIRTUALIZATION

Some articles at InfoStor (*http://is.pennnet.com*) deal with the subject of **storage virtualization**. The IBM Redbook *Storage Networking Virtualization: What's it all about?* is highly recommended. More and more storage virtualization products are coming onto the market, which means that an increasing amount of information on the subject can be found on the websites of the manufacturers in question. There's also a great technical tutorial booklet on storage virtualization available from SNIA.

APPLICATION OF STORAGE NETWORKS

With regard to the **application and use of storage networks** we unfortunately do not know of any comprehensive book. We can only refer you to the white papers of relevant manufacturers, various IBM Redbooks, to InfoStor, the SNIA and SNS Europe on the Internet.

NETWORK BACKUP

A good representation of the **components of a network backup system** can be found in Marc Farley's *Building Storage Networks*. For the interaction of network backup systems with storage networks and intelligent storage systems, we can again only refer the reader to the white papers of the products in question.

With regard to **NDMP** we can refer the reader to the same sources. In addition, the NDMP homepage *http://www.ndmp.org* and the standard itself at the Internet Engineering Task Force (IETF) on *http://www.ietf.org* should be mentioned.

DIGITAL ARCHIVING

The following sources have been used for the explanations of the **legal regulations**:

- Ulrich Kampffmeyer: *Compliance – ein Project Consult Whitepaper*, PROJECT CONSULT Unternehmensberatung GmbH, 2004
- Bundesministerium der Finanzen der Bundesrepublik Deutschland: *Aufbewahrungsbestimmungen für die Unterlagen für das Haushalts-, Kassen- und Rechnungswesen des Bundes (ABestBHKR)*
- Bundesrepublik Deutschland: *Handelsgesetzbuch (HGB) – Handelsgesetzbuch in der im Bundesgesetzblatt Teil III, Gliederungsnummer 4100–1, veröffentlichten bereinigten Fassung, zuletzt geändert durch Artikel 5 des Gesetzes vom 5. Januar 2007 (BGBl. I S. 10)*; online available at *http://www.gesetze-im-internet.de/hgb*
- République Française: *French Standard NF Z 42-013 – Electronic Archival Storage – Recommandations relatives à la conception et à l'exploitation de systèmes informatiques en vue d'assurer la conservation et l'intégrité des documents stockés dans ces systèmes*
- Armin Gärtner: *Medizintechnik und Informationstechnologie, Band 3: Telemedizin und Computerunterstützte Medizin*, TÜV Rheinland Group, Köln 2006

The theories of Jeff Rothenberg contributed to the **component-neutral archiving of data**. They can be found in his report *Avoiding Technological Quicksand: Finding a Viable Technical Foundation for Digital Preservation – A Report to the Council on Library and Information Resources*, January 1999.

For **data growth** we referred to John McKnight, Tony Asaro and Brian Babineau: *Digital Archiving: End-User Survey and Market Forecast 2006–2010*, ESG Research Report, January, 2006. This report is online available at *http://www.enterprisestrategygroup.com/ESGPublications/ReportDetail.asp?ReportID=591*.

David Bosshart's citation of about the **comparison of technological, entrepreneurial and social change** was found in *Allgemeine Zeitung* of 23.07.2005 in the article *Schöne neue Einkaufswelten am Horizont – Funk-Etiketten sollen Warenströme kontrollieren und den Kundenservice verbessern*.

The **Java Content Repository (JCR)** is specified in the *Content Repository API for Java Technology Specification, Java Specification Request 170 Version 1.0, 11 May 2005*. Information about the **eXtensible Access Method (XAM)** can be found at the SNIA (*http://www.snia-dmf.org/xam/index.shtml*) and information about the **Interoperable Enterprise ContentManagement (iECM)** can be found at the AIIM (*http://www.aiim. org/standards.asp?ID=29284*).

Business Continuity

Klaus Schmidt's *High Availability and Disaster Recovery* is an elementary introduction into that topic. Furthermore we used various IBM Redbooks and product white papers of various vendors.

MANAGEMENT OF STORAGE NETWORKS

Some articles on the **management of storage networks** can be found at InfoStor (*http://is.pennnet.com*). Some IBM Redbooks (*http://www.redbooks.ibm.com*) and white papers also deal with this subject. A detailed representation of the **Fibre Channel Generic Services** and the **Fibre Channel Methodologies for Interconnects** for the in-band management in the Fibre Channel SAN is provided by the pages of the Technical Committee T11 (*http://www.t11.org*). Information on **SNMP** can be found on the Internet pages of the SNMP Research Technology Corporation (*http://www.snmp.org*). A comprehensive description of **CIM** and **WBEM** can be found on the websites of the Distributed Management Task Force (DMTF, *http://www.dmtf.org*). Information on **SMI-S** can be found on the Storage Networking Industry Association website (SNIA, *http://www.snia.org*).

REMOVABLE MEDIA MANAGEMENT

The **IEEE Standard 1244** can be found at *http://www.ieee.org*. Related documentation and additional reading can be found at the homepage of the Storage Systems Standards Working Group at *http://www.ssswg.org*. There is an IBM Journal of Research & Development volume 47, no 4, 2003: *Tape Storage Systems and Technology: http://www.research.*

ibm.com/journal/rd47-4.html. Some interesting articles on the future of tape storage can be found here:

- Dianne McAdam *The Truth about Tape – Nine Myths to Reconsider*, The Clipper Group, Februar 2007 (*http://www.lto.org/pdf/truth_about_tape.pdf*).
- Dianne McAdam *Is Tape Really Cheaper Than Disk*, Data Mobility Group, Oktober 2005 (*http://www.lto.org/pdf/diskvstape.pdf*)
- Dianne McAdam *Tape and Disk Costs – What it Really Costs to Power the Devices*, The Clipper Group, Juni 2006 (*http://www.lto.org/pdf/Clipper_Energy_Costs.pdf*)
- Henry Newmann *Back to the Future with Tape Drives*, Enterprise Storage Forum, Dezember 2002 (*http://www.enterprisestorageforum.com/technology/features/article. php/11192_1562851_1*)

THE SNIA SHARED STORAGE MODEL

The SNIA provides a lot of material on the **SNIA Shared Storage Model** at *http://www.snia.org/tech_activities/shared_storage_model*. Tom Clark's *Designing Storage Area Networks* (*2nd Edition*) covers the model as well.

Appendix A

Proof of Calculation of the Parity Block of RAID 4 and 5

In Section 2.5.4 we stated that during write operations the new parity block can be calculated from the old parity block and the difference Δ between the old data block D and the new data block $\tilde{D}$. In the following we would like to present the proof for the example in Figure 2.16.

Mathematically speaking we state that:

$$\tilde{P}_{ABCD} = P_{ABCD} \text{ XOR } \Delta \text{ where } \Delta = D \text{ XOR } \tilde{D}$$

Taking into account the calculation formula for the parity block we must therefore show that:

$$P_{ABCD} \text{ XOR } D \text{ XOR } \tilde{D} = \tilde{P}_{ABCD} = A \text{ XOR } B \text{ XOR } C \text{ XOR } \tilde{D}$$

The associative law applies to the XOR operation so we do not need to insert any brackets.

We will conduct the proof on the basis of the values in Table A.1. The parity block will be calculated bit-by-bit by means of the XOR operation. The table therefore shows the occupancy of a bit ('0' or '1') from the various blocks.

The left part of the table shows the possible occupancies for the bits in the old data block D, in the new data block $\tilde{D}$ and the parity bit for the bits in the remaining data blocks (A XOR B XOR C). The values of the individual blocks A, B, and C is insignificant because at the end of the day the parity of these three blocks flows into the parity of all four data blocks. This proof is therefore transferable to arrays with more or less than five hard disks.

Storage Networks Explained: Basics and Application of Fibre Channel SAN, NAS, iSCSI, InfiniBand and FCoE, Second Edition
U. Troppens R. Erkens W. Müller-Friedt N. Haustein R. Wolafka   © 2009 John Wiley & Sons, Ltd

Table A.1 Calculation of the parity block for RAID 4 and RAID 5 by two methods.

A XOR B XOR C	D	$\tilde{D}$	P_{ABCD}	D XOR $\tilde{D}$	P_{ABCD} XOR D XOR $\tilde{D}$	A XOR B XOR C XOR $\tilde{D}$
0	0	0	0	0	**0**	**0**
0	0	1	0	1	**1**	**1**
0	1	0	1	1	**0**	**0**
0	1	1	1	0	**1**	**1**
1	0	0	1	0	**1**	**1**
1	0	1	1	1	**0**	**0**
1	1	0	0	1	**1**	**1**
1	1	1	0	0	**0**	**0**

The middle section of the table shows the calculation of the new parity block according to the formula given by us $\tilde{P}_{ABCD} = P_{ABCD}$ XOR(D XOR $\tilde{D}$). The end result is printed in bold.

The right-side column shows the calculation of the new parity by means of the definition $\tilde{P}_{ABCD} = $ (A XOR B XOR C)XOR $\tilde{D}$ and is also printed in bold. The two columns printed in bold show the same value occupancy, which means that our statement is proven.

Appendix B

Checklist for the Management of Storage Networks

In Section 10.1.2 we discussed the development of a management system for storage networks. As a reminder: a good approach to the management of a storage network is to familiarise yourself with the requirements that the individual components of the storage network impose on such software. These components include:

- *Applications*
  These include all software that is operated in a storage network and processes the data.
- *Data*
  Data is the information that is processed by the applications, transported via the network and stored on storage resources.
- *Resources*
  Resources include all of the hardware that is required for the storage and transport of the data and for the operation of applications.
- *Network*
  Network means the connections between the individual resources.

Diverse requirements with regard to availability, performance or scalability can be formulated for these individual components. The following checklist should help to specify these requirements more precisely.

Storage Networks Explained: Basics and Application of Fibre Channel SAN, NAS, iSCSI, InfiniBand and FCoE, Second Edition
U. Troppens R. Erkens W. Müller-Friedt N. Haustein R. Wolafka © 2009 John Wiley & Sons, Ltd

B.1 APPLICATIONS

B.1.1 Monitoring

- How can I check the error-free implementation of the applications?
- Which active or passive interfaces will be provided for this by the applications?

B.1.2 Availability

- What availability must I guarantee for which applications?
- What degree of fault-tolerance can and must I guarantee?
- What factors influence the availability of the application?
- What measures should be taken after the failure of the application?
- How can I guarantee availability?
- How can I detect that an application has failed?

B.1.3 Performance

- What data throughput and what response times will users expect?
- How can the performance be measured?
- What usage profiles underlie the applications, i.e. when is there a heavy load and when is there a less heavy load?
- What are my options for adapting to a changing usage profile?

B.1.4 Scalability

- How scalable are the applications?
- Can I use the same applications in the event of an increase in the volume of data?
- What measures may be necessary?

B.1.5 Efficient use

- Can applications be shared across business processes?
- Can one application handle multiple business processes?

B.2 DATA

B.2.1 Availability

- Which data requires what degree of availability?
- How can I guarantee availability?
- In case of a disaster: how quickly must the data be back online?
- What level of data loss is tolerable in the event of a failure?

B.2.2 Performance

- What data must be provided to the applications quickly and must therefore be stored on fast storage devices?
- How can I measure and check the data throughput?

B.2.3 Data protection

- Which data must additionally be backed up?
- How often must such a backup take place?
- How can I check the backup process?

B.2.4 Archiving

- Which data must be stored for how long?
- Which laws and regulatory requirements must be fulfilled?
- At what points in time must data be archived?

B.2.5 Migration

- Which data can be moved within the storage hierarchy from expensive media such as hard disks to cheaper media such as tape?
- For which data can a hierarchical storage management (migration) be used?
- How do I check an automatic migration?
- How do I check where the data really is after a certain period of operation?

B.2.6   Data sharing

- Which data sets can be shared by several applications?
- Where do conflicts occur in the event of parallel access?
- How can data sharing use be realised?

B.2.7   Security/access control

- Which users are given what access rights to the data?
- How can access rights be implemented?
- How can I check log and audit accesses?

B.3   RESOURCES

B.3.1   Inventory/asset management and planning

- Which resources are currently used in the storage network?
- Which financial aspects such as depreciation, costs, etc. play a role?
- When is it necessary to invest in new hardware and software?

B.3.2   Monitoring

- How can I determine the failure of a resource?
- Are there possibilities and criteria for checking that would indicate a failure in advance (for example temperature, vibration, failure of a fan, etc.)?

B.3.3   Configuration

- How can I view the current configuration of a resource?
- How can I change the configuration of a resource?
- Which interfaces are available to me for this?
- What consequences does the configuration change of a resource have?
- Can I simulate this in advance?

B.3.4 Resource use

- Which resources are consumed by which applications?
- How can I ensure an equable resource utilisation?
- How must I distribute the data over the resources in order to realise availability, efficient use and scalability?
- Which media, for example which tapes, are in use?
- Which media must be renewed, for example on the basis of age?
- How can I transfer the data on media to be replaced onto new media?

B.3.5 Capacity

- How much free capacity is available on which resource?
- Is sufficient storage capacity available for the capacity requirements?
- How many resource failures can be withstood with regard to sufficient storage capacity?
- Which trends can be expected in the capacity requirement?

B.3.6 Efficient resource utilisation

- How are the resources utilised?
- Are there unused resources?
- Where have I allocated resources inefficiently?
- Where can several resources be integrated in one?

B.3.7 Availability

- Which resources require a high level of availability?
- How can I guarantee the availability of resources?
- Which technologies exist for this?

B.3.8 Resource migration

- How can I – without interrupting the operation – exchange and expand resources?
- What happens to the data during this process?

B.3.9 Security

- How can I protect resources against unauthorised access?
- Which physical measures are to be taken for this?

B.4 NETWORK

B.4.1 Topology

- Which devices are connected together and how?

B.4.2 Monitoring

- How can I recognise the failure of connections?
- Are there criteria for predicting any failures?

B.4.3 Availability

- What level of availability of the network is required?
- How can I guarantee the availability of the network?
- Where are redundant data paths required and how can I provide these?
- Where do single points of failure exist?

B.4.4 Performance

- Where are there bottlenecks in the data path?
- How can I optimise the data path?
- Usage profile: when are which data paths utilised and how?
- Trend analysis: am I coming up against bandwidth limits?

Index

3rd-Party SCSI Copy Command 253–254, *497*
64b/66b-encoding 73–75, *497*
8b/10b encoding 73–75, *497*

A

Access control 172, 188, 302, 314–315, 322, 342, 425–426, 428–429, 431, 441–442, 450, 456, 460–462, *497*
Access control list (ACL) *497*
Access handle, *see* Removable media management
Access path 19, 177, 204, 257, 372–373, 424, 450, 459–461, 480–481, 483, *497*
ACL, *see* Access control list (ACL)
Active 19–20, *497*
Active/active 19–20, 209, 220, *498*
Active/passive 19–20, 220, *498*
Adaptability 219–230, 295–296
  archiving 295–296
  bandwidth 113
  connection of server to storage 62, 93
  device replacement during runtime 175
  limitations 167–172
  new applications 193, 197
  new end devices 197
  storage capacity 198
  user interface 195

Agent *498*
Aggregation *498*
AIIM 284, 319
AL_PA, *see* Fibre Channel
Appliance 150, 468–469, 471, 489, 492, *498*
Application server-free backup, *see* Backup
Arbitrated loop, *see* Fibre Channel
Archive bit, *see* Backup
Archive log file, *see* Databases
Archiving, 203
  access protection, 429
  access time, 300–301
  access to data, 292, 296
  access verification, 302
  adaptability, 295–296
  advantages of archiving, 283
  appropriate operational and organisational measures for, 342–343
  architecture of a digital archive system, 283–285
  archive management, 283–284
  archive storage, 296–299, 303–307
  archiving client, 328–329
  auditability, 292, 299
  backup and, 285–288
  backup copies, 286, 288
  cave paintings, 282

Storage Networks Explained: Basics and Application of Fibre Channel SAN, NAS, iSCSI, InfiniBand and FCoE, Second Edition
U. Troppens R. Erkens W. Müller-Friedt N. Haustein R. Wolafka © 2009 John Wiley & Sons, Ltd

Printed in the USA/Agawam, MA
April 9, 2013